THE UNHAPPY CONSCIOUSNESS

ANALECTA HUSSERLIANA

THE YEARBOOK OF PHENOMENOLOGICAL RESEARCH

VOLUME XIII

Editor:

ANNA-TERESA TYMIENIECKA

The World Institute for Advanced Phenomenological Research and Learning, Belmont, Mass.

EUGENE F. KAELIN

Dept. of Philosophy, Florida State University, Tallahassee

THE UNHAPPY CONSCIOUSNESS

THE POETIC PLIGHT OF SAMUEL BECKETT

An Inquiry at the Intersection of Phenomenology and Literature

D. REIDEL PUBLISHING COMPANY

DORDRECHT : HOLLAND / BOSTON : U.S.A.

LONDON : ENGLAND

Library of Congress Cataloging in Publication Data

CIP

Kaelin, Eugene Francis, 1926–
The unhappy consciousness.

(Analecta Husserliana ; v. 13)
Bibliography: p.
Includes index.
1. Beckett, Samuel, 1906– —Philosophy. 2. Phenomenology and literature. I. Title. II. Series.
B3279.H94A129 vol. 13 [PR6003.E282Z5] 81–12137

ISBN-13: 978-94-009-8524-7 e-ISBN-13: 978-94-009-8522-3
DOI: 10.1007/978-94-009-8522-3

Published by D. Reidel Publishing Company,
P.O. Box 17, 3300 AA Dordrecht, Holland.

Sold and distributed in the U.S.A. and Canada
by Kluwer Boston Inc.,
190 Old Derby Street, Hingham, MA 02043, U.S.A.

In all other countries, sold and distributed
by Kluwer Academic Publishers Group,
P.O. Box 322, 3300 AH Dordrecht, Holland.

D. Reidel Publishing Company is a member of the Kluwer Group.

Softcover reprint of the hardcover 1st edition 1981

To My Parents,
Who Knew,
Without Having Read of It,
The Truth of Consciousness

THE THEME

LITERATURE AND INTERDISCIPLINARY PHENOMENOLOGY: The Pessimism/Optimism controversy in the work of Samuel Beckett as viewed by Eugene Kaelin

Eugene Kaelin's book appears in the sequence of the research program: Interdisciplinary Phenomenology and Literature, which is being carried out by *The International Society for Phenomenology and Literature*, and follows the forthcoming Volume XII of the *Analecta Husserliana* series entitled *The Philosophical Reflection of Man in Literature*. Professor Kaelin plays an important role in this research program and we are happy that he has entrusted his book to us.

Proposing the philosophical program for our literature and interdisciplinary phenomenology research, I have attempted to highlight the main perspectives in which literature and philosophy meet.* The enigmatic situation of the human being, the vicissitudes of his existence, his ever repeated efforts to discover his "unique destiny" over against the common doom, present puzzling issues by which both are haunted. To evoke, differentiate and show the multiple variants of answers to these issues is the task of one, while to unravel and to determine their ultimate factors in the Human Condition is that of the other.

In some literary works the philosophical quest is intermingled with artistic literary pursuit; in others, philosophical reflection serves as artistic means. Yet, due to these common innermost concerns, even those literary works which appear as devoid of philosophical thought call upon philosophy in order to be appropriately appreciated.

Eugene Kaelin's philosophical analysis of Beckett's work shows that although on a primary analysis philosophical thought and literary artistry may blend, it still needs philosophically enlightened criticism to perform – as if counteracting philosophy's first impact – a secondary interpretation, so as to bring the significance of this blend to its maturity.

* Anna-Teresa Tymieniecka, 'Poetica Nova: The Creative Crucibles of the Human Condition and of Art / A Treatise on the Metaphysics of the Human Condition and of Art,' *Analecta Husserliana*, Vol. XII, forthcoming.

Samuel Beckett has visited upon the world one of the richest and most awe-inspiring of contemporary literary *oeuvres*. His life's work includes poetry, novels, short stories, plays, radio and television skits, a movie scenario – all marked by a high degree of self-conscious artistry which reduces an abstract thought to an electrifyingly concrete image, and the image to a pattern of sound that gives it body. Indeed, in the same *oeuvre* there are plays without words in which ideas are portrayed through bodily gesture, and films that are made to breathe in sight and sound. Never, in all this prodigious imaginative creation, is an idea pursued for its own sake as summarizing for our insouciant intellects the essence of our natural or social life-worlds, or the tragedy of our humanly projected individual existences therein, or how it is with the aesthete whose writing is performed in pain, and undertaken to relieve the pain of living, or even how it is with those of us who have never created a world or written a word the least pregnant with meaning. Ideas are for the intellect; art is made to wring the elevated element from our Human Condition.

Professor Kaelin has not written on all of Beckett's works. He begins his treatise with an overview of the criticism Beckett's art has already engendered: some good, some bad, some merely irrelevant. What makes criticism first relevant and then good is its fittingness to the corpus that has inspired it, its analytical acuity and perceptual clarity, and, ultimately, its effect in creating the desire to revisit the original artistic and creative mass. From the extant criticism on Beckett's art, Kaelin has chosen a dominant theme: Beckett as "philosophical" writer. What makes a literary artist philosophical? An attitude of resignation? Or of ponderousness? A penchant for analysis? Or merely a concern for the perennially unsolved problems of the human species? If so, the literary artist is like any other man or woman: humble before the awfully great, or prideful and a little pompous, as the case may be; too concerned with the accuracy of one's own categories, and foolish enough to continue stretching them to cover the imponderables of human existence, including the joys a well-turned phrase may evoke in the jaded hearts of the all-knowing but mysteriously unseeing cognoscenti.

The literary artist is philosophical to the degree that his works exhibit a philosophical structure; such is Kaelin's thesis. The philosophy of a work does not appear on that work's surface, nor in the literal meanings of the words whose tokens appear on that surface, but only in the plots, if the idea is metaphysical, or in the author's characterization, if moral. For this reason, Professor Kaelin attempts to read through the surface of the works, through the semantic constructions of the author's syntax, to whatever reality is depicted within the stratum of "represented objectivities," to use the term of Roman Ingarden, whether that is reality presented in a metaphysical or moral perspective. And since literature has been produced that is philosophical in this sense of the term – a genre of literature in which it is the literature and not the author that is philosophical, Kaelin has devised a method for the criticism of this particular genre of the literary art. We could call it "philosophical criticism," but he prefers the more specific label, "phenomenological structuralism." It is a way in which anyone proceeds who reads the work and then describes what appears therein, and which could therefore appear to any attentive consciousness.

The model for such a method had already been hinted at in the criticism of Beckett himself; the authors he criticized were Joyce and Proust. Such writers are "philosophical," claimed Beckett, in that they used a philosophical idea as a literary technique, merely as a convenience for giving structure to their works of art. Behind Joyce there stood Dante and Bruno and Vico; behind Proust, Schopenhauer and Bergson. This treatise attempts to gauge the extent to which Beckett himself is a "philosophical" author in the sense described; and behind him looms the shadows of a host of philosophers: Descartes, Berkeley, Schopenhauer, Hegel, Heidegger, and Sartre, to name only the most obvious.

The central question asked in pursuit of this method is not how this particular work of our author seemingly expresses the absurdity of the Human Condition, for that is at once a loaded and misguided question. The works do not express absurdity at all; absurdity is sometimes used, however, to express whatever it is the total work itself expresses. Misguided, in the same vein, is the question whether these works express pessimism or optimism or meliorism; for they express neither of the three. Beckett's vaunted "pessimism," like the philosophy of Arthur Schopenhauer Beckett has on occasion used to structure his work, is also a technical convenience. Where he has depicted the human being as completely caught up in the throes of inauthentic existence – either as living in the past in the flight from the present in order to avoid the necessity of making a decision to produce a better future, or surrendering one's own will to that of another, or merely monotonously repeating an action that is no action because it lacks significance even to the person whose spirit has become deadened by such a habit – others have found the tragedy of our contemporary age; if so, it is a tragedy in search of a theory, a Sophoclean phenomenon in search of an unknown Aristotle.

Aristotelian catharsis, we all recall, was developed for the humanity of antiquity, and necessitated a protagonist better than the average man or woman, albeit flawed by a particular human weakness, and could for this reason not fit the world inhabited by bums, weaklings, cowards, and the moral degenerates we all can be. Beckett's are no flawed characters; as artistic creations, they are characters perfected to produce the effect they do. Our perception of their fates does not shunt off our feelings of pity and fear as if into the atmosphere itself. Rather, it concentrates our feeling as if compressed into an ache in the center of our breasts.

Whether novelistic or dramatic, Beckett's characters are sublime. The more they suffer, the less significant their lives, the more inauthentic their struggles to relieve the hurt, the deeper into the mud their nameless essence is driven, the greater is the tragic lift controlled by an encounter with the literary or dramatic work which may thus be all the more authentic, as art, as the heroes or antiheroes of that art are themselves inauthentic as personalities. After all these years, the play is still the thing, not the idea; the novel, and not our reading of it, whose encounter produces the aesthetic effect.

For this reason, it is safer perhaps to aver that man's true nature is better known, because better learned, through our encounters with works of art than through any artist's particular vision of the truth of human nature. Such indeed is the assumption Professor Kaelin asks us to make in order to follow his account of Beckett's poetic plight. If you go along with his suggestion, you will be led through the labyrinthine ways in the imagination of a consummate artist.

Philosophically, that imagination has been variously shaped: from the earlier points of view that are Cartesian, Berkeleyan, Schopenhaurean, we move to the middle stages in which the Hegelian self-referring infinitude of absolute mind gradually changes into the ultimate stages in which existentialist themes dominate the literary and dramatic structures. And throughout, in both the novels and the plays, there is Beckett's constant concern for getting it right, for making the direct point indirectly, to *show* and not merely to tell. Always the aesthete, he has approached his artistic zenith in accomplishing the maximum effect with the minimum "natural" resources – a sight varying from blinding light to utter darkness, each concealing the litter of our natural environment, and constituting one variance interspersed with another: a field of sound constituted by a baby's cry out of a stark silence and alternating with an adult's sigh; the two sets of variances constitute a new phenomenology of consciousness unfolding in less than a minute's exposure, yet experienced as a single harmony of sound and visual sense.

In the later novels, as in many of the plays, the hero is the narrator. He may be nameless because unnamable, a voice struggling to control other voices, a ping, indeed; or he may be a groveler in the mud, or a personification unhappily divided into two parts, separated only by the nothingness of time, the one controlling the words, the other the music. But as always the puzzle to be solved entails the harmony of the music with the words, the rhythm with the image invoked. Indeed, it is in the novels of the creative voice and in the plays raised to the second exponential degree, which Professor Kaelin calls "dramaturgical," that the author's purpose becomes most obvious: showing how it is is perhaps the only way of relieving the pain of how it is, to live as a human being.

Separated from oneself as from others and from others as from any rational ideal, which Hegel calls "the Beyond," the literary consciousness is the unhappy consciousness *per se*, which, left to its own devices, may be allowed to languish in the pain of its own separation from the world and from others, or to assume the pain of its own self-deliverance, which is to create a reality of significance to itself. At their highest, i.e. in the slough of their despond, Beckett's literary and dramatic works illuminate in our every encounter with them a tragedy for our modern times; and they do this by evincing the sublimity of the human spirit.

Parts of this treatise were delivered at the second annual meeting of the International Society for Phenomenology and Literature operating under the aegis of the World Institute for Advanced Phenomenological Research and Learning, with its headquarters in Belmont, Massachusetts.

ANNA-TERESA TYMIENIECKA

TABLE OF CONTENTS

PART II

PREFACE

In the wake of so many other keys to the treasure, whoever undertakes still another book of criticism on the novels and drama of Samuel Beckett must assume the grave burden of justifying the attempt, especially for him who like one of John Barth's recent fictional characterizations of himself, believes that the key to the treasure is the treasure itself.

No one will ever have the privilege of the last word on these texts, since any words other than the author's own found therein must be referred back to the text themselves for cautious verification. Indeed, the words the author has used to create the *oeuvre* stand by virtue of their own creativeness, or fail in their pretense, and need no critical comment to be appreciated for what they have achieved or have failed to achieve. In criticism there is no privileged point of view – not even the author's own. He has consulted his knowledge and experience to make the work, and whoever would criticize his efforts would seem to owe him the indulgence of doing the same. If communication is mediated through the works, the author and his readers respond in reciprocal fashion to the expressiveness of their contexts.

For the philosopher of art, the challenge is extremely tempting – on a manifold count. Can the critic of criticism fulfill in practice what he demands of others in theory? Can he make sense of the non-sense embodied within an author's works? It all depends upon one's understanding of the words, both the creative and the critical ones. The sense_2 one seeks is the meaning of the author's words, themselves presented in sensuous_1 gestalt. Sense_1, then, is not equivalent to sense_2, even though in its patterns it controls what is understood as sense_2. Thus, in seeking out the sense of the words, one looks for the non-sense – the meanings and the meanings of the meanings they may intend.

Beckett's own criticism of the works of others is a handy procedural guide. Words suggest meanings in the images they create. The words are themselves perceived as real images (objects of sense), and their power is to suggest the images of a non-existent reality (the objects of imagination). But in their relatedness these images create within the imagination the picture of a structured, albeit unreal, world. Such a world may be like or unlike the one we all live in, depending upon its structure and the supposed structure of the real world – but whether or no, all the words spoken by the creative voice of the narrator betoken necessary lies that constitute the truth of his own creative world.

Descriptions of the structure of the real world are found in the theories of philosophers: 'metaphysical,' they are called; 'cosmological,' when they intend the synthetic wholeness of that world, or 'ontological,' when they intend the analytical elements of the universe described. For example, 'organic idealism' is a term describing the cosmological pattern of the real world when it is thought to be composed of ideas – percepts, concepts, or volitions – as ontological elements, and when reason has been subverted by a biological metaphor.

My criticisms are not for this reason to be taken as 'organic' or 'formal.' In the abstract, when critical reason grasps its subject whole, literary works are seen to be composed of words and their meanings, along with the relations they subtend, including the meanings of those meanings, should any be justified by the context in question. The tension between the words as perceived and the meanings as imagined and understood constitutes the expressiveness of the context. That is a fact of the critical consciousness.

The method used in the following analyses of Beckett's creative works may therefore be called a 'phenomenological structuralism.' The critic has focused his attention as closely as possible upon the context of the words (so that the work may appear as a phenomenon to his consciousness), and the structure of the meaning relations has been given as close a description as the state of his own perception, imagination and knowledge makes possible. Before such conscious attention, the patterns develop, between the words as signs relating to images as signified; and these in turn may become signs referring to still higher "signifieds," in levels of signification that offer a clue to the interpretation of the overall structure of the represented universe. One stops when the context seems clear. The clearness of the context should likewise be a fact of the critical consciousness.

Where necessary, hypotheses are formed, but these are always verified in a test of their fittingness against the perceptual and imaginative context of the words and images. The elaboration of such hypotheses is referred to as the 'hermeneutical expansion' of the sense of the words. The words as perceived are referred to as an 'aesthetic surface,' and they intend a field of "experiential depth significations." In reading from surface to depth the reader constitutes the significance of the stratified text, and the first reader of any text is its author.

Beckett is, moreover, a philosophical writer, and self-consciously such.

Not that his works argue for the truth of a cosmological or ontological point of view – that would make him a philosopher and his narrators, tellers of the truth instead of lies. Only absurdity lies in that direction. As a philosophical writer, Beckett uses the structures of a philosophical idea merely as a convenience, a device, to lend its relatedness to the field of experiential depth significations; that is what makes him a writer, just as in his own estimation the same technical literary device made philosophical writers of James Joyce and Marcel Proust.

The second challenge of this book, then, is to test the degree to which the philosophical structures of Beckett's works have changed throughout a representative sampling of his *oeuvre*. The central theme of the works is the disclosure of the nature of consciousness seeking the identity of full self-consciousness, either by acting or by reflecting on its actions. Taken as a whole, on the whole, they present a massive investigation of what it means to be human – what it means and how it feels to live a life one has never asked to be responsible for, and in living that life to assume the responsibility of disclosing the truth of the world in which one is responsible for living as one has. Existence exacts the fullest measure of personal responsibility, and a happy man is only an unhappy one, unaware of his basic existential condition.

For this reason, Beckett has often been called an "existentialist" writer. If that judgment is correct, and Beckett is self-conscious enough in his own existence not to vary from the principle he has found in the philosophical writings of others, existential ontology should lend the structure of its theory to the texture of Beckett's own works. In what follows I have tried to measure the extent to which this supposition is true.

It matters little whether Beckett accepts the "truth" of this philosophy or not; it would suffice, for my purposes, that his works exhibit the structures of the philosophical ideas in question. If he is unaware of the underlying philosophical ideas, the kindest thing to say is that he has discovered them in his own experiences and embodied them in the context of his works. From the point of view expressed in the philosophy, the same process applies to his own "disclosedness," his own openness to the world, as applies to his narrators and their respective worlds. States of mind must be brought to the full understanding of recorded speech.

Yet one should not be misled by the vaunted "pessimism" of the works. If that pessimism were Beckett's own, he might have come by it either empirically or theoretically; but however come by, it is attributable only to the works in which it is expressed. And the surface, ironic humor of its expression goes a long way toward mitigating its sting. Had he chosen the patterns of existential authenticity, instead of the prevalent inauthenticity of human consciousness, there would have been no pessimism, and, at the same time, no profundity of human feeling, no saving grace in the humor of its expression. In what follows, Beckett's "pessimism" shall be viewed only as a functioning counter within a total context of aesthetic significance. As such, it will supply the basis for a contemporary theory of tragedy, in which the value of the negative images becomes transmuted, through Beckett's artistry, into a positive affirmation of human dignity. Our knowledge of our own suffering is the source of all human dignity, not our capacity, through technology or ignorance, to avoid it.

The philosophical works most alluded to in this work are those of Hegel, Heidegger, Sartre and Merleau-Ponty. Should the names of Descartes, Geulincx, Malebranche, Berkeley, Schopenhauer, and others seem to stand out for their absence, it should not be forgotten that Hegel's *Phenomenology of Mind* traces the development of mind's evolution, and leaves ample place for the dualism of the rationalists and the idealism, subjective and voluntaristic, of the others.

The citations in the book reflect the conditions under which the manuscript was composed. Isolated in a corner of France, without library, and with only a shelf of Beckett's own works for my guide, I was forced by circumstances to adhere strictly to the demands of the phenomenological method: the books contained the reality to be described, and the philosophical idea is presented in the criticism much as it would occur to the reader of the text as the occasion arose. Documentation was prepared at a later date, when I had access to my own full library, along with that of the Florida State University, in Tallahassee.

The method used throughout the book involves at least two readings of each of the works: one for the formulation of the structural hypothesis, and another for its verification. I was able to perform the first readings of the majority of Beckett's works thanks to a grant of summer salary by the Florida State University Committee on Faculty Research Support awarded me in 1969. It was then I formed the hypothetical outline for the present work's structure. Then, in the Winter and Spring of 1974, I was awarded a second grant, this one by the Committee on Faculty Development Leaves, to continue my attempts to define in practice what I had learned from the theoretical study of phenomenological analysis. The time had come to accept the challenges offered by Beckett's work.

The first quarter of my second grant was spent in revamping the overall structure of the work – via the second readings; and the second, in preparing the manuscript, after a third, and sometimes fourth reading of the more difficult texts. For the plays I used a trilingual German edition published by Suhrkamp in 1963–64, and for each of Beckett's translations of his own works – from French to English and from English to French, I have availed myself of both versions. That is a problem in itself for any coherent criticism of Beckett.

I should like to express my thanks to my colleagues of Florida State University whose confidence in my ability to undertake the work successfully was expressed in awarding me the two grants mentioned above; and to my former colleagues of the Department of Philosophy at the University of Wisconsin, Madison, who in 1964 granted me leave with pay to begin my research in the philosophy of Martin Heidegger at the Albert-Ludwigs Universität in Freiburg-im-Breisgau. Without these three grants, this work would never have been brought to completion. Lastly, I am indebted to the encouragement of Professor Anna-Teresa Tymieniecka of the World Institute for Advanced Phenomenological Research and Learning for her encouragement to bring this work to publication. We share a profound respect for the pioneering work of Roman Ingarden in the phenomenological analysis of "the literary work of art." Readers familiar with Ingarden's work will perceive the added indebtedness. The method employed herein is concerned with the phenomenological problem of constitution; the results of the method display only one concretization – my own readings – of the individual works considered.

Tallahassee, Florida E. F. K.

ACKNOWLEDGEMENTS

To Grove Press, Inc., of New York, holder of the English language rights outside the British Commonwealth, and to Faber and Faber, Ltd., and John Calder (Publishers), Ltd., London, who retain the English rights to the works of Samuel Beckett within the Commonwealth, the author is pleased to express his thanks for their permission to cite and to paraphrase sections of Beckett's poetic, prose fictional, and dramatic works. The credits for the copyrighted works are as follows.

'Dante . . . Bruno. Vico . . Joyce,' from *Our Exagmination Round His Factification for Incamination.* First published in 1929 by Shakespeare and Co., Paris, and by Faber and Faber, Ltd. Copyright © 1929, 1972, by Faber and Faber, Ltd. Extracts reprinted by permission of Faber and Faber, Ltd.

Proust. Extracts reprinted by permission of Grove Press, Inc. All rights reserved. First published 1931. With concurrent permission of John Calder (Publishers), Ltd.

Whoroscope, from *Poems in English*. Extracts reprinted by permission of Grove Press, Inc. Copyright © 1961 by Samuel Beckett. With concurrent permission of John Calder (Publishers), Ltd.

Murphy. Extracts reprinted by permission of Grove Press, Inc. All rights reserved. First published 1938. With concurrent permission of John Calder (Publishers), Ltd.

Watt. Extracts reprinted by permission of Grove Press, Inc. All rights reserved. First published 1953. With concurrent permission of John Calder (Publishers), Inc.

Three Novels. Extracts reprinted by permission of Grove Press, Inc. Copyright © 1955, 1956, 1958 by Grove Press, Inc. With concurrent permission of John Calder (Publishers), Inc.

No's Knife, Collected Shorter Prose, 1945–1966. First published in Great Britain 1967. *Stories*, *Texts for Nothing*, *From an Abandoned Work*, and *Residua* (*Enough*, *Imagination Dead Imagine*, and *Ping*). Copyright © 1954, 1958, 1965, 1966, 1967 by Samuel Beckett. Translation of the *Stories*, in collaboration with the author, copyrighted by Richard Seaver, all other translations copyrighted by Samuel Beckett. *Stories, Texts for Nothing*, *Enough*, *Imagination Dead Imagine*, and *Ping* originally published by Les

Éditions de Minuit, Paris. *From an Abandoned Work* originally published by Faber and Faber, Ltd. Extracts of the translations cited by permission of John Calder (Publishers), Ltd.

Stories and *Texts for Nothing*. Extracts reprinted by permission of Grove Press, Inc. Copyright © 1967 by Samuel Beckett.

First Love and Other Shorts. Extracts reprinted by permission of Grove Press, Inc. Copyright © 1974, this collection by Grove Press, Inc. All rights reserved.

How It Is. Extracts reprinted by permission of Grove Press, Inc. Copyright © 1964 by Grove Press, Inc. With concurrent permission of John Calder (Publishers), Ltd.

Waiting for Godot. Extracts reprinted by permission of Grove Press, Inc. Copyright © 1954 by Grove Press, Inc. With concurrent permission of Faber and Faber, Ltd.

Endgame. Extracts reprinted by permission of Grove Press, Inc. Copyright © 1958 by Grove Press, Inc. With concurrent permission of Faber and Faber, Ltd.

All that Fall. Extracts reprinted by permission of Grove Press, Inc. Copyright © 1957 by Samuel Beckett. With concurrent permission of Faber and Faber, Ltd.

Embers. Extracts reprinted with permission of Grove Press, Inc. Copyright © 1958, 1959, 1960 by Grove Press, Inc. With concurrent permission of Faber and Faber, Ltd.

Words and Music. Extracts reprinted by permission of Grove Press, Inc. Copyright © 1962 by Samuel Beckett. With concurrent permission of Faber and Faber, Ltd.

Krapp's Last Tape. Extracts reprinted by permission of Grove Press, Inc. Copyright © 1957 by Samuel Beckett. With concurrent permission of Faber and Faber, Ltd.

Happy Days. Extracts reprinted by permission of Grove Press, Inc. Copyright © 1961 by Grove Press, Inc. With concurrent permission of Faber and Faber, Ltd.

Play. Extracts reprinted by permission of Grove Press, Inc. Copyright © 1964 by Samuel Beckett. With concurrent permission of Faber and Faber, Ltd.

Act without Words I. Paraphrase of action reprinted by permission of Grove Press, Inc. Copyright © 1958 by Grove Press, Inc. With concurrent permission of Faber and Faber, Ltd.

Act without Words II. Paraphrase of action reprinted by permission of Grove Press, Inc. Copyright © 1960 by Grove Press, Inc. With concurrent permission of Faber and Faber, Ltd.

Come and Go. Paraphrase of action reprinted by permission of Grove Press, Inc. Copyright © 1968 by Samuel Beckett. With concurrent permission of Faber and Faber, Ltd.

Eh Joe. Paraphrase of action reprinted by permission of Grove Press, Inc. Copyright © 1967 by Samuel Beckett. With concurrent permission of Faber and Faber, Ltd.

E. F. K.

CHAPTER I

INTRODUCTION: THE BABEL OF CRITICISM

A philosophical novelist or dramatist poses a special problem to literary criticism. Can the critic be accurate enough from the philosophical point of view and still remain faithful in his description of the artistic achievements of his author? The answer, of course, is, "That all depends upon the author and the critic."

Even before his Nobel laureateship of 1969, critical commentaries on the fiction and drama of Samuel Beckett were as common as the academic dissertations on the same topics. At age sixty, he was presented with a Festschrift, edited by John Calder,[1] containing essays by, among others, John Fletcher, Martin Esslin, Hugh Kenner, Harold Pinter, and Mary Hutchinson. Martin Esslin himself assembled a collection of critical essays entitled simply *Samuel Beckett*;[2] among his contributors were John Fletcher, Hugh Kenner, Alain Robbe-Grillet, and Ruby Cohn.

Ms. Cohn in her essay, 'Philosophical Fragments in the Works of Samuel Beckett,'[3] found traces of Descartes, Geulincx, Wittgenstein, Heidegger and Sartre, but none of Schopenhauer, perhaps because the world is not clearly anyone's idea unless that one be a writer of fiction, or because the primeval metaphysical Will of Schopenhauer creates only Ideas in the essential mode. Her own major work was the monumental *Samuel Beckett: The Comic Gamut,*[4] an explanation of Beckett's multiform mastery of that particular solution to the problem of pain, the cornerstone of Schopenhauer's ethics. Who knows what she might have found had she been looking for something other than "philosophical fragments" in her shorter critical article? Her fragmentary interests led her to emphasize the comic elements of Beckett's work at the expense of the tragic.

Another collecton of critical essays, likewise entitled *Samuel Beckett,*[5] appeared under the editorship of Melvin J. Friedman, who had assembled a new stable of contributors. Amongst them were Frederick J. Hoffman, Edith Kern, Germaine Brée, Robert Champigny, and Lawrence E. Harvey.

Professor Hoffman, for his essay, chose to abridge the theme of his earlier *Samuel Beckett: The Language of Self,*[6] viz. that within the twentieth century novelists have found epistemology more fruitful than metaphysics as a philosophical foundation for their fictional worlds. In the book he had written,

This is what I mean when I say that the philosophical ground of twentieth century literature has shifted from metaphysics to epistemology. Characters who were formerly maneuvered within an accepted frame of extraliterary reference are now presented as seeking their own definitions and their own language.[7]

There is nothing in the later Beckett, it is true, of any kind of extraliterary reference, but the shift referred to is not one from metaphysics to epistemology so much as from cosmology to ontology in the existential mode. Here there was perhaps too much implicit Schopenhauer, and not enough of the Heidegger Ms. Cohn had discerned.

Ms. Kern, on the other hand, went back centuries to find a striking similarity in the symbolism of the "contents of pockets";[8] it made no difference that the characters rummaging in their pockets were created by Jonathan Swift or Samuel Beckett. But the existential question is whether a man may be defined by what he possesses; is he a being, or a having? And in this Beckett differs from Swift.

Like Professor Hoffman, Ms. Brée looked where she should have been looking, at the whole *oeuvre* of Beckett himself, and recognized "L'étrange monde des 'grands articulés' de Beckett."[9] To call a world "strange," of course, is not to call it "absurd" – far too easy a gambit since Esslin's creation of that essentially critical fiction: all one has to do is to perceive that the created fictional world does not quite jibe with the one we all live in within the universal extraliterary frame of reference – but does suggest the uncanniness of the experience of consciousness turned inward upon itself. Here, too, the strong hint of a Heideggerian hermeneutical expansion goes unnoticed.

Discoursing upon "Les adventures de la première personne,"[10] however, Robert Champigny finds a base for the uncanniness of an inverted consciousness in the philosophy of René Descartes: starting from an initial skepticism, Beckett's characters are viewed as driven to a complete self-knowledge that is frustrated by the very thing to be known, precisely what was named by Professor Hoffman in "L'insaisissable Moi: Les M de Beckett."[11] But such a philosophical base produces a curious predicament for the philosopher whose "soul" was supposedly easier to know than its body since it was the certain measure of anything else that could be known; provided, of course, it was not forgotten in the continual recreation of God, in which case it would no longer be; or, not open to the evil designs of some malign genii, in which case it would most certainly fall into substantial error.

However that may be, Professor Harvey provides the overall context in which all the above philosophical theory assumes its characteristically Beckettian shape when he expatiates upon the initiation of the poet.[12] The majority of Beckett's heroes or antiheroes are in the throes of telling how it is, of showing in their halting attempts how difficult it is to say what we mean at the same time we mean more than what we say. For this same reason, Ms. Kern has, in a later book,[13] bracketed Beckett with Kierkegaard and Sartre as practitioners of the *Künstlerroman*.

Perhaps no collection of essays can do anything more than assemble various views of interpretation; but it is reasonable to expect, and we shall be forgiven our expectation and hope, that they all be taken of the same thing. In any collection each contributor gives his insight according to his own light, however bright or dim it may be. One can only describe what one has discovered. Some of these descriptions are pointed and penetrating; others are blunt and damage their subject; and no one, not even the perpetrators themselves, expect the several views to fund into any kind of unity. But the works of Samuel Beckett exhibit a very high degree of cohesion and unity. The critic's task is to discover it, and then describe it – with a great deal of caution.

When G. D. Barnard finds a new key to unlock the secrets of Beckett, we should therefore be all the more wary. His title, *Samuel Beckett: A New Approach,*[14] is promising, holding out as it does a hint of novelty and methodical precision; but in the end it, too, disappoints since it covers only one aspect of Beckett's fictional worlds. Commenting on what he takes to be inadequate in previous critical appraisals of Beckett, he states,

> ... they fail to deal adequately with the psychological and purely human aspects, and notably they entirely ignore the part played by schizophrenia in the novels.[15]

Indeed, he insists, "Without an appreciation of this one cannot fully understand the plays or the novels."[16]

Barnard is cautious enough, since his condition is stated only as necessary, and not sufficient, but what other conditions may be necessary for our knowledge to become sufficient remains totally obscure. We cannot diagnose a novel's condition, even if we can that of a patient, author or character; syndromes and recognizable patterns from an extraliterary frame of reference simply will not do as a substitute for critical appreciation, even when they are found to fit into some small niche of the literary frame, unless they throw some light on the structure of the works in question.

The schizophrenia of Beckett's characters, in fact, portrays the human content of a philosophical idea, that in searching out one's self-identity one becomes simultaneously subject and object of the search. One need only read Hegel on the phenomenology of mind to understand the structure of this process. What is needed is a critical literary method, such as the one developed by Beckett himself,[17] that is capable of showing the relationship between the words an author employs and the various levels of meaning they intend; for it is by their intention in the first place that the universe we call the literary work of art is created. Up to now our critics have limited themselves to describing archetypal patterns.

In this vein, two longer critical studies of Beckett's prose are worthy of particular notice. John Fletcher's *The Novels of Samuel Beckett,*[18] concerns Beckett's development of character that spans the several novels. Beginning with *More Pricks than Kicks*[19] Fletcher discovers the hero as "citizen," one Belacqua Shua; and *Murphy* is described as exemplifying the same motif. But Watt, the heroes of the *Stories,*[20] Mercier and Camier (the pseudo-couple, as they are identified in *The Unnamable*, from an early novel which was first published in 1970), Molloy, Moran and Malone are all viewed as heroes in the outcast's role. Finally, in *The Unnamable* itself, character has become entirely a voice, or congeries of voices; and these voices continue to be heard until *Ping,*[21] where sound finally gives way to silence.

Such a classification is useful since it tends to identify trends, but is open to Barnard's charge that it fails to capture the essential schizoid nature of most of Beckett's heroes from Murphy on, and in particular the division of Molloy's personality into three distinct voices. When one searches for a self-identity long enough one becomes aware of the fact that the searcher and the searched, being the same, but speaking in different voices, are both attuned to a third voice that summons the first to action and the second to reflection. That is Molloy. So what is there to be named?

Even with this recognition of the unnamability of that speaking voice which creates its world as it speaks, we should not be astonished to learn that it will go on to tell us how it is, even if its owner is forced to scratch his signals on the buttocks of the first character it finds crawling in the mud beside himself, 'Pim' or 'Bim' as you prefer to name him. Words are aimed at the intellect and may go astray; the direct anal etching may be more successful in hitting its mark.

Citizen, outcast, voice – that's the sequence of roles proper to "the seedy solipsist" for whom the world is but a personal idea. What Fletcher gives us, then, is a description of the basic continuity of the characters making their appearance in the successive fictional worlds that were indeed the personal idea of Samuel Beckett. The M's do recur, and they are unseizable, indeed unnamable, since they portray no essences to be grasped intellectually outside of their own limited worlds.[22]

The second study is Hugh Kenner's, entitled *Samuel Beckett: A Critical Study.*[23] Again progressing in sequence we may read "from *Murphy* to *The Unnamable*" as "the man in the room," "the rational domain," "the Cartesian centaur," "life in the box," and "voices in the dark." But here we learn that for Beckett setting is as important as character, that each tends to define the other, as Molloy himself notes that a region is not something to be journeyed through so much as what one already is. And should we be timid enough to ask what Beckett himself thinks about all this, we are told in Kenner's preface that Beckett knows no more about "his people" than what he has read in the books themselves: everyone is his own best critic, provided he has learned to read.

As for critical interpretations of particular works, they range from Robert Harrison's *Samuel Beckett's Murphy,*[24] an academic (eclectic) exercise in traditional analysis making use of such critical concepts as "form," "character," "symbol and motif," and "style," to Stanley Cavell's "Ending the Waiting Game," in a book of philosophical essays entitled *Must We Mean What We Say*?[25] Here we find the *Endgame* described as the final solution to the problem of *Waiting for Godot*; Hamm is Noah's son, and the setting is the arc coming to rest! Beckett's vaunted comic genius is displayed in this play, according to Cavell, as the "humor of literalness"; the characters are thought to be comic to the extent that they are absurdly unable to understand what is left unsaid in each other's discourse.

Obviously we do not always mean what we say, and then just as obviously we do not always say what we mean. Why should the dramatic artist speak in any other way? According to Cavell, Beckett's play gives a quiet, unintentional revelation of – an essence, the object of a reference to something other than what is directly said, in the sudden occurrence of what Beckett's mentor, James Joyce, always referred to as an epiphany: a revelation of character, a sense, something to be understood.

In collating only the above, of all the available literature, we find ourselves in an anomalous situation. There is one group of commentators with the opinion that Beckett offers nothing to the understanding but confusion, absurdity, or, what is perhaps the same, an infinity or quasi-infinity of layered insights into the human predicament; and another, with the incompatible view that his works give a precise understanding of implicit truths gleaned from an astute observation of that same predicament, but couched in terms that their author has happened upon, as if by trial and error, since that is the way we all use language anyway. No survey of secondary literature can yield anything else, unless there were prior agreement upon the method used to establish the truth claims made in the critical literature.

But, the truth is, our literary and critical specialists are still being trained in the least methodical manner possible. Either one looks for a "pattern" intuitively recognized; compares it to another found no matter where (e.g. the pocket contents of Gulliver with those of Molloy and Malone); and finds in this gratuitous comparison a unique significance. Shades of the collective unconscious! Or, one compares the structures of the real world with those found in the fiction and finds absurdity: a new significance in the irrational! Between the rational and the irrational only the rational man can decide, and he needs a method to make his distinction clear. Without some notion of the rules of evidence the "rational" mind is as caught up in his own world as the irrational characters of Beckett. Just as any rational man would, they too react to the awkward significances of their apparent worlds. And what would be the case, if, as Beckett's heroes maintain, they do not know what is true either about themselves or their worlds until they have succeeded in expressing their ideas in words? Surely we should be faced with a non-rational or pre-rational world of human experience where feeling, understanding and expression are not yet clearly delineated. To be human here is to create, and to fail in the act of creation is less damaging to our dignity than failing to accept the challenge.

For the writer's world there is no better authority than that found in the Johannine dictum, that in the beginning was the word; and fictional worlds are created by an assemblage of such words. That, too, was revealed by the beloved John's further context: and the Word was with God, and the Word was God. But who was God, but the creator of all things whose voice shaped the contours of that world in which we all live, and move, and have our being? 'Samuel' and 'Lemuel' are various Hebraic references to such a God; his name is Sam, or Lem!

It would be too easy here to suppose that we can know something of the act of fictional creation by what our coal miner's faith tells us of the divine; in truth, our basic religious faith is informed by what we should know of fictional creation, but which we are wont to express, in an act of primeval meta-criticism only in allegory, thereby illuminating the indirect by something equally indirect. But there is another way of handling this dilemma, i.e., by tackling it head on and illustrating the fact by a creative act of our own.

Such was the tack of Beckett, but it was not an immediate, intuitive decision. He too went through a period of apprenticeship, both as critic and as author with at first only a partial success; and he, too, learned from the success of others, primarily Dante, Joyce and Proust. As a student he had imbibed a thorough draught of philosophy and linguistics and has since never lost his awe at the power of the word or the eloquence of silence, where, according to Heidegger, conscience alone can speak. The Unnamable complains about Sam's auctorial intrusions in *More Pricks, Murphy*, and *Watt*; and Mercier and Camier, his schizoidal pseudo-couple, never made it to the press room until 1970.

For us, critics become self-conscious in our turn, there can only be one solution: total respect for the universe created by the author's words, beginning with their surface, i.e. their poetical chant, and proceeding along to what ever depth significance they may convey, and in as many layers as are open to clear exposition. The more such layers we find, the more dense the author's prose will prove to be; and the more dense the prose, the more forceful its tension. That is the way its expressiveness is felt. The very compression of Beckett's literary prose, we can be sure, is the source of the multifarious interpretations it has received in recent and contemporary criticism.

What is needed, therefore, is a critical method to decompress that prose, one that does not depend upon isolated pattern recognition or comparison with anything taken from another frame of reference, be it literary or extraliterary. To meet this methodological prerequisite, I shall propose a system of analysis that might be called 'structural phenomenology' or 'phenomenological structuralism.'

In short, the method will consist in "bracketing" out of relevance all reference to any theories, any previously established knowledge of the real world. What remains before the critical consciousness is the universe of the novel created as a unique frame of reference. This is the famous or infamous "phenomenological reduction" as found in the philosophical methodology of Edmund Husserl.[26] It implies a suspension of our beliefs in the existence of a natural world – the one we all live in, as described by the rational procedures of positive scientists and nomological theorists – in favor of an attitude of serious contemplation of the appearing phenomena, which then may be described as they appear to our attentive consciousness. The individual work of art is thus a structure of noematic elements constituted by the noetic acts of the reader.

With this reduction of the reader's attention to the noematic strata of an individual work, the judgments of rationality and irrationality, of sense and nonsense, of meaning and the absurd are cast aside as irrelevant. Where one novel makes a reference within its own context to that of another, that other becomes an integral part of the universe under observation, but only as intended within the referential context of the first. A literary work of art is an intentional object; its universe is significant only by means of its internal relations.

The fictional universe was, of course, created by the writer; he may or may not be aware of its significance; but if he is, his attitude of appreciation, as Kenner has suggested, is the same as any other attentive reader's; he too must read the words to discover the structures of the universe he has created. In this sense, an author may very well have created by discovering what he meant in the first instance to say; Molloy, the narrator, needs both Molloy the reader and Moran, his surrogate. By the dialectic of phenomenological experience the self is discovered in the other.

The structuralistic aspect of the method is linguistically motivated. The universe of prose fiction is a linguistic creation. As such, it entails a surface and a depth phenomenon. The surface of prose is the sensuous element of a sign: its visual and aural appearance as a phonemic structure. Assonance, dissonance, weight, measure and number of the liquid labials and the gasping glottal stops /caesura/ all work toward the determination of the sensuous qualities of the perceived sign, which, in proper morphological sequence, will convey a "signified," a meaning either connotative or denotative, an idea or a thing. And the system of these signifieds may themselves become signs for other signifieds at a higher (or lower) level of significance. If they do, they may be treated as "symbols," whose function in the literary context is to increase the density and compression of its expression.

At some level, however, the literary context will come to closure in a total structure of relationships between signs and signifieds. The closure principle itself may be only felt by the reader, or it may be raised to the level of understanding (and this, without distinction between the "rationality" or the "irrationality" of the content exhibited in the signifieds). For this reason, it should come as no surprise that the multileveled structures of Beckett's prose have been described as both "absurd" and "ultrarational," "absurd" when the content alone is considered, and "ultrarational" when the phenomena of linguistic communication are considered in phenomenological reduction. We may not always mean what we say, but we all are equipped to respond to the meaning of what we have said; the context is our clue, and the phenomenological reduction of the critic (or of the writer in the first instance) establishes the context of what has been said.

Beckett has merely complicated the issue by writing about writers, and these writers have compounded the difficulty by reflecting upon their own predicaments. So much for the "poetic plight of Samuel Beckett," where 'poesis' refers to the act of fabricating a story, and 'Samuel Backett' is a metonomy for the stories fabricated by Samuel Beckett, in which, generally, the narrator is doubly unhappy: first, in the separation of his mind (reflective consciousness) from his body (the prereflective, active self composed of mud; that forked, biped creature resembling nothing so much as a malfunctioning bicycle, and reducible by illness and the decrepitude of age into total immobility). How does one pump a bicycle without a functioning pair of legs? As every active sportsman feels in his bones, the leg reflexes are the first to go. And when they do, there remains only the vainly rational reconstruction of an habitual volitional intent. The game of life is for the young; old men and women are reduced either to passive contemplation, or to the inward and reflective attempt to recapture their past, lost mobility.

The second source of unhappiness stems from the eroding effects of time,[27] even for the reflective consciousness. The activity of the past is as far away from the reflective present as the preterite future of that past is from any given moment of the narrated incidents, which, if our grammar is to remain correct, occurred in some past anterior to the reflective consciousness's present. For the simple, even naive, past acting self, there was a birth and a constant anticipation of a death, two expulsions from place to place, from the comfort of the womb to the comfort of the grave, the one the mirror image of the other. In the meantime one is on the way, however one can proceed; but there comes a time when the bicycle balks, and there remains only the voice of the cyclist engaged upon a predetermined journey. What consciousness could remain happy surrounded by so much unconscious ignorance? Who am I? And what is my world? Only a view on the whole will give the answer to such questions, and no one still conscious enough to reflect on the matter is quite dead enough to give the correct answer. Beckett is a "philosophical" novelist and dramatist because his fiction and drama present this human existential predicament in its fullest, most concrete, symbolic form.

Our criticism will be philosophical in the sense of describing the structures of the worlds he has created. Even if the primary structure of those worlds were ethical, rather than cosmological, our task would still be ontological, and therefore metaphysical; for, as critics we intend nothing less than a description of the structures implicit in those fictional worlds, as strange or as uncanny as they both may be. For this reason, the "philosophical" statements of an author or of one of his characters may or may not be to the point; for, the philosophy of a novel is to be found only implicitly, within the structure of its events, one of which may be the dawning or the extinction of a consciousness.

If "The Unhappy Consciousness" rings a Hegelian bell for readers of *The Phenomenology of Mind*, and actually calls to mind the ultimate moment in the development of complete self-consciousness, so much the better; for consciousness to become supremely unhappy, Stoicism must give way to Skepticism, since what it knows or wants is the universal truth of its own free existence, and universals may be variously exemplified in the particulars of actual experience, about which first one, then another appearance seems to be the truth of the matter in question. Such indeed is the truth for the skeptic. Nor can self-consciousness lightly repudiate its consciousness of the other since one has become conscious of self only by virtue of recognizing the other. In its freedom and power the consciousness of the Master may lord it over that of the Slave; but without the work of his servants the lord is nothing (a Knott), and when the slave himself becomes aware of the lord's dependence upon himself, he too becomes free. Overcoming the dialectic of the master and the slave in the awareness of their own inner freedom, Marcus-Aurelius and Epictetus, the emperor and the slave, both became stoics. But Watt will be led to the madhouse.

The unhappy consciousness is skepticism maintaining its contradictions. That is why Beckett's skepticism is not Cartesian; Descartes was an optimist believing both in the immortality of the soul and its ultimate reunification with its Maker – another lord this, to be disarmed with the same dialectic of stoic renunciation. When both phases of the skeptical consciousness – one is and one is not this, that or the other – are united into a single affirmation, denying the true and affirming the false, it does nothing more than to affirm itself as a pure consciousness to which anything may appear as long as it is separated from its body (as in dreams, in imagination, or in death), experiencing itself as a drive to singular isolation from the domination of others (a condition satisfiable actually only by death), but ultimately conscious of itself as existing for itself in isolation from others.

Witness the "actions" of Beckett's dead: on one side of their double nature they seek solitude, only to be forced back to the other, to themselves in the past and hence as related to many others more, parents, friends, employers, charitable and civil institutions, submerged as it were in their desires and actions directed to the objects of this world, and, as such, inadequate to satisfy their own drives to a unique fulfillment.

Such is the culmination of the unhappy consciousness; and it must remain what it essentially is, striving for what it cannot obtain this side of its extinction by silence; and, consequently, it does remain doubly unhappy in that it can neither live nor die, telling us all along how it was and is. Living is the only way of dying, and dying is the end of the living. Dead or alone at last, it must talk to itself, but still using the words of others.

The unhappy consciousness may attempt to avoid the feeling of its own unhappiness by losing itself in practical desires and activities, but these measures can only exacerbate its essential unhappiness. As an essential condition, it must forever find itself between two self-contradictory drives: to attain to the unchanging status of a completed essence, and to do so submerged in a world of continual change, a phenomenon among other phenomena.[28] And in spite of its unhappiness, or perhaps because of it, it must persist.

In the meantime, persisting as it might, it can only serve as a symbol for Reason's highest development in the Absolute Mind, whose being and doing are properly and uniquely the synthesis of the many natural phenomena in the one constant law of Nature in whose knowledge uncertainty gives way to a certainty that comprehends all reality. For Hegel, only Reason can succeed in the projects of self-consciousness. Prior to the realization of its symbolic existence the unhappy consciousness must wait – for nothing that it may comprehend, since only the absolute mind possesses the concepts necessary to place the essentially relative and incomplete human expressions into meaningful relation to itself. Each relative being has its being in relation to another of like kind, and the whole of their system of relationships is the Absolute thinking itself. Indeed, the Absolute may encompass the relative because it thinks only itself; as such, it is the ideal of the limited, relative and solipsistic minds that must think of themselves in terms of something they are not.

Thus, having passed from the recognition of self in the other to stoicism, and from stoicism to skepticism, the purely relative mind of a man has become conscious that it is and it is not what it appears to be since it cannot avoid the attempt to maintain this unresolvable contradiction. Whether waiting passively (and telling themselves stories as a premeditated sedative to existence) or actively throwing themselves into feverish activity to escape knowing what they are only to have that revealed in a later act of reflection, the heroes of Beckett must suffer. But each of them suffers his condition in a way peculiar to himself. He is thrown from the outer world to the inner, where his suffering and his being are one.

Since they embody this philosophy, it is true that the individual characters of Beckett's novels and plays will seek their own definition and their own language to express that definition; not by discoursing upon that philosophy, but by living the condition it describes. And that definition, that language, must be discovered in the words and deeds of the characters themselves.

Their essential unhappiness (as fully self-conscious) cannot otherwise be understood.

But there are two ways of achieving this feat in a literary context: the narrator may take the position of the omniscient third person to describe how another character defines himself in action (*More Pricks, Murphy, Watt*), with all the attendant dangers of having the ventriloquist show himself in the process; or he may be a self-defining persona, the first person narrator reflecting upon his own predicament (*Stories*, the trilogy, *How It Is*). In his literary career Beckett has used both techniques, with mixed results. In making my judgment on the nature of these results, I shall lean almost exclusively on the compression of my author's prose. I shall attend to the phenomenal appearance of the universes he has created, relating them to the philosophical ideas or principles that give structure to their expression.

Hence, should I find a philosophical principle fitting the various ways Beckett's characters "define themselves" (a happy phrase that), it matters little if it were already enunciated in the abstract by a philosopher or by philosophers who were not the author of the events described. Nothing prevents a novelist from discovering for himself the truth of what a philosopher has already thought in the abstract; nor, it is hoped, for a philosopher to discover what the novelist has expressed in the concrete stipulation of his fictional universe.

Behind Joyce, we have Beckett's word, there was a Vico and a Bruno; and behind Beckett, the reader has mine, there was a Hegel and a Heidegger. And even if all these philosophers were essentially wrong, indeed wrongheaded, we human beings of this world still all perish, each of us alone, and each in our own way. Such is the experience that must be transformed throughout into a higher order, purely positive aesthetic value. Descartes' religious optimism must give way to a more rigorous demonstration; it cannot be founded on faith, but must be shown as a fact of human consciousness. The experience of a successful tragedy will allow us to contemplate how this inversion of values functions within our aesthetic experiences.

PART I

CHAPTER II

IN THE BEGINNING, THE WORD

But the word was with James Joyce. Beckett was twenty-three, already a published poet, but very brash, bent over as he was the translation of a section of his idol's *Work in Progress* (*Finnegan's Wake*):

> Now here am I, with my handful of abstractions, among which notably: a mountain, the coincidence of contraries, the inevitability of cyclic evolution, a system of Poetics, and the prospect of self-extension in the world of Mr. Joyce's *Work in Progress.*[1]

Even were the mountain real, rather than symbolic, it could be rightfully included in anyone's handful of abstractions, since what is called a "mountain" is the concept or essence unifying the multifarious perspectives that make their appearance to the consciousness that has bracketed one segment of the natural world: a rise on the horizon, a challenge or an obstacle, a climb to newer hights. The point to be registered is not that Beckett was already at that very early age a practicing phenomenologist, but that his thought was philosophical, and conditioned by the philosophy of the literary work he sought to translate. What is the relationship between philosophy and literature when philosophy is concerned with the nature of reality and our ways of knowing it, and literature remains a fine art?

The best way to go awry is to give the easiest answer and take philosophy as an abstraction that might be empirically illustrated alternatively in the real world of nature or in the fictional worlds of the literary artist. The philosophy of the literary artist, however, is not an abstraction that might be taken out of its context and judged to be true or false, and may not be able to be traced to a single dullard, whose thoughts run Germanstyle, "*tief.*" In this instance we know the latter, of course, because Beckett expatiates upon three original thinkers, none of whom is German; and his title says it all to the readers of those three gentlemen: 'Dante . . . Bruno. Vico . . Joyce.' As for the abstraction, let it be said here that the philosophy of a literary world is the very concreteness of its depth structures, the system of relatedness in meanings presented in the texture of its verbal surface.

"The danger," said Beckett,

> is in the neatness of the identification. The conception of Philosophy and Philology as a pair of nigger minstrels out of Teatro dei Piccoli is soothing, like the contemplation of a carefully folded ham-sandwich.[2]

One has to be hungry to find any pleasure in that. No doubt Beckett had something in mind like Professor Hoffman's "accepted frame of extraliterary reference,"[3] in which a clumsy author attempted to maneuver a character preformed in a creative imagination. The world and its character or characters must be created together, or forever remain apart; and when character is really created, there isn't much an author can do to force his will upon it. That notion too is an abstraction from reality, and the novel is nothing if not a concrete universe significant unto itself, i.e. an epistemological entity it pleased Vico to call "an individual."

Giovanni Batista Vico (1668–1744) is known, but only abstractly by those who may never have had the opportunity to read him, as the author of *The New Science* and *The Principles of the Philosophy of History*. The new science was indeed novel; much before Marx, the existence of individuals was explained by the force of historical circumstances: the essence is time, not a generalized conceptual idea, which could itself only develop in time through the human habit of making gestures with the vocal apparatus. Any human creature, such as Finnegan (albeit now dead), is an individual only in so far as he has lived in a social world, conditioned by the universal forces of history. But history is neither blind Fate, nor an ineluctable Chance.

The devout Italian saw Providence working on individual humans, but through such social institutions as the Church, marriage, and burial, each an institution made by and for humans. If this thought should shock devotees of the Petrine institution, let them not forget that Christ also said that his church was anywhere two Christians might be assembled in his name. It is the role of Providence to provide, and for humans to choose their destinies:

> This is not Bossuet's Providence, transcendental and miraculous, but immanent and the stuff itself of human life, working by natural means. Humanity is its work in itself. God acts upon her, but by means of her. Humanity is divine, but no man is divine. This social and historical classification is clearly adapted by Mr. Joyce as a structural convenience – or inconvenience.[4]

Thus Vico's philosophy of history (formulated a century before Hegel's, too) found its way into one of the classics of modern English letters, prophylactically.

A structural convenience for Joyce became, however, an intellectual inconvenience for the uninitiate, in particular Rebecca West, who had the temerity to write about her inability to perceive either the "content" or the "form" of Finnegan's world. Beckett was indignant:

> Here form *is* content, content *is* form. You complain that this stuff is not written in

English. It is not written at all. It is not to be read – or rather it is not only to be read. It is to be looked at and listened to. He is not writing about something: he is writing something.[5]

The logic of his indignation is clear; for, if form is content and content form, the so-called content-form distinction is inapplicable to a literary universe. It is, after all, only the Aristotelian ontology applied to the existence of a linguistic entity. In the literary arts, we have instead the distinction between linguistic surface, to be appreciated as much by the eyes and ears as by the mind, and a structured depth, where the structure is supplied, conveniently, by a philosophical idea.

That idea is, at best, a structural principle; and, without its relationship to a phonemic surface, that mirror through which we may sometimes perceive it, however darkly, it would remain an idea without a body – which is not to say it would compare favorably to a skeleton either; for surfaces are not added to depths, like muscle to bone and fat to muscle. Structure simply has nothing in common with an abstract or abstractable form:

By structural I do not only mean a bold outward division, a bare skeleton for the housing of material. I mean the endless substantial variations on these three beats, an interior intertwining of these three themes into a decoration of arabesques – decoration and more than decoration.[6]

Structure in its concreteness is the fusion of a system of stratified meanings into a single logos.

In search for a way to explicate what seemed simple to him, but which remained an enigma to Joyce's readers, Beckett proved himself, at age twenty-three, a phenomenological structuralist. He was forced to become one by the enormousness of his task, to translate what the literati of his time could neither hear nor understand.

It is a trite figure, of course, that refers to an aesthetic surface as mirroring an intellectual depth; and dangerous, too, for a warped mirror will show the universe askew. Was Finnegan's mirror too convex for a true picture of its world? No, Beckett found little reason for Joyce's ignorant reception:

On turning to the *Work in Progress* we find that the mirror is not so convex. Here is direct expression – pages and pages of it. And if you don't understand it, Ladies and Gentlemen, it is because you are too decadent to receive it. You are not satisfied unless form is so strictly divorced from content that you can comprehend the one almost without bothering to read the other.[7]

The distinction to be grasped here, however, is not that between form and content – old habits are hard to lose – but between surface and depth, between hearing and understanding the linguistic context. That was the poetics Joyce and Beckett had gleaned from Vico, now the philosopher of language.

Called a "scientific historian" by Beckett, Vico had presented a theory of society having developed in three stages: the theocratic, the heroic, and the human. Each stage is characterized by a distinctive function of language in human affairs. In the theological age, the written language was hieroglyphics and primarily sacred in its usages; in the heroic, linguistic expression was verbal, but metaphorical and poetical in its usages; and in the human, language became "philosophical," i.e., as the walking bicycle became more civilized in its reflexes, expressive of the abstract and general characteristics of human experience.

The poetics in such a scheme should be obvious. "Before articulation comes song; before abstract terms, metaphors."[8] Poetry, with its emphasis upon the sensuous surface of the linguistic vehicle, constitutes in this scheme the very antithesis of metaphysical abstraction. In epistemological terms, particular sensuous data are as different from the concepts used to refer to them as the real life deed is different from the pale cast of thought; but in genetic, and therefore historical terms, concepts, as a convenience for the inactivity of the intellectual, grow out of metaphorical resemblances, that sense or direction of the sensuous data toward closure which makes a metaphor an apt figure of speech (when it is). There is nothing materialistic in this process, as if the speaker possessed the materials for purely formal articulation; nor is it transcendental, or God-given. The process is rational, i.e., natural, an inevitable growth of the institution of meanings through historical time.

Historically, words have progressions and retrogressions as well as varying social uses. Beckett cites, as if to explain Joyce's love of multidirectional references, the progressive pattern of human living that evolved throughout Vico's three stages of culture: beginning in the forest man found his way to the cabin, the village, the city, the academy. The same progression may be perceived in the historical changes in the meaning of the Latin word "lex," which first referred to an acorn crop, and then, by metonymy, to the tree producing the acorns. The verbal form of the word, in the infinitive, "legere," meant to gather acorns; and the person performing the action was a gatherer. For this reason, an "aquilex" was a gatherer of waters, as the original word came to be used without reference to the specific objects gathered. With this separation allowed, a "lex" became a gathering together of peoples, as in public assemblies; or the principle by which these gatherings are organized, law. To understand the laws one had merely to behave in the prescribed way; and if not, one was forced to suffer the displeasure of one's peers. Since ignorance of the law is no excuse for one's breaking it, it was proper that laws be written; so it became necessary to read, and the new infinitive, "legere," meant to gather letters together into a word or to read – a process that still goes on in some literary academies.[9]

Real progress will have been achieved there when students are taught to respond to the fusion of surface and depth structures of linguistic utterances. As R. P. Blackmur put it so succinctly, "language is gesture,"[10] and literary works fairly beckon to our responses.

Joyce, claimed Beckett, made things easier, not more difficult. His aesthetic surfaces present in sensuous configuration the first level of their depth signification:

When the sense is sleep, the words go to sleep. . . . When the sense in [*sic*] dancing, the words dance. Take a passage at the end of Shaun's pastoral: 'To stirr up love's young fizz I tilt with this bridle's cup champagne, dimming douce from her peepair of hideseeks tight squeezed on my snowy-breasted and while my pearlies in their sparkling wisdom are nippling her bubblets I swear (and let you swear) by the bumper round of my poor old snaggletooth's solidbowel I ne'er will prove I'm untrue to your liking (theare!) so long as my hole looks. Down!' The language is drunk. The very words tilted and effervescent.[11]

This peculiar compression of the prose medium is what is meant by 'style' when that word is used by aestheticians to applaud a performer's gestures; and it would not be a difficult task to trace that word to the implement, or stylus, by which ancient hierophants carved their sacred messages.

Beckett called Joyce's style "purgatorial," borrowing a complex figure from Dante; but the compressed fusion of surface and depth is still Viconian:

This inner elemental vitality and corruption of expression imparts a furious restlessness to the form, which is admirably suited to the purgatorial aspect of the work. There is an endless verbal germination, maturation, putrefaction, the cyclic dynamism of the intermediate. This reduction of various expressive media to their primitive economic directness, and the fusion of these primal essences into an assimilated medium for the exteriorisation of thought, is pure Vico, and Vico applied to the problem of style.[12]

That's the description of an essence resulting from Beckett's contemplating the universe of *Finnegan's Wake*.

For a further exegesis of the "purgatorial" epithet, however, we must move on to Dante, the philosopher-poet "intoxicated with the number three."[13]

No wonder; he too needed a fusion between surface and depth. Hegel's vaunted dialectic is only a pale abstraction therefrom. But then Hegel was no poet, and wrote in that civilized age when humanity had already lost sight of its past linguistic achievements. But out of this purgatory; on to Dante.

Other than his preoccupation with the number three, Beckett notes Dante's enormous task of almost single-handedly "inventing" a vulgate Italian in the composition of *The Divine Comedy*. In France Descartes later shifted from Latin to the vernacular to reach a wider philosophical audience; and although he was a master of it, his language was there, already made. Although Beckett's "Mr. Joyce" cannot be said to have invented a new English, as Dante (together with Petrarch and Boccaccio) did in establishing a modern Italian for their disunited country, his *Work in Progress* did create an expressive vehicle suitable to the demands of Finnegan's world, composed on those three dominant beats: the Church, marriage, and burial. They were the institutions that formed a life; but a language was lacking to create the effect of those institutions in this fictional context. If these institutions are called 'divine' or 'ministerial,' it is no doubt for reasons of their human failure. But whatever, they turn the mind to immediate theological speculation.

Dante had theological reasons for the conical shape he gave the realm of souls expiating the sins committed during life; when the punishment was fulfilled, it had to be possible for the cleansed soul to escape to that higher realm of the Church triumphant, heaven, where its everlasting reward was the Beatific Vision itself. But Joyce structured Finnegan's new abode in circular shape; there was no escape here, for reasons entirely human. According to Beckett, Dante's purgatory is a mean between two static, and therefore lifeless, conditions, the one all virtue and the other all vice, between "unrelieved immaculation" on the one hand and "unrelieved viciousness" on the other.[14]

Purgatory, on the contrary, is full of constant vacillation, all life, " . . . a flood of movement and vitality released by a conjunction of these two elements."[15] And in Joyce's purgatory, when the scene is of drunks the words get drunk; when sleepers sleep, so do the words. Beckett himself puts the question, "In what sense, then, is Mr. Joyce's work Purgatorial?" And answers, "In the absolute absence of the Absolute."[16]

Had he replied with "in the absolute sense of the Absolute" the meaning would have been the same, for the Absolute is nothing if not the locus of all things relative. It all depends upon whether we continue to look for a literary form as an abstract frame for the content of a life or as the concrete structure of the human events determined by the institutions affecting the life of the individual. To arrive in Purgatory, however, the individual must first be dead. Like Mary's virginal motherhood, Dante's privilege to be there alive was a special dispensation, and one Beckett will readily grant his own deceased narrators.

Death and divinity are not separable concepts. That may be the only conclusion to draw from our experience of the Church as the fulfillment of man's eschatological needs (some would say "fears"), since it first unites parents, should their own love fail to do so, and then baptizes, confirms, shrives, marries, and stands by, always at the ready, to bury their children. Yet, for some, the Church holds out hope for a new beginning where there can be no further error.

Joyce, however, looked for a more theoretical explanation of man's eventual return to the fount of being. He found it in the philosophy of Bruno, one of the first martyrs of modern science.

Giordano Bruno (1548–1600) was burnt at the stake for his opposition to Aristotelian scholasticism and for his rather mystical notion that all things of the universe participated in the divine substance. Spinoza's "natura naturans" and "natura naturata" had their source in Bruno's vision, according to which all things sprang from God, as if from a fountain, only to return there for another spewing forth. In this process, because of the eternal return, anything may be identified with its opposite. Beckett's explanation of this system of absolute relativism is as follows,

> There is no difference, says Bruno, between the smallest possible chord and the smallest possible arc, no difference between the infinite circle and the straight line. The maxima and the minima of particular contraries are one and indifferent. Minimal heat equals minimal cold. Consequently, transmutations are circular. The principle (minimum) of one contrary takes its movement from the principle (maximum) of another. Therefore not only do the minima coincide with the minima, but the minima with the maxima in the succession of transmutations. Maximal speed is a state of rest. The maximum of corruption and the minimum of generation are identical: in principle, corruption is generation. And all things are ultimately identified with God, the universal monad, Monad of monads.[17]

Vico had adopted this view of successive transformations in his account of the generation and corruption of societies in history, and of signs within a language. Joyce, true to his dedication to the maximal compression of the word, applied it to both the surface and depth of his prose.

Thus Dante, Bruno, Vico come to rest in – Joyce. And Beckett found his first solution to the problem of how philosophy relates to literature.

His second scrutiny of the same problem was published two years later under the metonymical title, *Proust.*[18] He explains the metonymy in a foreword to the seventy-two page essay, by indicating his intention to consider only Proust's monument of prose fiction, *A la recherche du temps perdu*, which, because of the anticlimactic nature of the ultimate epiphany of *Le temps retrouvé*, he thinks improperly named.[19] English readers of Scott-Moncrieff's translation, by the way, are equally misled by the English title of the whole work, *Remembrance of Things Past*; not for its association with John Milton, but for its passivity. Although the "remembrance' captures one aspect of the novel's intent – the function of "unconscious" memory in Proust's view of the human psyche – it is entirely too passive to convey the active search of Proust's narrator to compose a work of fiction that would embody that psychological view.

Proust's first-person narrator is called 'Marcel,' like Proust himself, and his novel exhibits the structure of time revealed to Marcel in *Le temps retrouvé*, a coincidence that has led many to suppose an identity between the author and narrator and between the two novels. But Beckett wisely declined to chase that rabbit: the identification of author and narrator could give no insight into the structure of the work, where their identical view on time could. Here, once again, the essence is of time. If the author and narrator were indeed the same, the true beginning of Proust's novel would occur at the end, where the ultimate significance of time is revealed to the narrator: flashback, and on.

The connection between human consciousness and time as a means of measuring temporal events has been clear to philosophy since Augustine. In modern epistemology, time itself became subjectivized, as the succession of conscious events came to be interpreted as the genesis of its very concept. And more recently that concept has formed the cornerstone for the philosophical edifices of Hegel (and through Hegel, Marx), Bergson, and most recently, Heidegger. But Proust was no student of the *Phaenomenologie des Geists*, nor of *Matière et Mémoire*, and he died before the publication of *Sein und Zeit*. His theory of the connection between consciousness and time is more unique than derived, and thus poses a problem to criticism not apparent in Joyce.

While Joyce used the philosophical ideas he found already formulated by others, Proust formulated his own philosophical ideas as he expressed them within his fiction; but since Proust's narrator is a novelist worried about his own lack of productivity, those ideas are developed within the texture of the larger novel itself. What *A la recherche* shows, then, is the manner in which an author, in the person of the narrator, Marcel, comes by the philosophical idea that gives its structure to the novelistic depth. As we shall see later, Beckett makes use of the same device; and the two sets of ideas are not totally unrelated.

Whether the idea has been formulated before by a recognized philosopher or whether the novelist himself is the source of that idea, it is still a creative convenience. And whichever is the case, the task is the same for the phenomenological critic: to bracket the real word from the linguistic edifice attended to as a single context of significance, and then attempt to describe the way in which surface and depth are perceived to fuse into that significance.

Beckett began this task sensibly:

For the purposes of this synthesis (of Proust's 'perspectivism') it is convenient to adopt the *inner* chronology of the Proustian demonstration, and to examine in the first place that double-headed monster of damnation and salvation – Time.[20]

He begins this way to save some time. Although the hypothesis of the temporal structure of Proust's own novel has been clear from the beginning, his narrator is led to an understanding of it only at the end, in the last sections of *Le temps retrouvé*, when the idea of a book (similar in all respects to Proust's own) dawns upon him, as if by accident:

The scaffolding of his structure is revealed to the narrator in the library of the Princesse de Guermantes (one time Mme. Verdurin), and the nature of its materials in the matinee that follows. His book takes form in his mind.[21]

Thus, at the end of Proust's novel the narrator must begin his, in search of a time past that has already been recounted. Beckett begins with this idea as the "hermeneutical hypothesis" to illuminate both stories at the same time.

What does one find, when, as a novelist, one looks for a time that, naturalistically considered, has passed away and continues in the present to recede further into an ever enlarging past anterior? Only an account given by human consciousness that suffers the separation from its own past could possibly provide the answer. Whence the need of a narrator. But what does he know?

Subjects and objects in perspectival projection, that's what; a more or less well-connected series of impressions that may or may not correspond with the physical or psychological scientist's view of real things. Such a correspondence could only produce a "naturalistic" novel, like those of the infamous journalistic reporters, Jules and Edmond de Goncourt, whose *Journal* Marcel has just read prior to finding himself in the Guermantes' library. At the time of his epiphany, Proust's narrator had been musing on the style of the naturalists, just prior to taking down a copy of George Sand's idealistic *François le Champi*, when his past becomes mysteriously revived: all that past was merely a prelude to his salvation in art. (This last idea, borrowed from the aesthetics of Schopenhauer, is retained even by the narrator of Sartre's *Nausea*.) But the novel had to be written, as indeed it already had, by Proust in the very pages and various subdivisions of *A la recherche du temps perdu*.

Beckett's job was to explain the philosophical ideas that afforded Proust the elaborate structure of that monumental work. I shall take the liberty of reversing the order of his own demonstration, and begin with his description of the Proustian style.

Proust has been declared a romantic impressionist for his style bears out the judgment, here made by Beckett:

> The Proustian world is expressed metaphorically by the artisan because it is apprehended metaphorically by the artist: the indirect and comparative expression of indirect and comparative perception. The rhetorical equivalent of the Proustian real is the chain-figure of the metaphor.[22]

Human figures appear, for the most part, in botanical images; "And this human vegetation is purified in the transcendental apperception that can capture the Model, the Idea, the Thing in Itself."[23] For, if humans seem to be vegetal, they give the essence of the living thing as it appears to a willess contemplator, who merely records the impression.

Beckett explains,

> He (Proust) is almost exempt from the impurity of will. He deplores his lack of will until he understands that will, being utilitarian, a servant of intelligence and habit, is not a condition of the artistic experience. When the subject is exempt from will the object is exempt from causality (Time and Space taken together).[24]

Proust, of course, was infected with the Schopenhauerian virus that compels the view that the artistic contemplator be a pure, willess subject of knowing. His narrator finds reality (time without space) in the musical compositions of Vinteuil, where the Idea of Schopenhauer's primal metaphysical will is expressed immediately, without the aid, as in the literary and graphic arts, of concepts or representations.

But he might just as well have been working under the influence of Henri Bergson, whose epistemology gives further hints for the interpretation of the Proustian style.

Bergson's *Introduction to Metaphysics*[25] goes to some lengths to explain the difference between the reality of images and the pragmatic constructs of scientists. These latter, from the metaphysical point of view, are never strictly true. They are stated in terms of concepts which are at least twice removed from the reality they supposedly depict: first, by the distance of sign from referent; and secondly, by that of the generality of the concept to the particularity of the events referred to. So, while scientific statements are assuredly useful, like habit and memory, to allow adjustment to the practical conditions of life, they cannot be true.

Artistic statements, however, and metaphysical ones as well if Bergson is right, are true, but certainly not useful. The images of the artist allow a subject to surround its object in an aura of expressiveness, and permits an eventual fusion of subjective feeling and objective appearance in an act of representationless intuition. The artist and metaphysician, who are likewise constrained to use metaphorical images (Bergson's own prose is a case in point), are said to "know" by direct sympathetic intuition, where the 'direct' here refers to the lack of symbolic mediation in the activity of contemplation.

Whether Proust were inspired by Schopenhauer or by Bergson, however, Beckett recognizes the importance of intuition in the Proustian epistemology; indeed, he claims, "Proust is positive only in so far as he affirms the value of intuition."[26] A closer analysis might have revealed that Proust's philosophy resembles the epistemology of Bergson in the surface of his prose, and Schopenhauer's metaphysics in its depth.

That a direct process of knowing should be described as "indirect" by Beckett causes only a moment's reflection. A metaphor, as an implicit comparison, does not state directly what it means; but what it states, the attribution of the like between the unlike, is immediately and directly understood. The differences disappear before the appearance of a striking similarity that is expressed in paradox. That is the obliqueness of the philosophical author who would remain a literary artist; and Proust's images are direct presentations of this "enchantment of the senses" that occurs as a new phenomenon, linguistically controlled. "Correspondences" Baudelaire had called them, since in the literary image the things of experience are reduced to symbols by which the intellect becomes unified to the newly created object. Seated in the intellect, these correspondences are the poetic images in which the intellect and its object fuse and interpenetrate in a mystical, but adequate, union.

Beckett chooses to ignore the Swedenborgian idealism of Baudelaire's poetics, and limits himself to an intellectualist criticism of the manner in which the Baudelairian correspondence is constructed:

The Baudelairian unity is a unity 'post rem,' a unity abstracted from plurality. His 'correspondence' is determined by a concept, therefore strictly limited and exhausted by its own definition. Proust does not deal in concepts, he pursues the Idea, the concrete.[27]

But the concrete idea cannot be found outside the linguistic image itself: "For Proust the object may be a living symbol, but a symbol of itself. The symbolism of Baudelaire has become the autosymbolism of Proust."[28]

In referring in this way to themselves, as a "poetical" presentation, the images of Proust's prose continually enliven the surface of the writing, without merely prettifying the philosophical depth: "For Proust the quality of language is more important than any system of ethics or aesthetics. Indeed he makes no attempt to dissociate form from content. The one is the concretion of the other, the revelation of a world."[29] And should we replace the terms 'form' and 'content' in the above by 'surface' and 'depth,' we should have stated the Joycean definition of a literary style, linguistically conceived.

If the more or less tense relationship of linguistic surface and philosophical depth defines the unique significance of the novelistic context, it remains, for the case of Proust, to trace out the philosophical ideas expressed in that depth. I referred to them above as "subjects" and "objects."

Unfortunately, however, there can be no quick summary of Proust's non-naturalistic philosophical view. For if objects appear in various guises at different times even to a single subject, as impressionists (and phenomenologists) have always maintained, the same is true of another subject, such as Albertine, the object of Marcel's love.

We see her because he sees her as a member of a young band of cyclists, a floating nymph as it were, then in succession as a solidly appearing woman with the inflamed temple, a beauty spot on her chin, and finally as the slang-talking lesbian who could be true to no man's love. The narrator, of course, moved in, from the surface global appearance, to the central figure, to her own sensuous makeup; in, as it were, to the quick of her quickness. But she lies, not only in her appearance, but in speech as well. How can such a person be the object of a sane man's love? The tragedy of their relationship was that she couldn't, and the reason is simple:

The person of Albertine counts for nothing. She is not a motive, but a notion, as far removed from reality as the portrait of Odette by Elstir, which is the portrait not of the beloved but of the love that has deformed her, is removed from the real Odette.[30]

The real Albertine is known, and badly at that, only by Albertine herself, because her consciousness, too, undergoes the transformations of time, locked up in her own self before, during, and after her imprisonment in Marcel's rooms.

Readers of *Swann's Way* have recognized in the Marcel-Albertine love-jealousy identity the same pattern of events as those experienced by Swann and Odette – love as frustrated desire; but the alternative to having the experience is boredom – or ignorance. Why must the project of love always fail? Because, as Sartre has explained, almost in Proust's own words, to love is to wish to be loved, and no one has any effective control over the beloved. Beckett reverts to Proust's words for an explanation of the inevitability of the failure:

'We imagine that the object of our desire is a being that can be laid down before us, enclosed within a body. Alas! It is the extension of that being to all the points of space and time that it has occupied or will occupy. If we do not possess contact with such a place and with such an hour we do not possess that being. But we cannot touch all these points.'[31]

Our very will to be identified with the object of our desire gets in the way of the union; and since the object is spread out for others to appreciate, our love degenerates into jealousy.

Beckett will later depict the identification of a person with a region when his own narrator, Molloy, finds out, in search of himself, that his own reality has been regional, created by a stretch of his conscious temporality. But, more existential in conception, his solution of the literary problem of recapturing the past is strictly his own.

For Proust, the nub of the problem is the multiplication of the narrator's own personality, situated in the beginning of the novel between two "ways," the road leading to his neighbor, Swann, and that leading past the Guermantes' estate. This physical placement allows him to enter into the sociological spaces of the bourgeoisie and the aristocracy (true and false), whose members all undergo the same sort of temporal transmutation as himself. The reason: the nature of human consciousness, which substitutes the matrices of habit (a second nature) for the vibrant coloring of the present moment's awareness in which the phenomena of the present may appear as a flower, a friend, or a friend in guise of a flower.

Our habitual reactions, established in the past, represent an imposition of a past order, that of yesterday, on the tasks of today. Without this continuity of orders and the ability of the conscious memory to recall what a past consciousness had conceptualized, there would be no "character" as such. But the truth is, all our characters suffer the diremptions of successive moments:

"We are not merely more weary because of yesterday, we are other, no longer what we were before the calamity of yesterday."[32] Where then is to be found a significant future? Not even the existentialist's projection of a goal to be attained will guarantee the continuity of a human personality, without the resoluteness of a continual repetition of a past choice: "The aspirations of yesterday were valid for yesterday's ego, not for to-day's [*sic*]."[33]

This separation of yesterday's ego from today's produces two individuals, only one of whom is operative at any given moment. Although the attainment of a present goal would be celebrated by the future union of subject and object, with the passage of time

> The subject has died – and perhaps many times – on the way. For subject B to be disappointed by the banality of an object chosen by subject A is as illogical as to expect one's hunger to be dissipated by the spectacle of Uncle eating his dinner.[34]

All our conscious attempts to fill in the gaps suffer the results of the Bergsonian dilemma: if they work, they are not literally true. To get at the truth an individual must get past his second nature, suspend his habitual responses, and, in a word of Heidegger, let his world "world."

For the existentialist, of course, the individual's world is organized around a context of involvements, ultimately the assignment of value to things as useful to achieve other things (shades of Bergson's *Homo faber*!); and these ultimate things are themselves of value only in so far as they serve the interests of the developing transcendence of the individual.

But this solution of the existential problem was lacking to Proust; he could only appeal to the mechanism of "unconscious" memory, those precious moments when the old ego dies, hard: "Such as it was, a minister of dulness, it was also an agent of security."[35] But,

> When it ceases to perform that second function, when it is opposed by a phenomenon that it cannot reduce to the condition of a comfortable and familiar concept, when, in a word, it betrays its trust as a screen to spare its victim the spectacle of reality, it disappears, and the victim, now an ex-victim, for a moment free, is exposed to that reality . . . [36]

In these moments, the things of nature appear in their autosymbolic, self-referring, clarity that is nothing if not a mystical enchantment.

Thus, if our habitual and conscious memory has created a system of significations that in essence can only hide the Idea in a maze of concepts, the liberated subject finds himself locked into it, face to face:

> . . . when the object is perceived as particular and unique and not merely the member

of a family, when it appears independent of any general notion and detached from the sanity of a cause, isolated and inexplicable in the light of ignorance, then and only then may it be a source of enchantment.[37]

Not being consciously stored in a memory bank of prefabricated concepts, these moments, as Bergson had said, may only flower in images.

Beckett's *Proust* is itself an imaged account of the Proustian metaphysic. The critic expounds the breakdown of Marcel's habitual self; depicts the anguish of the loss of conscious security, the despair of the attempt to possess what cannot be possessed, the object of one's love; and lays out, as if in chronological order, the list of Marcel's encounters with the enchantment of reality from the most famous, the taste of a madeleine dipped in tea, to the last, his perusal of George Sand's *François le Champi.* All in all, it is a masterful piece of criticism that might have been entitled, had its author seen the connection between Bergson and Proust, *Schopenhauer . . . Bergson . . . Proust.*

But such a title would have emphasized a different facet of the critic's problem. He had chosen to divulge the structure of his author's novel, emphasizing that author's uniqueness.

I have read that criticism to extract its lesson, that the description of a novel's depth structure is also a description of a world, based in this case too on the structure of a philosophical idea. As long as one remains cautious and refuses to interpret the philosophy as the "meaning" of the prose, there can be no violation in criticism of the novel's own aesthetic intent. In the beginning was the word; the idea remains a device, a convenience, used in the discovery of phenomenal significance, here as in any linguistic context, measured by the tension between sensuous surface (the signs) and experiential depth (the various levels of signifieds).

Philosophically considered, Proust, who had described the structure of his narrator's phenomenal world, was already a phenomenological writer. And Beckett, having bracketed the context of Proust's prose and described the linguistic context of that world, did a masterful job of phenomenological criticism.

In both cases the test for success is the same; it may be put in the form of a question, Has our vision become clearer for the description? If so, like any scientist testing his hypothesis by judging it against the facts of nature, the author and critic must see the structural principle clearly embodied within the concreteness of the fictional universe.

As always, that universe is composed of signs and signifieds in relation. Its signs are apparent to our senses; and if its field of signification remains for any reason obscure to us, we need only form a hermeneutical hypothesis (one after another if need be) to order the most apparently divergent details of the phenomenal world it reveals.

Many times the hermeneutical device is already available, clearly expounded by some philosopher; and if not, since we are always confronted with a literary context, moving between sense and non-sense, there is always the critical concept of "the literary world." For better or for worse, that is a metaphysical notion, and structural phenomenology is only a critical tool for expositing the content of that notion as befitting the particular case.

Beckett has twice performed this critical task with an eminent degree of success. He had long been the sorcerer's apprentice; so it seems meet now to consider the results of his apprenticeship in sorcery, as he strives for the words that will create his own fictional worlds.

The true beginning is there, and the very first attempt to depict the ravages of time in his own works happens to be a poem the persona of which is none other than the philosopher René Descartes himself.

CHAPTER III

APPRENTICESHIP IN SORCERY

Since criticism is easier than art, it is less difficult to explain the tricks the master has learned than to perform them. The true master's presence is rarely visible on the face of the conjured world that is made real principally by the sounds of his creative voice. To learn these tricks, it should come as no surprise, the apprentice may stammer, stutter, or lisp; his words remain sounds, audible to the ear but intractable to the mind, transmitting static more often than announcing the miraculous birth of a world. But with time and practice and patient reflection on the still resources of the inner voice, the miracle becomes easier, and the voice will talk itself out until even its own stillness has found expression, in the silence surrounding that one last creative ping.

Beckett was a very bright young man who did not himself become a master of the craft until he had forged a personal style following the examples of his predecessors. It was not easy to follow Joyce, and Proust himself needed an alter ego to make a successful fictional world of his manifold of impressions. Beckett's mastery first becomes fully apparent in the trilogy, *Molloy, Malone Dies*, and *The Unnamable*, written in French and later translated into English by their author (*Molloy*, in collaboration with Patrick Bowles). In order to reveal the signs of his developing mastery, it is my purpose in this chapter to examine a selection of works from Beckett's early period, when he was still working toward the full command of his art and craft.

The first is his "Cartesian" poem, *Whoroscope*, itself an exercise in translation – of thought to image; and there will follow the three novels (*More Pricks than Kicks, Murphy,* and *Watt*), all of which, according to the Unnamable were " . . . clumsily done, [as] you could see the ventriloquist."[1]

Although the Unnamable's condemnation may be applied equally to *Mercier and Camier*, since its schizophrenic hero, that infamous "pseudo-couple,"[2] likewise makes its appearance to him, the novel was not published until 1970, and was therefore not generally available for criticism surpassing the obvious: that it was rejected for a time by Beckett, as well as by his own alter ego narrator.

Whoroscope is a poem of ninety-eight lines arranged in thirteen stanzas, and a series of seventeen reference notes attached to particular lines, following a three sentence general explanation of the poem. Without a reading of these notes, or the biography of Descartes from which the events of the poem were culled, the poem itself would remain white noise.

Descartes, it appears, would eat no egg that was not sat upon by a brooding hen from eight to ten days; otherwise, "stinking fresh," it could be given to Gillot, his valet, who was actually given the mathematical problems Descartes himself had neither the time nor the inclination to brood over. And as for the title, one of the general notes explains: "The shuttle of a ripening egg combs the warp of his days."[3]

If one image fails to fuse with another, it may also fail to explain another; but the object of all these images is the philosopher René Descartes reflecting on the events of his life, whose actual beginning he always kept a secret because he wanted no one to throw a hex on it by casting his horoscope. If not by the stars, why then not by the ripening of an egg? What a challenge for the poet to play astrologer in the guise of resident ovologer! Had not the ancient hierophants sought the favorable and unfavorable omens for their clients in the entrails of animals? Why not, then, in the fruit of their loom? The Unnamable was right, we can hear the voice of the ventriloquist, lisping, trying too hard to imitate the inimitable Joyce.

It is general knowledge, of course, that Descartes usually spent his mornings in bed, never arising before noon. Was this merely another astrological superstition? Or, was it true as he claimed, that a bed is the only proper place for philosophical meditation? Why not, since there the mind would naturally become more active as the body is sunk in repose? In sleep indeed the very soul of man becomes separable from its corporeal dressing, and is free to indulge its most secret yearnings. How vile of it, then, to be thinking of food when it should be thinking its loftiest or profoundest thoughts. And all those insults from the mouths of idiots! How petty to have noticed them at all! And how envious, to see only the weaknesses of the strong! What pretention, what hypocrisy, in making it appear that the philosopher made appeal only to the souls of his followers!

A soul detached from its body is a naked whore, and a whore's destiny can only be cast in a whoroscope. But it is a neat touch to let that brazen bitch, Descartes' soul, lay herself bare. Descartes lies abed, and his immediate desire is for food, since the body, once awake, refuses to be denied; his ultimate hunger, however, is to be reborn after death in a "second/starless inscrutable hour."[4] Here, too, the essence is of time: Descartes waits for the egg to ripen, and, waiting, reflects on the course of his life's events.

Alternating stanzas present a movement from the annoyance at his current wait to reminiscence of the past, until the final image that reveals his expected future rebirth.

The avowed purpose of Descartes' published *Meditations* was to prove the existence of God and the immortality of the soul. If his soul were indeed immortal and the hour of his death could be kept a secret as was that of his birth, his death, being nothing but a rebirth, would indeed constitute a second starless hour, inscrutable to the most habile astrologer.

Descartes actually died at the hands of his treating physician, more through the ineptitude of the good doctor's treatment than the malice of his will, though that too was a fact of their relationship. The physician, Weulles, was an intellectual enemy of Descartes at the court of Christina of Sweden. Descartes had accepted the invitation of the blue-stocking queen to visit her court for the purpose of introducing her to the mysteries of mathematics. He accepted, not knowing the queen was an early riser, and would ask for those daily lessons at five A.M. The climate was cold, and Descartes' physical condition, long pampered by whole mornings in bed, could not withstand its rigors. A pulmonary infection led that tiring body into the hands of his enemy, and death. Descartes' mind, the old whore, could then think no more.

Such are the grand outlines of *Whoroscope*. Had Descartes been a materialist, like another of his enemies, Gassendi, Beckett might even have written a proctoscope. As it was, however, he forged the whoroscope as an instrument for penetrating Descartes' mind, panting in ecstasy in anticipation of his omelette, that "slim pale double-breasted turd,"[5] which he "will eat with a fish fork" / "White and yolk and feathers."[6] How's that for brooding, you image-famished critics: a philosopher of the mind on the pleasures of ingurgitating the broodings of a barn fowl, as he waits for the beginning of a new day!

In the first stanza of four lines

> What's that?
> An egg?
> By the brothers Boot it stinks fresh.
> Give it to Gillot.

the outstanding feature is the oxymoron of "stinking fresh." The conjoined incongruous notions can only be ridiculous – or reveal the tastes of a peculiar individual; when we discover that the persona of the poem is Descartes, reflecting not on the existence of God and the immortality of the soul, nor even on the separation of soul and body, but on the prospects of his own second coming, the peculiarity of the individual is affirmed. Indeed, the prejudice the reader may have, to view the philosopher as a lover of wisdom detached from all corporeal cares, may give an oxymoronic tinge to the total structure of the poem.

When that poem's puzzles have been solved, with the aid of the references, it can be read as the exposition of the absurd conjunction of irreconcilable opposites. But then that was Descartes' definition of a man, that mystical union of material body and spiritual soul. The effect may even be comic or tragic, depending upon one's expectations from life, but the tragi-comic effect is the hallmark of Beckett's genius.

The medieval scholastics were just as humorous in their definition of man as a "featherless biped"; they merely carried to its absurd limits Aristotle's ("the master of those that know") method of definition: first to locate the species to be defined in its proximate genus (biologically, in the bipeds) and then to distinguish that species from all others of the same genus by naming the differentia (still biologically speaking) as "possessing no feathers." Individuals, of course, could not be defined by this procedure, since they were the logical result of partitioning a species into its members; whence, since the essence of anything was what its definition referred to, the essence of Socrates would be the definition of Socrates if only Socrates had a definition. Thus, already in the scholastic Middle Ages, there was a confused notion of the absurdity contained in an individual's attempt to define himself in terms of preformed concepts.

Even if it cannot be determined by a reading of the stars, what a man is, is determined by his actions. So the existentialists have maintained; but in 1930 Beckett was not an existential philosopher, not even an existentialistic one. That was to develop later, when he had achieved greater insight into the relations between the mind and the body of man. Here he is content to show the actions of Descartes' mind as he waits for his breakfast.

As if to show the subtlety of that mind, the poem renders a Cartesian solution to a theological conundrum posed by Antoine Arnauld: if mind and body are two separate substances, how can we explain the miracle of the Eucharist in which the substances of wine and water are "transubstantiated" into the blood and body of Christ, united as a single person possessing both a divine and a human nature? Beckett puts this solution into the mouth of the exulting Descartes:

> They don't know what the master of them that do did,
> that the nose is touched by the kiss of all foul and sweet air,

and the drums, and the throne of the faecal inlet,
and the eyes by its zig-zags.
So we drink Him and eat Him
and the watery Beaune and the stale cubes of Hovis
because He can jig
as near or as far from His Jigging Self
and as sad or lively as the chalice or tray asks.
How's that Antonio? [7]

If one remembers the association of spirit with the volatility of air, the images fall into order. Air is moved hither and thither by the solid objects displacing it; so it can come into contact with all our senses, nose, ears, eyes, and anus, where, as sweet as it was, it becomes fetid and foul. It's volatility allows it to zig and to zag, in such a way that a pure spirit may be said to jig as far as necessary to give life to its body. "The watery Beaune" is an implied judgment on the quality of the eucharistic wines served to the faithful in the chalice, and "the stale cubes of Hovis," the same, on the quality of the unleavened bread served on the tray. And "Antonio" is Italian for "Antoine."

For all its cuteness, however, or perhaps because of it, the poem fails. At any rate, the texture of the poetic images needed the support of their notes, as the allusions they contained were too recondite even to be recognized without them. But it fails, too, because the images, even annotated, fail to fuse on their own first level significance. Musing, for example, on Harvey's discovery of the circulation of the blood, following a reminder of his illegitimate daughter's death, Descartes' persona asks:

Oh Harvey beloved
how shall the red and the white, the many in the few,
(dear bloodswirling Harvey)
eddy through that cracked beater? [8]

The "red" and the "white" pose no problem, where "the many" (red corpuscles) and "the few" (white) might; nor does "the cracked beater" when it is a question of the heart. And if the beater being used to prepare that omelette is also cracked, we might be able to understand the delay in its serving. But it was the heart all along, and not the chef's whisk, for the persona of the poem alludes to Descartes' presence at La Flèche, the Jesuit lycée he had attended, when the heart of France's Henry the Fourth was brought there for interment. The cadence is as crisp as any summative image by T. S. Eliot:

And the fourth Henry came to the crypt of the arrow.[9]

One need only translate 'la flèche,' and recall the gruesome idolatry of the dead king to make the connection. But the apparent link is a reference to hearts and the swirling of blood. In the one case the heart is an idea, of the organ in its quick state, pumping the blood through the body; and in two others, still, as Descartes remembers the death of his daughter as the associative spring to recalling his disagreement with Harvey on the physiological functioning of the heart. Then this stanza, the seventh, is brought to a close by his recall of Henry the Fourth's burial at the school. The living and the dead are united only in thought, the measure of the passing of time.

Failing to fuse at the first level of interpretation owing to the obscurity of the references (whence, the necessity of the notes), the images are forced into unity by the philosophical idea structuring the whole poem, the psychophysical dualism of Descartes applied to the case of Descartes by the persona of Descartes himself. The passage in time from the quick to the still is the fate of all bodily parts; but the soul will live on as it is – thinking its thoughts, measuring the time of a single significant life.

Indeed, the succession of images, spoken by the persona, constitute the stream of consciousness that is Descartes' mind. Each image is an individual *cogitatio*, and the stream of *cogitationes* shows not the least hint of a substantial ego, the idea of which allowed Descartes "to prove God by exhaustion."[10] The vainness of that proof is exposed in the description of the idea upon which it is based, Descartes' own vain wish for a personal immortality.

Whoroscope is indeed, therefore, a Cartesian poem: its soul is separable from its body. The apprentice was trying too hard, and missed the lesson of his master. Beckett had tried to keep himself out of the poem by conjuring up the persona of Descartes, but his words revealed his presence.

More Pricks than Kicks was first published in 1934, four years after *Whoroscope* and three after Beckett's *Proust*. The title, from a philosophical point of view, is more ethical (even Biblical) than ontological, but its ethical import is supported by the same Schopenhauerean foundation Beckett had found in Proust. If "the world is my idea (representation)," a Proust might very well make use of that philosophical pronouncement to structure his universe using a technique – impressionism – that reflects this basic ontological noncommitment to the real world, or one might, off on another tack, try to create a universe that is actually someone's idea, without respect to the truth or falsity of that idea and without respect, as well, to the impressions of others.

Technically, of course, as Proust had demonstrated, this would require a narrator whose unique point of view orders this solipsistic universe, the first person speaking in awareness of itself. That, of course, was the function of Descartes' persona in *Whoroscope*.

In structuring *More Pricks* Beckett made his first mistake in choosing the omniscient narrator's stance, placing himself astride all the events of his hero's solipsistic world; and once again, we detect the moving lips of the ventriloquist:

> My sometime friend Belacqua enlivened the last phase of his solipsism, before he toed the line and began to relish the world, with the belief that the best thing he had to do was to move constantly from place to place.[11]

The original Belacqua, of course, sat in Dante's antepurgatory folded back in upon himself in the foetal position, and not even his friendship for the living Dante could induce him to make a move up the winding roads of purgatory itself; his punishment had to last as long as his impenitence on earth, doing nothing here as there. Molloy will learn the futility of moving from place to place, and will indeed decide, for philosophical reasons, to remain where he was.

Belacqua Shua, on the other hand, is the solipsist turned peripatetic; but like his namesake he continues to deny one external commitment after another. Beckett's narrator describes him as living "a Beethoven pause,"[12] whose silence promises to give way to a thunderous climax and final reminiscent coda.

The movement of the whole novelette is prefigured in the first instant, in which Belacqua ponders the death of a lobster "lepping fresh,"[13] caught as it was that very morning. It dies instantly, unlike those immortal souls in the superheated waters of Hades, as it enters the boiling water to become food for some epicure's already sated body. Whereas the parboiled souls of Hell served Dante's intellectual and moral purpose of urging the avoidance of sin, Beckett's Belacqua seems preoccupied with searching it out.

The quick transition from things corporeal to things spiritual, and from the living to the dead, takes place in Belacqua's mind, as a Dantesque Italian line strikes his consciousness: "qui vive la pietà quando è ben morte."[14] To die suddenly in the state of grace does not quite do justice to the oxymoron of the living dead, but a life of unrequited love might turn the trick. Belacqua, as the sequel of the work shows, lives such a life, so his thoughts turn inward: "Well, thought Belacqua, it's a quick death, God help us all. It is not."[15]

For a solipsist to fall in love would be as quick a death as that of the lobster, but, as the saying goes, blood is thicker than water. For it to achieve the necessary temperature, the solipsistic lover must first fall out of love with himself, and that may take a very long time, and many trials and errors. And, as will be seen, all of Belacqua's trials seem to have been errors, since he was looking for more kicks than pricks. In any life of too little joy no death is quick enough, even allowing for the ambiquity of the "quick," as referring to living and to the shortness of time.

Thus *More Pricks than Kicks* not only names the novel describing Belacqua Shua's misadventurous life, it also makes a comment on the character of that life, which is short and replete with more pains than pleasures. The additional suggestion that the anticipated pleasures are to be felt in the quiver of his erectile organ is not, however, gratuitous, since in his peregrinations Belacqua runs though a series of doxies, all of whom, of course, die in the end, whatever their state of grace.

One of these partners is a Ruby Tough, with whom Belacqua has arranged a double suicide plot. She is dying of natural causes anyway, so her sin might not be against the Holy Ghost (in Catholic doxology, the only unforgivable sin, that of final despair). But the plot fails:

> When the first shock of surprise had passed and the silence spent its fury a great turmoil of life-blood sprang up in the breasts of our two young felons, so that they came together in inevitable nuptial.[16]

and the obtrusive narrator makes his own comment: "With the utmost reverence at our command, moving away on tiptoe from where they lie in the ling, we mention this in a low voice."[17] The voice of the ventriloquist, however, is never low enough.

Yet, however ironically, the truant couple, having had their despair uplifted by a surge of life's blood, accomplished their mission in disguise; for, the narrator comments, "l'amour et la mort – caesura – n'est qu'une mesme chose."[18] Whether love and death are a single phenomenon, they both may be considered a kind of nonpoetical caesura. In this non-Beethoven-like pause, however, the narrator again inserts his own presence: "May their night be full of music at all events."[19]

The title of the episode above is 'Love and Lethe,' a double reference of some note: first, to the oblivion of the solipsist's past desire through the union with his own idea; and, second, to Dante's drinking from the Lethe before being reunited with Beatrice on leaving the earthly paradise on his journey to the Beatific Vision. But a *pas de deux* of any sort is unthinkable for the consistent solipsist, unless his partner is compliant enough to remain forever an idea, rather than the living (quick) refutation of his own ultimate creed. Dante, of course, is reproached by Beatrice for his faithless inconstancy, and when he pleads innocence she reminds him that he has drunk of the forgetful waters.

Ruby, who was moribund of natural causes all along, "died duly,"[20] of those causes, and remained an idea of Belacqua's furtive felicity; or, as he preferred to express it in mockery of the Mass, his "sursum corda." And if he is to remain constant to that idea, he must be unfaithful to anyone who is unfortunate enough to serve as the occasion of its recall. Like the meditating Descartes' hungry body, his flesh will continue to make itself felt, as his mind serves up one beloved after the other.

On the spiritual plane, when during the Mass the priest intones his "sursum corda," the congregation of the faithful is signaled to lift up their hearts in joy; the miracle of transubstantiation is about to occur. Belacqua's joy is centered somewhat lower in the anatomy, and the miracle happens only too often. He uses the expression twice in his conquests. The first time is with Winnie, in County Fingal. The narrator sets up the sequence of events that begins with a reminiscence of a past amour:

The last girl he went with, before a memorable fit of laughing incapacitated him from gallantry for some time, was pretty, hot and witty in that order.[21]

But Winnie on this occasion shows herself witty, pretty and cold, in a newer order of feminine profiles. She and Belacqua are on a trip during which he steals a bicycle from the Portrane Lunatic Asylum, when, in the countryside, the presence of her wit is displayed as follows:

'The lovely ruins' said Winnie 'there on the left, covered with ivy.'

The ruins are of a church, and two small fields further on is a square "bawnless" tower, where Dean Swift is known to have kept a paramour, named Stella.

'That,' said Belacqua 'is where I have sursum corda.'
'Then hadn't we better be getting on,' said Winnie, quick as lightning.[22]

Two shady proposals in one afternoon's outing are too much for a girl in full possession of her wits, and only Belacqua feels the sting of the prick. He leaves her to a Dr. Sholto, there being no sursum corda that day. Winnie, in effect, is left to "die in decency."[23] Love and death for the solipsist constitute the same threat. His loves are quick, but his death is, alas, all too slow.

After Ruby there is Lucy, and after Lucy, the Smeraldina. Neither of these latter is dupe of Belacqua's request for a surge of heart: Lucy because she is as witty as Ruby, and the Smeraldina because she knows that love and death are one. Lucy, however, is the first to fulfill the Smeraldina's intuition. Belacqua moves in as follows,

'Then I thought' he said at last 'that the best thing to do was to go to the wood for a little sursum corda.'[24]

And the move is parried: " 'Corda is good' said Lucy."[25] Here, too, more of a prick than a kick.

In pressing his suit Belacqua promises Lucy a cicisbeo to fulfill those moments of her heart's surging desire when he should himself be incapacitated. But, as it turns out, she will need no paramour, and both become incapacitated as they happen upon a Mr. Tanzherr lifting up his own heart to a loved one. The dancing master catches up to and throttles Belacqua, and Lucy falls from her jennet, crippling herself for life. She marries another, and actually dies on the second anniversary of her accident.

So Belacqua marries Thelma, née bboggs, she with the muddy stammer in the name. To the wedding is invited one Walter Draffin, Thelma's own paramour, just to provide competition to the new husband, who remains unaware of this reversal in his relationship with members of the fairer sex. But Draffin runs off with Alba Perdue, Balacqua's partner at a party earlier described in *A Wet Night*. Could it have been a dream? At that party Alba's lost whiteness came to symbol in the dress she wore; it was red, and buttoned up the back.

The best man at Belacqua's wedding is one Hairy Tiny Capper Quin, and he earns title to the epithet contained in the name of his role as he ends up in possession of the Smeraldina. Another friend, James Skrym, leaves the festivities in tow of Hermione, who walks on crutches. How far can they get without falling down, the better to lift up their hearts?

The whole wedding scene is entitled by Beckett *What a Misfortune*. The wrong girl at the wrong time? Not at all. Married, the solipsist has lost his reason for being.

If "More Pricks than Kicks" were not cuter, and did not suggest the ethical structure in a solipsist's world, the work might have been entitled "The life and loves (or death) of Belacqua Shua." The multiplication of couples at Belacqua's wedding gives ample expression to Beckett's ironic humor, and what follows the climactic misfortune of his marriage is both dénouement of the many knots tied that night and fulfillment of the analogy between love and death. Married by misfortune, Belacqua is to die by mistake.

His untimely end is foreshadowed in the love letter he receives from that other object of his heart's desire, the Smeraldina, whose "billet doux" suggests the bitter truth: "Der Tag will kommen und die stille NACHT."[26] A night without music is a night without love, and the Smeraldina will form a new couple with Belacqua's best man, Hairy, who thus fulfills (a second time) the cicisbeo theme struck up in Belacqua's relationship to Lucy. Although the narrator insists that the two sets of three lovers offer "a most foully false analogy,"[27] it does take three to make a triad, if not a trinity, and all triads are in some respect alike.

Belacqua actually dies on the operating table of a hospital, thereby allowing the other pairs of lovers to rest in peace. On the way down to surgery he recalls the punch line of a parson-actor joke:

'By God! I'm shot!' [actor's lines]
'Mercy, Upon my word!' [parson's emendation]
'Oh' he cried 'oh . . . By Christ! I AM SHOT!'[28] [lines delivered]

Belacqua, of course, was shot with the sedative preparing him for surgery, and he needed mercy upon anyone's word. But the narrator has none for the victim:

By Christ! he did die! They had clean forgotten to auscultate him![29]

Even though his readers are forced to submit to this auctorial intervention, Belacqua's heart could not stand the surgical one. And with his death, the world of the solipsist has disappeared.

There remained only *Draff* for an explanation of the solipsistic hoax. With Belacqua's passing all his ideas must have ceased to be, but there is the miraculous exception of Hairy and the Smeraldina. Dead are all the loved ones of Belacqua's tumultuous past. "Thelma née bboggs . . . perished of sunset and honeymoon that time in Connemara" (hence of consummated love).[30] And the others in solemn procession: "Lucy of course long since, Ruby duly, Winnie to decency, Alba Perdue in the natural course of being seen Home."[31]

The coda expected from all these musical passages in Belacqua's life comes in the form of Nick Malacoda, who measures Belacqua for his casket. The day did come, and the still night, in which no music could be heard.

The hero is buried, and sent over to his just reward. But where? Perhaps he will have attained that eternal abode which is the right of piety, of all those who have died in the state of grace. His death was gratuitous enough. And then perhaps not, and for the same reason. Being gratuitous, his death was an accident; consequently, ill prepared for, much less resolutely anticipated by a sincere act of contrition. Is that why, on their return from Belacqua's funeral, Hairy and the Smeraldina find their home in flames, the maid raped by the gardner, and the firemen not there? Perhaps. But Old Nick had already done his job.

The moral of the tale of Dante and the lobster is that no lobster ever has a quick death. Belacqua's opening thought was thus a mistake; and since, as a solipsist, his thought constitutes whatever real world there is, that too must have been a mistake. After all, it takes time to fall in love, and more time still to have that love fully requited. Moreover, no one can claim to have experienced death, since that event in the real world is excluded from one's conscious life, except in anticipation, and thus in fact always happens to someone else. So what is there to do? Nothing, nothing at all, in the stillness of the night.

If his death is a mistake, so is his last judgment. For in repeating to himself the words of the actor-parson's joke he has found the bribe of salvation. Calling upon the Holy Name and asking for mercy, he has certainly achieved enough to resume his position in antepurgatory where Dante had left, and Beckett found, him. He thus experienced the eternal return as the circle of his world closed, and with it, one elaborate literary pun.

If one could be as Dantesque in the criticism of this pun as Beckett was in writing *More Pricks than Kicks*, it would be tempting to show the relationship between the literal meaning of Beckett's words and the allegory they intend. That has already been done: the quick dying, or living death, of the lobster in relation to the dead living of Belacqua. These two levels of significance would, of course, yield a superintendent moral sense, that life is full of more pricks (pains) than kicks (pleasures). That is the hedonistic paradox, which states that anyone seeking more kicks than pricks in life is destined to feel more pricks than kicks. Check. There would remain, then, only the supersensible, "anagogic" meaning of the whole linguistic structure. And that too is there in the Beckettian reversal of Dante's exhalted spiritualism: love is of the body and is equivalent to death, at least for the solipsist.

Dante's own method of hermeneutical expansion is therefore not incompatible with the methods of phenomenological structuralism, and his *Divine Comedy* was still the most complete example possible for showing how philosophy and literature come to the same thing in the structure of the literary world. Dante, of course, was under the instruction of Virgil and Beatrice, the latter counter-balancing his love of the wisdom promulgated by Plato, Aristotle and St. Thomas. Beckett had only Dante and Joyce and Proust. That should have been enough; unfortunately, however, it wasn't.

The apprentice had failed to see that a novel structured in solipsism needs no third person, omniscient narrator. And the presence of such an agency in Belacqua Shua's world was one cicisbeo too many – like the God who left the lazy and the negligent in antepurgatory, where, turned inward upon himself, wrapped up in the foetal position, Belacqua protects himself against the sun, under the shade of a rock on those unscalable cliffs.

Belacqua will be reborn again, many times: first in Murphy, then in Watt, and then again in Molloy and Malone. But Beckett did not realize that his quasi-omniscience was insufficient to write a novel, until he let Molloy tell his own story.

Beckett's *Murphy* reincarnates the dead Belacqua about the year 1938. Murphy has no Christian name, since he was not baptized; but the signs of the reincarnation are clear. For the purpose of explaining the events in the life of the unnamed body that looks insistently for peace in the possession of its own mind, the author still employs an omniscient narrator, who depicts the essence of that mind in three different ways.

First of all, we are led to understand Murphy's mind by the order of events in his life that are recounted as the reflective search for personal identity under the handicap of a failing heart, which at the most unpredictable moments physically skips an essential beat, leaving its owner gasping for breath amidst the onslaughts of a hostile world; a symbol, as it were, of its inability to attach itself to a single object of love outside those defined by the projects of its own mind. The heart, it has been said, has its reasons that Reason cannot know.

The second depiction is found in the former, as Murphy's superstitious dedication to the astronomical determination of the events of his life. He will go nowhere, do nothing, without the aid of his horoscope, which is duly laid out for the reader and learned "by heart" by Murphy himself.

And, finally, the narrator gives a quasi-philosophical description of the workings of the solipsistic mind united as if by miracle to the body with the failing heart.

Murphy's experiences were either actual or virtual, and the difference was not of intensity, or of logical possibility, nor even of quality, such as a pleasure or a pain, but rather the coincidence of the external with the mind's internal conditions:

He distinguished between the actual and the virtual of his mind, not as between form and the formless yearning for form, but as between that of which he had both mental and physical experience and that of which he had mental experience only. Thus the form of kick was actual, that of caress virtual.[32]

How clever of the narrator to inform us in this way that Murphy's life had more pricks than kicks, where the prick is an actual, physical kick and the psychological kick is only a virtual caress! Although the physical kick felt in the body is a pain, the psychological kick that is a pleasure is registered mentally as a physical caress, which, in Murphy's case, can only be mental, i.e., virtual, because physically Murphy is as repulsive to the rest of humanity as it is to him.

The narrator strives hard to picture such a mind, and to show how the two sorts of experiences may be thought to overlap, even to imagining the Platonic solution:

... he felt his mind to be body tight and did not understand how the two experiences came to overlap. He was satisfied that neither followed from the other. He neither thought a kick because he felt one nor felt a kick because he thought one. Perhaps the knowledge was related to the fact of the kick as two magnitudes to a third. Perhaps there was, outside space and time, dimly revealed to Murphy in its correlated modes of consciousness and extension, the kick *in intellectu* and the kick *in re*. But where then was the supreme Caress?[33]

If Murphy had remembered his existence as Belacqua, he should have been on the lookout for this easy reversibility of pricks and kicks. After all, Plato's souls are reincarnated, and know by reminiscence.

But Murphy is the complete solipsist, whose experience goes little beyond the present moment, except for the vague intimation of an unhappy future. Perhaps the solution was supernaturally determined, as the dualistic occasionalists, in particular Arnold Geulincx, maintained. But that was a problem of little interest:[34]

Any solution would do that did not clash with the feeling growing stronger as Murphy grew older, that his mind was a closed system, subject to no principle of change but its own, self-sufficient and impermeable to the vicissitudes of the body.[35]

Unfortunately, however, no one is an island unto himself, as Murphy learns; and his body will come to its end, as accidentally as Belacqua's, distributed as ashes on the barroom floor, only to have "been swept away with the sand, the beer, the butts, the glass, the matches, the spits, the vomit,"[36] no longer caressing, nor even caressable.

The peripeteia of Murphy's existence develops in a story framed by his relationship to the prostitute, Celia Kelly, who plans to marry him in order to escape from the streets. Immersed of necessity in the physical aspects of love, she aspires to a life of decency such a marriage would represent. But Murphy and his stars are opposed to the union, and Celia is forced back into the streets.

Murphy, it seems, was the student of the Cork County Irishman, Neary, who possessed the power of voluntarily stopping the beat of his heart. Not that Murphy's own heart left anything physically to be desired, were it not for its occasional self-stoppage, and not that Neary was a master of regulating his own pulse in tune with his romantic ideal of the moment. Indeed, this latter was frequently the victim of unrequited love; instead of the normal *quid pro quo* that ends in *sursum corda*, his love was returned to him in a circuitous chain of lovers.

Neary's latest interest is a Miss Counihan, who does not refuse his advances outright, but insists that she is engaged to Murphy and can do nothing for Neary's palpitating heart until she has proof that Murphy is dead, or has been unfaithful to her, or has otherwise given (written) testimony of his having forsaken her.

Neary then hires a private gumshoe, Cooper, to trace Murphy in London, since he cannot go there, where he is wanted for bigamy. So Cooper, Miss Counihan, and another of Neary's students Wylie, are off in search for Murphy; and they are eventually joined by Neary himself, when Cooper discovers that Neary's second living wife has died. Each is in pursuit of his own intention: Neary to use Murphy as a means to obtain Miss Counihan's affections; Miss Counihan to force him into marriage; Cooper to earn his pay; and Wylie to pick up the pieces of Miss Counihan's broken heart.

In the search, only Cooper's project remains unchanged, and it is he who is responsible for Murphy's dust's returning to the sand and garbage of a barroom floor.

Everyone needs Murphy; but Murphy needs no one except himself, and so continues his pursuit of the objects created by his own mind. The emotional balance he sought, but never felt, at Neary's feet can be achieved only symbolically in his rocking chair, where, as his body is appeased, his mind is released for pleasures indescribable:

> He sat in his chair in this way because it gave him pleasure! First it gave his body pleasure, it appeased his body. Then it set him free in his mind. For it was not until his body was appeased that he could come alive in his mind. . . . And life in his mind gave him pleasure, such pleasure that pleasure was not the word.[37]

Trussed up in his chair, as naked as the day he was born, he could pursue his "Belacqua fantasy" without interference of the outside world.

This reference to Dante's impenitent sinner is the narrator's,[38] and would suffice to connect *More Pricks* with *Murphy* even should we have missed the coincident structures of the two life-worlds. Belacqua Shua is incapable of loving, given the solipsistic nature of his experiences; and Murphy, himself described as a "seedy solipsist,"[39] suffers a similar malady. In looking for a reason to explain Murphy's appeal to women, Wylie even suggests that it must be his "surgical" quality:[40] clearly a reference to Belacqua's end, to Murphy's incisiveness, and to the notorious feminine sympathy for all suffering humanity.

Murphy's incisiveness is not a figment of the narrator's imagination. Where he himself is concerned, he sees clearly indeed. Celia insists that he get a job so that she may leave the streets; and although he would like her to do just that, he cannot see how getting a job would be good for either of them. How is it possible to love someone and then wish him to change into something else? That surely is to wish the death of one's love. Since he possesses only his body, his mind, and her, should he find a job, he will certainly lose one or all of these possessions. His prediction, of course, comes true; and Celia is forced back onto the streets, with her last and sole tie to the world, her aged, chair ridden grandfather, whose chief ambition in life is to fly his kite out of sight. Celia's relationship to her grandfather provides the frame for Murphy's own sad search for a personal identity.

Celia and Murphy are reconciled after one argument, and she disturbs one of his sessions in the chair with the news that she has procured the horoscope he sought. There was only one system external to his own mind in which he had the least confidence, and that was the signs of the zodiac; and Beckett's narrator affects a similar disposition by recounting the temporal sequences of his narrative in terms of the positions of the heavenly bodies. Earthly bodies are not a matter of indifference to Murphy, but he cannot put his trust in them or in his reactions to them. Neither Miss Counihan nor Celia can permanently dissuade him from the pursuit of his mind's self-reflecting pleasure. After all, kicks and caresses must be ordered by the mind lest the pains stemming from the realm of the actual dominate the pleasures of his purely virtual world.

In accordance with his horoscope, Murphy accepts a job on the first Sunday to occur on the fourth of the month, an October; the year was 1936. A penurious Irish "pot poet" was looking for a replacement in his job as male nurse in a psychiatric hospital, the Magdalen Mental Mercyseat. As if by miracle Murphy is offered a job that fits his horoscope; a "great Magical

Ability of the Eye, to which the lunatic would easily succumb"[41] and a custodial career[42] are clearly indicated by the fake astrologer.

As a custodian of lunatics, Murphy has never had a peer. He is not hampered by textbook descriptions, since he needs none: he looks upon all his wards as having achieved Nirvana on earth, as fully satisfied Belacquas. The alienated personalities on this earth are those on the outside, writing the texts and vainly trying to adapt themselves to the impossible demands of an external world. And when Murphy looks into the eye of his favorite patient, the chess playing Mr. Endon, he sees only his own reflection; he himself remains unseen, although their faces are close enough to approach the position of a "butterfly kiss" (the brushing eyelids), because Mr. Endon's gaze is fixed too far into the "blooming buzzing confusion" (Neary's adaption of William James' description of the world of pure experience) to have any figure detach itself therefrom in the formation of the least significant gestalt. Neary's psychological instruction was always couched in terms of formal gestaltism, but even Kurt Koffka himself had refused to answer his query as to why such a pretty figure as Miss Dwyer's refused to form on the background of the world after she has finally accepted Neary's physical attentions.[43] Could it be true that love destroys its subjects and objects?

Mr. Endon, pursuing his *idée fixe*, looks upon Murphy as the occasion for a game of chess, so whatever affinity Murphy finds for him goes as unrequited as Celia's love for Murphy. Mr. Endon proves his indifference by escaping from his cell, and registering a crazy pattern of visitations on the other patients. Even a suicidal schizophrenic does not lack for a sense of humor.

Murphy's end, since all things must come to an end, is as accidental as Belacqua Shua's. He chooses to live in the garret of the mercyseat with only a gas fire whose source is an unused jet in a water closet on the floor below. He sees by candle light. The gas pours into the garret and Murphy is burnt beyond recognition, except for his telltale birthmark, recognized at the inquest by Celia. His body is incinerated, and the ashes turned over to Cooper to be disposed of in accordance to his last wishes. But Cooper enters the saloon and gets into a fight, during which he throws the packed ashes at one of his assailants. And so Murphy's round is complete: dust unto alcohol sodden sand and vomit. Celia returns to her grandfather and the streets.

The comparison of *Murphy* with *More Pricks* is compelling. Although the philosophical structure is the same – the solipsism of the heroes, the latter novel pushes one step further the "moral" meaning of the written text. Murphy, being more self-conscious in the intellectual sense than Belacqua, is capable of ruminating on the hedonistic paradox. In pursuit of himself, he becomes alienated from the world, which he holds as of no account and is capable of sympathizing with his patients liberated, from his point of view, by their own alienation. His mind, after all, was not an instrument of adaptation, but a place,[44] just as later on Molloy will discover himself to be a region. And what are the events of the "big world" in comparison with those of the little, that mind which remains the measure of all things?

The solipsist, himself rejected by the world he repudiates, must continue his drive to be united with the only object he can know, his own mind. Even Geulincx's *Ubi nihil vales, ibi nihil velis*[45] constitutes only half a truth for the solipsist's ethic:

> But it was not enough to want nothing where he was worth nothing, nor even to take the further step of renouncing all that lay outside the intellectual love in which alone he could love himself, because there alone he was lovable.[46]

All this knowledge, even with the soothing aid of the rocking chair, could only predispose him in the right direction, not clinch the object of his search.[47] Thus, the life of little joy may soothe the consciousness of the anchorite and give the practical solution of the hedonist's paradox; but it cannot yield up the object of the solipsist's love. Life is still more pricks than kicks.

Murphy finally liberates himself from the enslavement to the fake Indian's horoscope, when he realizes that even those stars which have supposedly determined his ways are of his own creation. Existential in import, the realization is moving:

> Between him and his stars no doubt there was a correspondence, but not in Suk's sense. They were his stars, he was the prior system. He had been projected, larval and dark, on the day of that regrettable hour as on a screen, magnified and clarified into his own meaning. But it was *his* meaning. The moon in the Serpent was no more than an image, a fragment of vitagraph.[48]

And he saw the fruit of this discovery in the liberating alienation of his mental patients. Their very salvation was in their present alienation from the macroworld. Scrutinize such a person, as Murphy does Mr. Endon, and you will find a reflection only of yourself.

Such is the new moral to be gleaned from the solipsist's predicament, and it applies just as faithfully to the reader who looks as closely at Murphy as the latter had at Mr. Endon, the suicidal schizophrenic with the sense of playful humor. The veritable end of it is that we are all schizophrenic, being minds tied to a decaying body, and capable of reflecting on our predicament;

so no further moral discussion seems to be in order, beyond the statement that Beckett's pessimistic world-view is brought into sharper focus. It is not yet clear, however, how this pessimism may be transformed into real tragedy. That is a demonstration that must wait upon a fuller comprehension of the human transcendence. The pessimism of this stage is mitigated only by the comic wit of its narrator, all too plainly here Beckett himself.

With Murphy's liberation from the Hindu's drawn horoscope, Beckett became liberated from Dante's superstitious world view. From here on in, his narrators will be concerned with an understanding of human self-alienation. But we are to suffer yet one more third person account, that of the narrator, Sam.

Watt was published in 1953, eight or nine years after it had been completed, by the Olympia Press in Paris. Such a time span between completion and publication indicates some diffidence on the part of the world's publishers, even considering the fact that when the author dated his manuscript World War II was just coming to an end.

The name "Watt" figures almost elusively in the story of Murphy when this latter uses his "surgical" quality to bilk Vera, the waitress, out of a second cup of tea. The narrator comments on the scene:

"Murphy had some faith in the Külpe school. Marbe and Bühler might be deceived, even Watt was only human, but how could Ach be wrong?"[49]

Whether psychologists and schools of psychology are being referred to or not, Watt's being only human sticks in the mind.

For Watt is all too human, crushed by external circumstances, and misunderstood by his fellows as clearly as he is by himself. His story is told by a narrator, who calls himself 'Sam' in one passage, and who claims that all he knows of Watt was told him by Watt himself. But Watt had communicated the essentials of his story in a curious fashion.

In turn, he inverted the order of letters in words, words in sentences, sentences in periods, and the possible permutations and combinations of these inversions, as it were, simultaneously.[50] But, by sympathy and repetition, he and the narrator communicate. They once inhabited the same "pavilion," but became separated when Watt was removed to another. They met in the garden when the weather was both windy and sunny, since one preferred the wind and the other the sun, and neither sought out the other. Hence they met accidentally, by virtue of their tastes; and Watt's story is told the narrator after the events had already transpired, when both have been institutionalized, the one as teller and told, the other as told and teller. In order to make it easier for the reader, the narrator takes the liberty of straightening out Watt's tale. Although all stories have a beginning, a middle and an end, they do not necessarily, Aristotle to the contrary notwithstanding, come in that order.

The story concerns Watt's travel to the house of Knott, to assume the tasks of a house servant; his service as first-floor orderly; his ascension to the duties of the second floor; and his return to life outside the Knott household. Given his propensity for inversions, however, one is not surprised to learn that Watt prefers his own ordering of beginning, middle and end: two, one, four, three in the four-part scenario. Although the narrator understands, and approves the novel ordering ("Heroic quatrains are not otherwise elaborated."[51]) in which Watt's service on the first floor precedes his arrival at Knott's, and his leaving, his service on the second floor, we are given the story in naturalistic order, in which his coming and his going, both reflecting the inability of society to accept such a creature, serves as a frame for his life of service in the Knott household. Watt's second-floor service, however, is an event which is recounted to Sam after Watt has left Knott and become institutionalized.

The time reference for this service must be left obscure, since Watt himself, who recounted the events to the narrator, was forced to deduce how long he remained in the service of Knott. The procedure of replacement of the servants, however, was clear. When a new first-floor orderly arrived, the previous keeper of that floor was removed to the second floor and he who had previously served there left the service. When Watt arrives, Arsene leaves and Erskine moves to the second floor. When Arthur arrives, Watt takes Erskine's place. And when Micks arrives, Watt makes his return to the outside world. There is thus a serial order of servants, all fitting two general categories of physical appearance; and the service periods in rotation seem to be equal, perhaps a year on each floor.

In leaving, Arsene gives Watt a long harangue on the little hope for any change to occur in his new domain, and consequently, on the inanity of wanting anything or speaking about it at all; for once a new servant arrives he shall discover

> . . . at night rest in the quiet house, there are no roads, no streets any more, you lie down by a window opening on refuge, the little sounds come that demand nothing, ordain nothing, explain nothing, propound nothing, and the short necessary night is soon ended, and the sky blue again over all the secret places where nobody ever comes, the secret places never the same, but always simple and indifferent, always mere places, sites of a stirring beyond coming and going, of a being so light and free that it is as the being of nothing.[52]

Although the newcomer to such an existence may feel happy, because for the first time in his life his tasks are assigned, and in a definite order, he also loses his freedom of choice. That, after all, is what is so difficult about life in the phenomenal world, to choose a stand in the give and take of social circumstance. Here, had the word been used, close to the "Absolute," everything is determined; and the fit between the circumstances and the personalities is made to order.

Once Arsene thought he had perceived a change, but that was a slip, and only he had noticed:

> To conclude from this that the incident was internal would, I think, be rash. For my – how shall I say? – my personal system was so distended at the period of which I speak that the distinction between what was inside it and what was outside it was not at all easy to draw. Everything that happened happened inside it, and at the same time everything that happened happened outside it. I trust I make myself plain.[53]

Arsene's mind is structured as was Murphy's, if indeed he was not a reincarnation of Murphy himself. The trick, of course, as *Murphy*'s narrator knew so well, is to express what has happened, and that is never easy. Impressions won't do, for they are of the external, phenomenal world where, as in Proust's fiction, things and persons, objects and subjects continually change aspects in time; and in entering Knott's house the servant has entered the province of the negative, the seat of the mind.

It was Hegel, of course, who stated that mind is negativity; we need only refer to his *Phenomenology of Mind* to discover what that means, and Beckett will later inform us, as if by way of explanation, that No's knife is Yes's wound: the mind's negativity cuts into the positive features of the phenomenal world. So we should pay little attention to his last caveat to the reader: "no symbols where none intended."[54] Symbols are where they are to be found, and they virtually invite interpretation. We should, in avoiding them, prove ourselves to be as inept literary critics as Watt was connoisseur of abstract paintings and foundational physics.[55]

He caught himself wondering what the painter had meant to say instead of continuing to look at what had been painted – a black circle on a white ground, with a blue dot in the upper eastern quadrant and a missing arc at the bottom. That's the part of mind's circuitous return to itself that is forever left open, the object of one's search for self-identity. What was there but a circle and a center in search of one another, but pulled by the force of perceptual tension, all positive this, into a boundless space and endless time? Why, then, was Watt called ignorant of physics? His impression was correct, and instead of interpreting the symbols as being of a mind attempting to order itself, he shows himself an inept psychologist, refusing to see in that design the configuration of his own mental state.

The periphrasis of his statements, his manner of walking, the elaborate rational orderings of simple phenomenal fact all find substance in that one painting. Had he been able to make the interpretation, of course, Watt would not have been Watt, and he would not have been put under the care of Sam, who understood enough to reorder the events of Watt's encounter with the nothingness of existence in the house of Knott.

Watt lived the experience and went mad, and Sam tells us how it was. In the future, Beckett will allow a first-person narrator to tell us How It Is. A journey, yes; but unless one can locate the center of the direction, it is a return to oneself, where nothing is clear, unless by hypothesis, because nothing is fixed until the word has been found, or created, to fix the nothing.

Thus Sam feels free to change the order of Watt's story because in telling Sam that story Watt has placed himself in Sam's charge; and if Sam uses the elaborate constructions of Watt's own mind to express what Watt has experienced, that gives us a clearer picture of *what* is being talked about: the search for a missing order in which the hero substitutes his own negative order for the positive facts of the world.

Arsene had felt the difficulties involved, because he couldn't "eff the ineffable."[56] That too should have been a warning to Watt; as indeed it was, without his knowing it. Watt's attention was drawn to the moral of Arsene's discourse, viz. that any apparent change in the situation was surely false, being only an appearance:

> What was changed, if my information is correct, was the sentiment that a change, other than a change of degree, had taken place. What was changed was existence off the ladder. Do not come down the ladder, Ifor, I haf taken it away.[57]

The free interchange of "f's" for the "v's" in the last sentence indicates a Germanic intonation of the English language, and at least one critic has found an allusion to Wittgenstein,[58] who likened philosophical language to a ladder, which was to be discarded when one has used it to adapt, via a first-order language correction, to reality. For the being of nothing or the nothing of being, however, there is no appropriate language; Watt shall find that ladder veritably taken away. No one can eff the ineffable.

Without the support of a language, the hero is thrown back upon a life with no expectation of any change:

> To hunger, thirst, lust, every day afresh and every day in vain, after the old grog, the old booze, the old whores, that's the nearest we'll ever get to felicity, the new porch and the very latest garden.[59]

Stoic renunciation, and not the hedonistic expectancy of Belacqua, is the essence of the new morality. The lobster does not die a quick death. And when one renounces the positive features of the world, whether in fact or in theory, there remains only the freedom, the negativity of the mind.

Belacqua, whose conduct indicates that he had found the transition from 'monad' to 'gonad' irresistable, confused the functions of his mind and heart, and therefore failed to make this renunciation; and he died by error. Murphy did make the renunciation, but he died in the same way, a victim of his stars. Apparently, then, it makes no difference whether one renounces or not; for the human task is to discover what life means, with death as its only end, and to make this discovery one must continue living. To continue living, however, is to make the next step up the ladder of experience, and to react to this paradox with skepticism. So the dialectic runs.

The German words *meinen* and *weinen* are more than feminine rimes. In the service of a superbly knotted negativity there is no meaning, not even a mean, golden or otherwise, with which to discover a happy denouement to the travail of living, for living is the same as dying. So Watt will be forced to live out his existence looking for a meaning he cannot discover without the aid of his alienist, who must interpret the Rorschach-like painting for him.

That Watt begins and ends in alienation is quite clear. Already on his voyage to Knott's he gives ample demonstration of his malcoordinated personality. His body is not like that of others, attuned to its tasks by the order of nature; and since it is not, he must walk as if directed by what is left of his mind, the reasoning part, separated from the imperious forces of the id. The narrator describes this walk as the dread movements of some Frankensteinian monster:

> Watt's way of advancing due east, for example, was to turn his bust as far as possible towards the north and at the same time to fling out his right leg as far as possible to the south, and then to turn his bust as far as possible to the south and at the same time to fling out his left leg as far as possible towards the north,
>
> . . . .
>
> and then again . . . , and so on, over and over again, many many times, until he reached his destination, and could sit down.[60]

No wonder, then, that self-satisfied "normal" people will feel obliged to throw rocks at his head, as the daily traveler, Lady McCann, does. That is a lady?

At the entrance to the train station, we become privy to the utter meaninglessness of the "normal" daily living in the persons of the Hackett's and Nixon's, and, on the train, to the equal vacuity of spiritual palliatives offered by Mr. Spiro, the editor of the Catholic periodical, *Crux*. There can be no salvation either in living the social life, or in renouncing it in favor of any supernatural concern.

After his stay at Knott's establishment, Watt returns to the same station for his return journey (destination unspecified, the nearest or the furthest terminal, since, in Hegel's world, all movement outward is a return to self anyway), but not before being first ignored, then knocked unconscious by a swinging door, and finally brought to by the contents of a slop bucket, to which is added the liberal sprinkling of his own blood, as the porter lets the bucket fall along with its contents. So much for the external world's treatment of the uniquely human.

The crux of the story is revealed in Watt's service to Knott. The downstairs orderly occupies himself with slop buckets, being careful, under orders, not to mix Knott's slops with those of the servants, and with disposing of Knott's uneaten food. For this there is a dog, kept emaciated to insure final disposal of the food; and a family, the Lynch's, to take care of the dog and his slops. But dogs, like servants, need replacement periodically; and families need continuity. Watt tells us how he supposes this to be done, in proper mathematical matrices. During his first stage of service his commerce is with the Lynch's; Mr. Graves, the gardener; and Mrs. Gorman.

Her day is Thursday, and Watt serves her stout. They spend hours kissing, taking turns sitting on each other's lap, but go no further on the way to Cythera because Mrs. Gorman lacks the time, and Watt the force, to liberate the id.

The appearance of the Galls, father and son, to repair the piano, disturbs Watt's newly ordered life and infects him with reflective curiosity:

> Thus the scene in the music-room with the two Galls, ceased very soon to signify for Watt a piano tuned, an obscure family and professional relation, an exchange of judgments more or less intelligible, and so on, if indeed it had ever signified such things, and became a mere example of light commenting bodies, and stillness motion, and silence sound, and comment comment.[61]

He had lost, we are told, the ability to symbolize, to make an interpretation, since the age of fourteen or fifteen.[62]

As a result he had adopted the only attitude left him, that of taking appearances for reality: " . . . whatever it was Watt saw, with the first look that was enough for Watt, that had always been enough for Watt, more than enough for Watt."[63] For the substance and forms of real things Watt had no concern. His awareness was therefore reduced to the first stage of consciousness' development, if Hegel was right, that of sensuous certainty, where being is empty meaning, the *meinen* of our feminine rime. But Hegel, too, was a punster; *meinen* also meaning "mine."

The person having an experience may say "here is a tree," but in a passage of the moment when the subject reorients himself inward "Here is no tree" is the truth of the statement. To change the bare subject of attribution into a thing, such as a tree, there must be a universal, and for the universal a name; and so by the dialectical swing of everything into its own opposite the mind's experiences must grow until the consciousness of bare sensuous certainty evolves into self-consciousness, and this eventually into the Spirit, where the Absolute mind, like Knott, knows only itself. And, indeed, Watt's stay in the house of Knott would follow this tortuous dialectical path, if only Watt could achieve the state of self-consciousness. But Watt's manner of walking was as close as he could get to the swings of the dialectical consciousness.

Beginning with his reflection on the meaning of the visit by the Galls we find:

> But what was this pursuit of meaning, in this indifference to meaning? And to what did it tend? These are delicate questions. For when Watt at last spoke of this time, it was a time long past, and of which his recollections were, in a sense, perhaps less clear than he would have wished, though too clear for his liking in another. Add to this the notorious difficulty of recapturing, at will, modes of feeling peculiar to a certain time, and to a certain place, and perhaps also to a certain state of health, when the time is past, and the place left, and the body struggling with quite a new situation.[64]

The solipsistic mind is caught in its present moment; as for the past, even Proust needed volumes to reveal its essence, and Molloy will experience the same difficulties.[65]

Words, indeed, are of little help. They name the universals, and particulars have a way of escaping their net. Watt thus becomes anguished before the realization that universal names leave a meaning only partially fulfilled:

> Looking at a pot, for example, or thinking of a pot, at one of Mr. Knott's pots, of one of Mr. Knott's pots, it was in vain that Watt said, Pot, pot. Well, perhaps not in vain, but very nearly. For it was not a pot, the more he looked, the more he reflected, the more he felt sure of that, that it was not a pot at all. It resembled a pot, it was almost a pot, but it was not a pot of which one could say, Pot, pot, and be comforted.[66]

The pun of the comfort in the pot is perhaps indelicate, but the Hegelian meaning is clear. Caught since the very first stages in sensuous consciousness of things, Watt is inevitably thrown back on his own act of meaning; and all of his attempts to explain his experiences will appear both true and false with the passage of time. After his stoic renunciation, he moves to apparent skepticism – the two stages of the dialectic of experience just prior to the attainment of the stage of self-conscious unhappiness.

In these attempts he will use analogies, whose affirmations are always simultaneously negations. How to explain that nothing happens at Knott's?

> For the only way one can speak of nothing is to speak of it as though it were something, just as the only way one can speak of God is to speak of him as though he were a man, which to be sure he was, in a sense, for a time, and as the only way one can speak of man, even our anthropologists have realized that, is to speak of him as though he were a termite.[67]

For a moment, the seeming suffices, whether it be of phenomenal appearance or intellectual explanations. Indeed, all explanations, which exorcise the doubts of a moment, are but specially concocted hypotheses, which can be changed like worn-out clothes that in spite of their present condition may yet be used, as rags, to cover a lost meaning:

> . . . generally speaking it seems probable that the meaning attributed to this particular type of incident, by Watt, in his relations, was now the initial meaning that had been lost and then recovered, and now a meaning quite distinct from the initial meaning, and now a meaning evolved, after a delay of varying length, and with greater or less pains, from the initial absence of meaning.[68]

And when the mind loses its suppleness and can no longer adapt old hypotheses to new meanings or old meanings to new hypotheses, it must fall back on the ready-made rationality of the mathematical matrix, finding truth in the universal connections of things, but in abstraction from the things themselves. The symbol of Watt's madness is his preference for the rational scheme. And his narrator never tires of tiring us with the rationality of Watt's mathematical schemes.

The exorcism of doubt occurs when the word has been found, but it in turn gives rise to new doubts. Where does all this search for meaning stop? If Hegel is to be our guide, with the Absolute. And, for Watt, the Absolute is the master, Knott, who, without his servants, is nothing, a mere not. That's the tragedy of it all. A master whom all servants serve, Knott speaks not; and his risings are the same as his retirings. Like the phenomenal appearance of things, his outward condition changes without the least change in his substantial constitution. This is what Watt discovers in the second stage of his service.

In the following passage, descriptive of Knott's movements, the narrator's style succeeds in creating the content of the narration:

> Here he stood. Here he sat. Here he knelt. Here he lay. Here he moved, to and fro, from the door to the window, from the window to the door; from the window to the door, from the door to the window; from the fire to the bed, from the bed to the fire; from the bed to the fire, from the fire to the bed; from the door to the fire, from the fire to the door, . . . [69]

So the Absolute moves in its dialectical swings, on and on, in relentless, unchanging, change of attitude. The epiphany, albeit a bit hard to read, comes as no surprise to us, armed as we are with the hermeneutical hypothesis that sets all in reasonably clear order, like one of Watt's own lost hypotheses.

To be united with the Absolute is to be lost in Knott, who, according to Watt, needs only two things: first of all, to need nothing, since he or it is the center around which all things spin; and, second, to need a witness of his not (Knott) needing.[70] Absurd, you say? Yes, the unhappy consciousness has embraced both ends of a self-contradiction. To become successful dialecticians, as Hegel has said, we must learn to think contradiction.

Knott needed his servants, but even in their madness they became masters of their master. For Watt did understand, even if now he had begun to walk backwards; and he told his story to Sam, who recounts it to us in a story whose very style mirrors on its surface the depth world of continuous, successive negation.

Sam notices Watt's peculiar walk:

> Continuing my inspection, like one deprived of his senses, I observed, with a distinctness that left no room for doubt, in the adjoining garden whom do you think but Watt, advancing backwards towards me.[71]

Since this quotation, in the jagged juxtapositions of its syntax, gives some indication of the growing instability of Sam's own mind, the refutation of the false rationality of the human mind is left to Arthur, whose coming into the household of Knott sent Watt upstairs into the inner sanctum.

Arthur refutes it by telling the story of the futile academic committee bamboozled by the student, Louit, who has trained a peasant to recognize the roots of cubes and cubes of roots merely by identifying the last digits of numbers on a prefabricated list. Thus, mechanical recognition is both as true and as false as the false and the true can be of the "Fruchtbare Bathos des Erfahrung."[72]

The fruitful bathos of experience is both Watt's and Hegel's in a common expression, entitled by Beckett, *Watt*, whose hero, like Murphy, possessed no Christian name.

After repeated attempts to achieve a style that creates a world, the sorcerer's apprentice finally pulled off the trick. In *Watt*, Beckett, the former servant, was able to overcome the power and the influence of his masters to become a true philosophical novelist in his own right. His auctorial intrusions are, for the most part, relegated to the notes appended to the novel, where any interested reader may easily find them. In a still later performance he will discover the first person, but that will be in his first French short story, "Premier amour" (1945). The first person is, of course, the only authentic witness of human suffering, because it alone can suffer, and be aware of its suffering, in solitude. Sooner or later the solipsist will die – and tell us that story, too, in the three selected French short stories first published in France under the title *Nouvelles* in 1958,[73] although they too were written in that first year after World War II.

We will, then, see more of Watt. The solipsist gone mad will reappear to recount his first love, in a language that is its author's first love; and in the *Stories* he will appear again expelled from his rooms (like the previous lover) in *L'Expulsé*, telling a story to calm his awareness as the surviving consciousness of a dead body (*Le Calmant*), and as the author of his own death in *La Fin*. He will be revived again, and make an appearance in his own name, to explain his plight to Mercier and Camier in a novel by that name written in 1946. But that story is complicated, and deserves a chapter to itself. What began in a vision of the separation of mind from body, and thus of the self from the other, will continue as the author discovers how self and other live and move through a dialectical relationship toward fuller self-consciousness.

CHAPTER IV

THE GIFT OF TONGUES

Watt was written in France, in English, and published there in the same language. And when the character Watt makes his appearance in Beckett's first full-length French novel, *Mercier et Camier* (written, 1946; published, 1970), he informs the peripatetic "pseudo-couple" that his own fame will never reach as far as London. Hardly ever has such a clear autobiographical reference been given in the fictional creations of this author – not even the veiled statement of his preference for the French language in the title "First Love." Obviously he resented being ostracized by the English publishers; so, in a characteristic gesture, the Irish poet turned novelist rejected his native tongue, if not to take his revenge, at least to create his own characteristic means of expression, like Dante, the better to sound the depths of the creative word's source.

He had not yet discovered that the names for universals are mere motions of the air (expiration, if not an idealistic inspiration from above), the *flatus vocis* of the medieval nominalists; nor that the essence of man's being could not be named at all. But when the words used to name the least existent or nonexistent thing all reduce to the breath exhaled to express the ideas of them, the expiration will continue until finally there is no more breath to be exhaled.

The breath, the tongue, the lips constitute so much bodily equipment for producing the sounds that break the silence and, so, organize the surface of the linguistic expression. If an author is led to accept the gift of tongues, the purpose is to discover who and what its possessor might be; he need not lose himself confronting the conventions – and deadening habits – of the guardians of his mother tongue. Thus, if it is the spirit that must be made soul in the exteriorization of thought, the self that receives that spirit must pass outwards unto the Other, before returning back unto itself in full cognizance of its being.

Writing in a foreign tongue places the poet in the location of the other, and writing successfully in a foreign tongue allows a deeper significance to the return. For those readers not capable of following the poet on that voyage through the outlands, Beckett will oblige with English translations of the works that allowed him to discover his essential self-identity, just as he will offer French translations of his English works to permit his French readers at least that pale reflection of a voyage through unknown fields. What is important is the experience, and Beckett has ensured the basis of an artistic rendering of self-alienation by meeting the problem on the level of language itself. He knows whereof he speaks because in speaking he has created the experience, risking all in the pursuit of his first love.

He had, of course, chosen to spend the war years in France, and he had been English reader at the *Ecole Normale Supérieure* in Paris, as well as French reader at Trinity College in Dublin. So the language was not new to him when he returned to Paris, in 1945, to continue his literary career. Then, having chosen French as the means of his artistic expression, Beckett exhibited the predicament of the writer whose struggle with words is as visible on the surface of his prose as it is in the understanding of its depth signification (since the first-person narrators of his French novels are all concerned with the necessity and the simultaneous impossibility of finishing a story).

An edition of *Murphy* had been destroyed in a London air raid, and English publishers were slow in reacting to his works – so many circumstances contributing to an eventually successful artistic achievement. His decision to write in French was passed off by Beckett as the desire to impoverish himself still more;[1] but that "impoverishment" conveyed a typical Beckettian irony. There would be fewer readers of the French works, perhaps; but even with the diminished fortune that would cause, the artist would have deliberately impoverished his means of expression, purifying it, as it were, for a leaner, tenser expression. For the words of the foreign tongue are more clearly words: i.e. sounds, at first difficult to articulate, and then meanings; and not just a preformed stylized pattern of meanings alone. A challenge to the poet, yes; but those foreign sounds provided still another means for artistic exploitation. As his future efforts will attempt to establish, the most lengthy story can be reduced to a breath, to a nonrational ping that announces the birth of a new image.

A choice of the new was a temporary refusal of the old – the obstinate assertion of the author's right to achieve authenticity according to his own lights. Now the pre-Joycean English literary culture need no longer be satirized; it could be roundly ignored, in a gesture similar to the solipsist's rejection of the universally accepted meanings and patterns of conduct in favor of the few, personally chosen significances provided by his own inner experience. Of such stuff are Beckett's "monsters" made, and French supplied him with the tool for making them.

Let us retrace the steps that brought him here, beginning with the solipsists, both "peripatetic" and "seedy," already created in his English prose.

Belacqua, the peripatetic, succumbed at the point of the surgeon's knife; and Murphy, the seedy one, died dully, by his own obstinacy in the pursuit of the pleasures of the mind. Already wielding No's knife,[2] Watt was turned out of Knott's house, not knowing why, even though subsequently he was capable of telling his story to Sam, his keeper, inverting all the orders conceivable within rational discourse. And in listening to Sam's account of Watt's life the reader learns the truth of the latter's expulsion: too long a service in the house of Knott would allow the servant to become aware of his own creative freedom, to be expressed in a judgment on the character of Knott himself. The servant's power is felt in his own freedom to negate, and to do so as the spirit moves him.

Self-consciousness, in Hegel's account of it, begins with the dialectic of the master and the slave, and proceeds to stoicism and skepticism before it affects itself with the inner contradictions that constitute its own undying unhappiness. How can these negations be avoided? Why should they be? Watt follows the dialectic of this negative consciousness, retaining his sense certainty (which is, after all, his), his perceptions, and the force of his understanding (albeit divorced from the conditions of the actual world) until he is brought to the house of Knott, who seems unlimited in the power of his negation. When he is released to his freedom, Watt is resigned to being misunderstood, mistrusted, and mishandled by the average, normal people of this world; and there the story ends, with Watt unaware of the power of his own negation.

That he has continued on his dialectical way, however, is clear. By the time he has met Sam to recount his story of service to Knott in his own quixotic way, he has achieved the ultimate state of an unhappy consciousness, then capable of maintaining the truth of any falsity and the falsity of any truth simultaneously, with the result that he has become institutionalized, in which state he shares his story with Sam. The transitional stage is there by implication; for prior to the simultaneous maintenance of the positive and the negative (of the world and his own mind), he has, perforce, passed the stage of skepticism in which nothing is accepted as true because everything phenomenal is only an appearance.

Appearances, of course, are ordered into law by the force of the mathematical understanding. And it is here that Watt's mind no longer functions, except abstractly. He can understand the relations and can perceive the things, but he can't bring the two together into a coherent representation of the world. His unhappiness is expressed in accepting both the appearances and their formal schemata as real, and he consequently loses all connection with reality. His mind now functions like his body, having a rational direction, but no connection with the means for attaining its ends, swinging from right to left in order to move forward in a straight line.

Can the same story be told in the first person? That is the sorcerer's task if his ventriloquist's lips are not to move; or if moving, to remain unseen. Ultimately, the *Stories* of *No's Knife* will open the way. Although they are a prelude to the monumental structure of the trilogy (*Molloy, Malone Dies, The Unnamable*), they are not the first attempt to solve the author's problem. The first person narrator was introduced in *First Love* and the solipsistic character will be bifurcated into the famous pseudo-couple of *Mercier et Camier*, who meet Watt in person, but whose story is related by a narrator who claims to have been along on their voyage. Chronology is not the surest of the guides to interpretation.

Here once again the concept of literary structure is more illuminating. The Hegelian dialectic of Watt, which is repeated in the title of the collected *Stories*, will in the trilogy be replaced by the existential ontology of Heidegger. The reasons are both philosophical and literary.

Hegel's dialectic of experience is an ideal, sequential process of the return of consciousness to itself. At first aware of an object (any kind at all, an in itself), consciousness then becomes aware of itself as aware of that object, existing for itself, and may return to itself, where as fully self-conscious it then exists both in and for itself. That, of course, is the solipsist's goal, but it is constantly obstructed by the existence of others. *First Love* will give an account of the obstruction to self-sufficiency experienced in an aborted act of love. Since the dialectical movement of self to other back to self constitutes an ideal schema for an understanding of all conscious experience, time in the process becomes unreal, hidden by the appearances of objects in the world.

Historically considered, Hegel was the original phenomenologist whose ultra-rationalism (satirized in Watt's characterization) spawned, by way of reaction, the existentialist movement, although no one seems to be willing to have invented the name of that school of thought. The phenomenologist's explanations change when time is viewed as the very stuff of consciousness, when the temporality of consciousness and the spatiality of the world are taken, in analytical, phenomenological ontology, as the "existentials" to replace the dialectician's "universals."

We, of course, know of the change from our reading of the history of philosophy; the literary artist, for his part, is forced to make his discoveries in the context of his words – not in contemplating the structure of the "real" world as the philosophers sometimes pretend, but in forging the structures of the pretended worlds lived by his characters. There may still be a dialectic of the other, but it will be embodied in the interchange of voices as character dialogues with its creator, always suspecting, yea, fearing, that the latter will bring him to an untimely end; or, what is perhaps worse, to no end at all.

The question will occur time and time again, Who am I? Where, and why? How can I put it? In what tense, since time is so important? Is there a way out? Such questions as these are as important for the author, as they are for us; they verily bristle with ontological concern, but we and the author are only two figures of the communication that must be mediated by the words used by the character to enunciate that concern for us all.

Molloy, Malone Dies and *The Unnamable* will leave no room for doubt; but the narrator of the *Stories*, and still later, that of the *Texts* must come to their own understanding by finding the proper words. The saving grace is there, in the gift of tongues, first received by Beckett through his education and experience, then liberally showered upon his narrators.

It will be convenient, for the purposes of my demonstration, to consider the structures of *First Love* (1945), *Mercier et Camier* (1946), and the *Stories* (1945) in that order.

Without the vantage point of the *Stories* that look forward to the trilogy, Beckett's first two attempts at prose creation in French might appear as simple repetitions of the self-other dilemmas found in *More Pricks, Murphy,* and *Watt*. How can the solipsist love? Under what conditions be intimately united with another? We already know the answers from the earlier English novels: unhappily, contingently, suffering the contradictions of internal inconsistency. For if every lover carries with himself the seeds of his own destruction (whence the adult's fear of the child) and every self, his own other, the project of solipsism must go for nought. The later narrators, beginning with the one of *Texts for Nothing*, will come to know this, since the words to which they give tongue are the words of others, even if, as the Unnamable claims, there were no man on duty, no man in charge. Since there are no private languages, there can be no solipsists fully aware of themselves, only would-be solipsists haunted by the existence of others.

In the abstract, there are two ways out of the would-be solipsist's predicament: either to come to grips with the existence of the other, and thereby to give up any claim to being alone in the world; or to achieve one's self-identity by developing a personal style in the use of the power of speech that everyone shares.

The first way is that of *First Love* and *Mercier et Camier*, but it leads to a philosophical failure in that the narrator of the former and the heroes of the latter are driven back into their respective solitudes. Yet the telling of those stories was more than a literary exercise.

Beckett was led to the second formulation of the surcease from solipsism, and eventually to an enriched basic philosophical structure for his fictional worlds. Buying time, as it were; going back to apprenticeship in a new language for the development of a style that will allow his future narrators to borrow his own words for the telling of their stories, Beckett would try the old themes (self-other-world) once more, before inventing the new (human consciousness as existential disclosedness).

The first-person narrator of *First Love* has no name, Christian or otherwise. His narration is composed in the past tense around the nucleus of a single act, his "marriage" and separation from the girl whose presence to him is necessary so that he might be freed to think of other things. Her name was 'Lulu,' pronounced as in English 'Loulou.' As he got to know her better, the narrator exercises his right of authority by giving her another name, 'Anne.'

The affliction that goes by the name of love came upon him at about the time his father died. His father had always protected him from the other members of the family, and had provided in his will that the son inherit his own room when he (the father) was no longer there to protect him. The son, of course, thought it would have been better to inherit the whole house; that way it would have been harder to dispossess him of his inheritance. The money he was given at that time he kept for future contingencies.

Often the son went to the cemetery, not only to visit his father's grave, but to find himself alone among the dead. He was not depressed there, ate well, and eliminated his wastes wherever he pleased. On one trip he had noted the date of his father's death – since it was marked on the stone, along with that of his birth. He had a reason to remember the death date, for his solitude began with the passing of the old man.

Curious, one day, to know how old his father was at the time of his death, he had to return to look up the date of his birth. With both dates noted in his little book, he was capable of calculating his own present age– about twenty-five. The union of father and son is curiously related. To compute his own age, the son would have to know how old his father was at the time of his own birth. That day was not engraved on his father's tombstone, but was indelibly marked on his own inner psychic structures: the date of his first expulsion, from his mother's womb into the world, the beginning of his own conscious time, announced by his cries as he began drawing his first breath in a foreign world. If, by sympathetic projection, this day was translated into a date within his father's life, the calculation of his own age at the time of his father's death would be possible; and allowing for error, the answer would be given approximatively: he was about twenty-five when his father left him to fend for himself.

He was dispossessed of his inherited room under the most peculiar circumstances. Always severely constipated, one day he took too long a time discharging his waste in the bathroom – was it really constipation, or actually diarrhea? A verbal nicety of difference, at best – and upon returning to his room he found the door locked with all his belongings piled in front. A hint? Yes; all the other rooms were locked too.

He was expelled into the world once more, in spite of the hypocritical past he shared with the other members of the family. The son who was living – and thus dying – could not restrain himself from musing over the relationship with his already dead father:

> Pauvre papa, il devait être bien emmerdé ce jour-là, s'il pouvait me voir, nous voir, emmerdé pour moi je veux dire. A moins que, dans sa grande sagesse de désincarné, il ne vit plus loin que son fils, dont le cadavre n'était pas encore tout à fait au point.[3]

It was the son's bowels that had so reluctantly moved, and the father's will that had been shat upon.

The narrator met Lulu on a bench facing a canal, built between two dead trees. Facing away from the canal, at least he could not be surprised by an unexpected and unwanted visitor from that side. He was lying on the bench when she approached and asked for room to sit down. At first tempted to move on, but too tired to do so, he made room. She sat, and ignoring him, sang songs in a voice never on key, but nonetheless pleasant to hear. Nothing happened between them that first night, except the pleasure he had experienced in hearing her voice.

Since it was raining the next night, he thought he would not be disturbed on the bench again. He was mistaken. He bluntly asked her if it was her intention to disturb him every night. – "Really, am I disturbing you?" – "Yes, I can't lie down on the whole bench when you are taking up part of it." That, it turned out, was a mistake; for in answering her first question, he allowed her to ask another. – "You really insist on lying down?"

They solved the problem by her sitting, and his placing his legs across her knees. That, too, was a mistake. He could feel the plump flesh of her thighs under his own calves. The thought occurred to him that he might kick her in the privates, just to make her go away, but the thought was vain. Instead, he reflects on how, when hearing the words "lying down," one immediately has the image of a body spread out in its full length. That, after all, is the trick of the literary conjurer; but this conjurer was at grips with his own mind, trying to keep straight the idea of itself as distinct from what is not itself in the foreground of the world.

Murphy and Celia, all over again: the solipsist whose mental comfort is more than slightly deranged by the physical exigencies of the body. As Lulu caresses his ankles, he can feel the tumescence of his sexual organ; for her part, being a woman, she can flair it out – which, in his estimation, she would do even at a ten kilometers' distance. That was the night of their union, and consequently of the narrator's "exile":

> Ce qu'on appelle l'amour c'est l'exil, avec de temps en temps une carte postale du pays, voilà mon sentiment ce soir.[4]

Exiled first from his warm enclosure within his mother's womb, then from the patronage of his father, then from his inherited room, he is now exiled from the very center of himself. Not even his ideas are any longer solely his: Lulu's image gets tied up with the bench.

What happens to the bench during the day not even Bishop Berkeley could tell; for the narrator is never there during the day and Berkeley had to invent or evoke the deity to preserve the existence of any object under the conditions of non-perception. During the days, the narrator supplied himself with food. But such a tenuous arrangement could not be maintained. Lulu disturbed him on the bench physically, and, psychologically, her image disturbed him even in her absence: a new, no-win situation.

Taking the dilemma by the horns, he tells her he has had enough. That was too vague, and offered another opening for his adversary. – "You don't want me to come by any more?" He then answers by going through the horns. – "You can come by from time to time, but less often." That solution will give him time to reflect on his pains. If he were only pains, if life were only pricks and there were no kicks to be delivered against them, the problem would be easy. Each part of the body, beginning with the hair on the head and ending with the nails on the toes could in reflection upon them be placed in a meaningful order. Indolence, like the original Belacqua? No, that's the coward's way out.

While he is lost in his ideal world of pure pains, she has not been idle. His last answer was still vague. – "How often?" – "Not at all if possible; and if not, the least often possible." And if she didn't get the message from that, he could settle the issue by taking refuge in an abandoned barn. And that he does.

But he had reckoned badly with the other horn of his dilemma. Away, she continues to haunt his consciousness. He finds himself scratching her name on the dried "cow-pies." That must have been love. But what kind? Physical? No, he was not a satyr. Platonic, then? No, for why should he be writing her name on dry cow turds if his love were pure and disinterested? One way of ridding himself of his obsession would be to give her another name. He does: 'Anne.' After all, baptism exorcises the devil until the loss of innocence lets him return to reclaim the fallen as his. The expected gift of grace fails to happen. Now he writes 'Anne' on those cow-flops, and licks his fingers when he is finished.

To put an end to this suffering he retreats to the first horn of his dilemma: to suffer her presence. At least that way he is relieved of the necessity of thinking of her; when she is beside him he is freed to think of other things. So he returns to the bench one day, only to find that she will not show up. The next night he comes by earlier, and is not disappointed. The frost had already made its annual appearance, but that was not enough to dissuade her. – "Could she explain what she saw in him?" – "No." He cries, in desperation: not surprisingly, for he had always cried easily. – "Did she come by just to see him?" – "Yes." – "Well, here I am." – "And you, did you come by to see me?" ––"Here I am, Here I am."

They sit down next to one another, and he feels the urge to get up as quickly as they had sat down. The bench was hot, or at least seemed to burn his flesh. Would she sing a song? The words of the song escape him, as they tend to escape all true poets, in the beginning; but upon reflection he can recall something about lemon or orange trees.

His departure that night prefigures his final departure from her rooms, where she is giving birth to their child. As he leaves, her voice becomes more faint, and he can no longer make out the words or the tune of the song she sings. He comes back to settle the uncertainty. The voice comes and goes, and when the song has finally been finished, he leaves her without being able to determine what song she had been singing.

Some time later, they meet on the bench once more. It is too wet for them to sit down, so they stroll along the path. Feeling that he might have some pleasure in taking her by the arm, he does, but withdraws his support from her as soon as he notices that he feels no pleasure. He loses himself in the inspection of her features, and is disturbed by her question: Could she sing him a song? No; she could say something instead.

She reveals that she has an apartment: two rooms, a hall, and a kitchen, where she installs him in one of the rooms. As per habit, she gets undressed as quickly as possible, even before showing him the house. He will have none of that, and asks to see his quarters. The room is full of furniture, which he moves out into the hall. The couch is put in a position facing the back of the room, close by the door, but that position allowed a weak light to pass through the frosted window. So he turns the sofa, back side to the center of the room, and crawls over the top into his protective niche: back to the womb within the room. There being no chamber pot, whose name (*vase de nuit*) sends his mind off in a fanciful poetic flight, he must settle for the equally symbolic "fait-tout," a large, covered, stewing pot, used for "doing" everything. He doesn't need the cover, so he refuses it, and goes to sleep still fully clothed.

The next morning he wakes up disheveled, the chamber pot mysteriously empty, but with his bed filled with the nude corpulence of his hostess. If only his sex could talk, it might have told the story of what happened. She brings him his food, and allows him to mark the time of his own existence.

But not without some uncomfortable, yea, distressing accompaniments. She lives by prostitution, and the giggles and groans of her clients disturb the narrator's peace. He tries to arrange the situation by shaming her: So you live by prostitution? But she retaliates in kind: No, *we* live by prostitution. And the statement is true in a double sense, for they both have prostituted their authentic selves, and he lives off her ill-got earnings.

Although she feels no shame before this Other, he feels shame before himself. What brings about his final decision to leave her is her badgering him with her pregnancy. Already the child is leaping in her womb. But if it leaps, he says, he could not be the son of this Belacqua. Then, as she begins crying out in the pains of travail, he leaves the apartment, "hearing" the sounds, knowing they are in the air. He tries to control them by coming and going, towards and from the source. When he walks, he cannot hear the sounds because of his footfall; but as he stops they fill his ears. Since it is important that they stop, he continues to walk away.

Had he stayed there, he would only have heard the cries of the new born child announcing a new expulsion, a new life instead of the death to which his own life has been committed. Therefore, instead of the symbolic death his love had been, and the symbolic death his own fatherhood would have been, he has chosen to pursue his own real death, reunited once again with himself: alone, and on the road. Ironically, the narrator's first love has always been himself.

The structure of the story may be said to go against the grain of nature, yet the events are determined in their pattern. Refusing the real world to live in the comfort of his own, the narrator may structure his tale on the same basis as the life recounted has been lived. There were a series of expulsions: from mother, from father, from family – to self; from self in his fateful union with Lulu, where he becomes lost in the Other; and, finally, expulsion from that union with the Other, back to Self. That is the dialectical path on which every consciousness is thrown, if it is to achieve its fulfillment in total self-consciousness.

And for those readers who cannot perceive the structure, Beckett has his narrator reveal the secret in that climactic moment on the canal-side bench when he muses on the discomfort caused him by the girl. His task, then, as it became later when it fell to him as narrator to tell his story, was to give order to the ideas – of himself, and of the other, and "by laziness," of the world.[5] And once this private world has been structured as it is, the least perceptive reader can understand those relations – other than the coincidence of times – by which the death of the narrator's father and his own "marriage" have come to be associated.

Each marked an expulsion, but in relation to the direction of the expulsion, in an inverse pattern. With the death of his father, he was expelled into selfdom; and with his union with Lulu, from self-dom to a necessary relation with the Other. The two expulsions, then, form a chiasmatic figure in the come and go of conscious existence; and it is only the indolence of the solipsist that allows him to stop the dialectical process with the achievement of a self on the other side of that Other, for there are still others to come.

In *Mercier et Camier*, the philosophical structure of the dialectical process is maintained, but there is a familiar strain in the relationship between the narrator and the events narrated. These latter concern the attempts of Mercier and Camier to undertake a voyage together (symbolizing a life's project), but how they transpire is recounted by an omniscient, third-person narrator, who claims to have accompanied them on the trip. Since there is no justification for this claim, other than the opening statement of the narrator, one can be sure that Beckett has fallen back into the errors of his former ways, playing the ventriloquist, but allowing his lips to be seen as he breathes the words for his puppets.

It may have been the length of the composition, or the impossibility of deciding between Mercier and Camier for a central point of view. Whatever were the case, we can be sure that the unnamed narrator bears the name 'Sam,' like Watt's keeper; and his claim to having been there on the voyage of his characters could bespeak an admission that the novel is in some sense autobiographical.

Watt will appear, towards the end of the journey, to disturb the habitués of a bistro with his rising defiance of life; and he will claim to have known Camier's mother as a saintly woman (to have been cursed with such a son). In return, Mercier will claim to have known one Murphy, who has disappeared from this earth under mysterious circumstances, without leaving a trace of his physical existence; but these curious facts will go unexplained, in context, until the Unnamable lays bare the true connection between the creator of a fictional character and the creature created in that impersonal voice.

In this respect, therefore, *Mercier et Camier* was a mistake, and we can understand Beckett's hesitation in publishing it in the year it was finished. It's actual date of publication, moreover, suggests that he had been persuaded to capitalize upon his Nobel laureateship (1969), since the *Editions de Minuit* produced the first full version of it in 1970. That was one way the author had of repaying his editors for taking the chance of losing their hats on the publication of *Molloy* in 1950.[6] As of this writing (1974), the text has yet to appear in English translation.

But let us return to the internal evidence.

The narrator begins his account with a prologue that defines the conditions under which the heroes will make their trip. It will be relatively easy: they never really leave their home region; and although the weather will often be inclement, they are acclimated to it; besides, the zone is temperate. Though they are not rich, they have what it takes to finish the voyage without begging; and the voyage will be undertaken determinedly, following frequent consultations between the two voyagers. Except for the duality of the protagonists, that is a description that fits the life of a single, reflective person. But to pursue that symbolism too soon will remove some of the pleasure of the epiphany.

Mercier and Camier had taken an appointment in a public square to begin their trip. Mercier gets there first, doesn't find Camier; waits for five minutes and takes a stroll. Meanwhile Camier arrives, doesn't find Mercier; waits for five minutes and takes a stroll. Each strolls for about fifteen minutes and returns for another five minute wait; then each returns to find the other gone, and leaves for another stroll, this time for ten minutes. Since Mercier had arrived at 9:05, he returns from his last, ten minute stroll at 9:40, just as Camier is departing after his five minute wait following the fifteen minute stroll. Mercier waits five minutes again, departs at 9:45, and returns after a five minute stroll to find Camier returning from his last ten minute stroll. They come together at ten till ten, like the hands on the clock.[7]

The process is almost Wattian in conception, and indeed the narrator comments that such a beginning "stinks of artifice."[8]

In their first sortie (for the voyage is a series of returns to the city), it is raining. As if to celebrate having found one another in that out-of-phase meeting, they begin to count cadence – one-two, one-two. The rain drives them into the doorway of a public display house. There two dogs are copulating, and a bicycle is tied to a fixture. The guardian orders them to get out with "their" dogs and "their bicycle." To stay the night there, they would have to bribe the guard with a half-crown. They consult after hazing the guard, and decide to leave.

Mercier will take the bicycle and the raincoat and finish the first leg of the trip. When he has done everything necessary, Camier will join him, with the umbrella. But as this decision was taken, the sun comes out, just in time to go down. They continue their trip together, Mercier taking the handlebars of the bicycle, and Camier the saddle-seat: mind and body, then, in proper alignment for the moment; but the bicycle flaps against their thighs as they walk. It is not built for two.

Walking toward the canal (where the entire voyage will end, much later), Mercier is given to his thoughts, "about the horrors of existence,"[9] and reflects that one could bring the whole voyage short by jumping into the canal. They stop at a bar to refresh themselves, but are not permitted to enter because of the bicycle. They chain it to a tree, and go into the bar across the street, where they enter into another conference. They consider ten propositions, the first of which is that it is useless to go on for the moment; and the ninth, that only one thing counts, to leave. The tenth comments on the whole situation: "Merde!" (Shit).

In the street they cannot decide which direction to take, so they throw the umbrella in the air, to allow the direction to be decided by chance. As they take off down a street, they meet a man in full evening dress, leading them to the home of a prostitute, Hélène. She has only a divan and her own bed for the three of them (including herself), and Mercier opts for the divan. Camier accepts the prostitute's bed, if the love-making could be short; but she will have none of that, and he is forced to make love to her until four o'clock in the morning, when she serves him onion soup.

The next day they leave Hélène's, but discover that they have misplaced the bicycle, their sack of provisions, and the umbrella. Deciding that they do not need these appurtenances, Camier bilks Mercier into taking the first train South (that's the closest station). But instead of the express Mercier expected, the train is an omnibus. On the way to the station, they take turns wearing the raincoat, since the two of them have only one. On the way, they feel hungry; and since Camier is then wearing the coat, he goes for a cake. Aside from their night of sleep, this episode is their first separation.

While Camier is away, Mercier is accosted by his own children, whom he drives away, cursing the day he had inadvertently prolonged the life of the species, instead of bringing it to an abrupt end, as he would have, had he been in full possession of his faculties when he had procreated them. Camier finds Mercier in tears, but they are both cheered by the sight of a woman run down in an accident.

In the train, they are entertained with the life history of a traveler, Mr. Madden, expert gelder of cattle and butcher, who was forced to flee the home of his adoptive parents, as a tender boy, for having knocked up a milkmaid. The rest of his not uneventful life was spent in caring for cadavers. Mercier curses their having taken an omnibus. At the station following the one where Madden had gotten off the train, the two travelers get down to look for food. Camier explains that, given Mercier's condition at the time he had driven off his children, they had to leave, but without leaving; and this way, they could always return, either by tramway, bus, or train.

The inn where they seek their food has been sold out of what they want, and they risk getting nothing, unless they take a room. They go up, with two bottles of scotch, drink and sleep the night away.

In the meantime, a Mr. Conaire (an unbelievable name, since it suggests a person looking like a cunt) appears looking for Camier, who had the day before given him an appointment at this very inn for this day. Camier, it appears, was a gumshoe, a searcher for lost persons. Mr. Conaire had lost his dog, and hired Camier to find it.

Mercier and Camier arise early to take up their travels. Now in the fields beyond the inn, Camier throws away an envelope in which he had stored some clues for his work: several buttons, two samples of hair (head and body), an embroidered handkerchief, several laces, a toothbrush, a piece of rubber, a garter, and some odds and ends of cloth. He won't work anymore, he won't need them. What's worse, he has a cyst on his private parts that has been acting up. Mercier consoles him in his pain. Think of it this way: when you fear for your cyst, think of your fistula, and when that makes you tremble, think about the chancres. The system works as well for those who think they are happy. After all, if you had no pains at all, neither in the body nor in the head, you would have nothing to think about. That's nature's way of smiling, if not laughing, at you.[10]

So much philosophy forces them into reflecting about their present situation once again; and in the reflection they decide to return in order to recuperate their lost possessions. Camier's thoughts remain on the exigencies of the trip, while Mercier loses himself in his dreams.

Camier goes back to the inn for sandwiches, where he is accosted by Conaire. Explaining his decision to quit his profession, Camier gives Conaire a lesson in Berkeleyan idealism: yesterday he had thought a lot about his client, but then he had stopped thinking of him. At that moment, it was as if the client had never existed (hence, there was no appointment to be kept). Well, almost; it was as if the client had ceased to be – or better, as if he existed, but unknown to Camier.[11]

When Camier returns to Mercier with the sandwiches, he finds him queasy in the stomach. Mercier vomits, and feels better for it. They decide to throw away the raincoat, since there was only one for the two of them, but forget to empty its pockets: a few punched entrance tickets, used matches, some ends of newspaper margins scribbled with forgotten appointments, the last tenth of a used pencil, a few sheets of dirty toilet paper, some leaky condoms, and a bit of dust.[12] Would that be missed? How could it be, since it contains all that is left of a life? To continue their trip they need a slogan: "lenteur et circonspection" (slow and easy).

Proceeding in all circumspection, slowly, they arrive some time later back in the city. It is raining, and they hole up once again at Hélène's place, where they pass a quiet evening. The next day it is still raining, so they spend the day with the prostitute, making love before the fire on the rug. The day following they are in the street again, without having found their sack of provisions. They go to a bar for another conference on the conduct of their trip, and return to Hélène's. No love that night, even Camier wanted to conserve his strenth for what is to follow. They leave at noon the next day, but have forgotten the umbrella. Camier returns to retrieve it, and takes so long that Mercier knows his corporeally minded friend has had another go at the prostitute.

Should they throw the umbrella away? Not if it should ever be useful. But it is raining now, and cannot be opened. Camier takes it back to Hélène for repair, staying longer this time than the last – it always lasts longer the second time! While he is gone, Mercier is left to the content of his inner thoughts on the slowness, and the insufficiency of time. His task was to find the sack and the bicycle.

In Camier's favor it should be stated that he would have preferred to look for the sack and bicycle rather than face a second round of Hélène's favors, but Mercier had insisted.

When they come together again, Camier is sitting in a bar, and listening in on a conversation about the sin of artificial insemination. He threatens to report the interlocutors to their archbishop. Mercier finds him, and they are served by a giant waiter, each looking at the other in turn until nine images have been created: a personal one for each of the three, and one for each of the others, multiplied by the three persons in the round of gazing. In addition, the three of them are the object of all the other gazes, from the other clients of the bar. They are served, and try to drink to the success of . . . but they forget of what. Each is for the moment drawn in upon his own failure to succeed in the last mutually decided mission. Camier had broken the umbrella, by making an abrupt movement as he was helping Hélène with the repair; and Mercier had found the bicycle, but with all its essential parts stolen. The sack he never saw.

Out of the blue, Mercier says "Yes." And Camier wants to know what he is affirming. Mercier denies having said yes, since the last time he remembers having said it was to take himself a wife. And that was disastrous. What a funnel! Ever try making love to a hole in the ground?

Where are we going, without umbrella, sack, and bicycle? Who should be caring, as long as we are going? The answer is Mercier's, since he has long since understood that all voyages of life end in the grave. They join their steps, and continue along until they meet a policeman, from whom they ask directions to the nearest, cleanest whorehouse. He tries to correct them of the error of their ways, by grabbing Camier's arm and twisting it. But Mercier, with an awkward but effective kick to his groin, brings him down; and Camier finishes him off with repeated blows to the head with the victim's own billy-club. To show the sympathy connoted by his name, Mercier likewise gives a single blow, once the head has been covered, and suggests that the final blow might have been the one which actually killed that guardian of the peace. For refuge, they wend their way back to Hélène's.

The next day, walking through the heath, Camier has taken cold, and they both have the impression of being seen. Conscience? Whispers of guilt for the murder they had committed? Perhaps. But equally probably, their awareness of the narrator's presence; and hence, the awareness that their story is being told. But that would be the ventriloquist caught with his lips in motion again.

Since they had spent about eight consecutive nights at Hélène's place, they decide to pass this one under the open stars. Mercier tells Camier it would be good to let his privates sink into the moisture of the peat bogs; certainly they have been overworked, recently, what with the cyst and all. After a short conference, in which it was suggested that they use half their energy for a short walk and the other half for a return and sleep, they continue to grope their way through the darkness until they reach the walls of a ruin. There they sleep.

But before bedding down Camier moves along at a slight distance, and in the morning takes off on his own. When Mercier awakes to look for him, he finds Camier gone. Following at a distance, he sees Camier take the right fork of a road; and when he reaches that point, he deliberately takes the left, leading back to town. The mind will no longer follow that body. Such is their fourth separation of the trip.

In the summer, they are both back in the city. Camier leaves a house, walking with his head resting upon his breast. To his surprise, he feels his head lifted by the hand of a person who asks if Camier has seen that person's friend, Mercier. Camier says that Mercier is somewhere behind. The stranger is none other than Watt. Watt, it seems, has known Camier's mother, and has known Mercier since this latter was a boy. Watt re-introduces the estranged couple, who insist on using the polite form of the second-person address – such has been the result of their last parting.

As Watt leads them to a bar, they meet another policeman. But it is Watt who avoids another catastrophe. They were walking, all three, arm in arm, when the policeman told them to separate in order to allow more honest citizens to pass. Watt cajoles the policeman into ignoring the aggressive attitudes of Camier and Mercier, now being brought together once again before the disapproving glance of the symbol of society, by telling him that they are both a bit off their rocker: Mercier takes himself for John the Baptist; and Camier takes himself for either Julius Caesar or Toussaint L'Ouverture; while he, Watt says, is content with the destiny nature has laid on him. (As we recall from reading his story, he was himself a bit crazy, maddened by his vision of Knott's nothingness, and rejected by the world.)

In the bar, Watt directs them to a table and orders double whiskeys. They would have been suspect, ordering beers at the bar itself. They drink to Mercier and Camier's reconciliation, and to Watt's knowledge that from one of them, there will be born, indeed there has been born, the one who, having nothing, wishes for nothing.[13] That, we recall, is the message of Geulincx's *Ethics*. But who is that one? Watt himself indeed, but he is only the reflection of the union of Mercier and Camier: a man born of flesh and spirit, but who always has had the greatest of difficulty in bringing both halves of his parentage into a significant union.

Having united them only for the moment, Watt suddenly goes berserk. He disturbs the whole bar by shouting "La vie au poteau!" (Life to the gallows, or Let there be death). The barkeep comes over to throw them out, and now rôles become switched. Symbol becomes signification; it is Mercier and Camier who talk the keeper of the bar into serving another round. Watt, they say, is suffering badly. His wife has just died; he has just lost his twin brother! They get their drink, but Watt persists, with "La vie aux chiottes!" (Life to the shithouse, or Life down the drain). As if affronted by the success of his friends' ploy, he had taken up Camier's cane and broken it on the table of a beer drinker sitting at a table nearby, overturning table, beer, and drinker. Despairing of saving the situation again, Mercier and Camier take to the air, as they hear Watt's last war-cry: "Vive Quin!" Who is or was Quin, they reflect, not being able to say who? Belacqua's best man? Hairy Tiny Capper Quin? Why not? For the narrator was at that hoedown too, and the first hero to have realized the wisdom of desiring nothing, because he was nothing, was none other than Murphy.

Together they approach the canal once more, where they stop on the *Pont des Ecluses*: the bridge over the locks. Here is the end of their journey. What are they to do now? Sigh? Mercier has been here before, under the guidance of Watt. Camier descends, twice, to look into the water. They both see the lights of a hospital – for diseases of the skin, but it is too early for them to hear the cries of the inmates. Before they can be heard, Camier takes his leave, with a genial "good bye," while Mercier returns the gesture with a sincere "Good night." Later he will hear the human cries from the hospital, along with the patter of the rain on the water of the canal. Off the bridge, the waters flow on at independent levels. So the voyage ends, at the bridge of sighs.

The clue to the interpretation of this tale is given by the Unnamable, who refers to Mercier and Camier as a "pseudo-couple."[14] Reflection will show why. A couple can be false in one of two ways: either the couple is not one because its members fail to coalesce; or the couple is one, and not a composite at all, but only suffering an apparent split into two. According to the first interpretation, Mercier and Camier never achieve the status of unity. They decide to meet, to make a trip together; but each pursues his own habits and inclinations, until they finally part forever. That is the surface, literal, reading of the events in the story.

But there is another interpretation, which corresponds with the symbolic significance of the two personalities. Mercier, the name suggests, stands for pity, grace, detachment, in a word, the soul; he is the merchant of ideas, always ready with a general explanation of Camier's and his own attachment to their voyage. Camier's name suggests, on the other hand, the driving physical force (a cam), lost in hunger for food and sex; incorporating, in a word, the body. By profession he is a gumshoe, a searcher for lost persons: not insignificantly, for his own as well as those he has been hired to find. United with Mercier, from moment to moment, he need no longer look for his own; for with Mercier he forms a complete human being. But it is Mercier, in his function of philosophical interpretant, who knows that they each have reached the completion of their essence by being united to the other. When they part for the last time, their story must end, for neither can live without the other. They form a pseudo-couple in this interpretation because they have always been one, and not two persons.

Nor is it a matter of paying one's money and taking one's choice; for both interpretations fit the context. The first level of the events' happening points to a second, that of their significance. And both levels are united in a third.

The self-other-world complex that exists between Mercier, Camier and the various symbols of society (a museum guardian, numerous inn-keepers, and two policemen) follows the pattern of a dialectic structure. Their conversation tends to be contradictory, and after each consultation they wind up in skepticism or in resignation to the effects of chance. The mind becomes what it is by its relation to a body, and vice versa. What tears the two apart is the presence of a woman (Mercier's wife, and Hélène). Coming together and parting, only to come together and part once more, their trip is dialectically determined.

But whether together or apart, they are both essential functions of a single person. That single person may be dominated at one moment by one of his functions and at another by another; but in each period of dominance, that function never loses complete contact with the subdominant. Camier knows that he causes himself pain by his physical excesses, and counts on the future presence of Mercier's meditations to soothe them over. For Mercier has lived long enough to know that pains are necessary in order to have something to think about; and when one has thought long enough one knows that Geulincx was not far from right: he who has nothing will want nothing – if not to retain the nothing that he has.

Thus, once again, we find a tri-level literary universe in the structure of *Mercier and Camier*: the literal level of significance is the dialectical pattern of the events in the lives of two distinct personalities, Mercier and Camier; that passes over into an allegorical interpretation of the real unity of these personalities into a single essence of a human being; and that illustrates the moral meaning of a life of suffering; salvation exists in wishing for nothing when one has nothing.

No Dante scholar could stop here, however, The anagogical meaning is that God has died. Camier refers to God as our little "omni-omni," the impassive one.[15] Mercier had begun the give and take by cursing the rain: "Quant à toi, je t'emmerde!"[16] (As for you, I'm shitting all over you!) Camier misses the aptness in this reversal of roles between Creator and creature (as the liberated mind, may, in its freedom, play havoc with divine providence; and the structure of the novel bears out the degree to which the will of Mercier is uncontrolled by the well ordered progress of nature), and tries to correct his friend. It is the omni-omni who shits on you, he claims.[17] But as for the Almighty's vaunted passivity, that only reminds Mercier of his wife, who, in her own freedom, has always remained completely impassive to his every expressed desire. And he begs Camier to leave her name out of the discussion.

Later, when neither Mercier nor Camier can open the umbrella in the rain, Mercier wants to know what happened to God, and Camier replies that they had earlier denied him.[18] True, says Mercier, but that is no reason for Him to be so vengeful! Thereafter, all their actions are self-determined, and their suffering self-endured. Man is alone in this world, for as much time as the temporality of his consciousness shall itself endure. The significance of each such life is its self-transcendence. To wish for anything more is to make oneself more unhappy than one already is. Thus, in this early novel the basis of Beckett's pessimistic vision shows a Janus-faced configuration: looking backward toward the unhappy separation of mind and body, and forward to the full self-consciousness of human reality separated from the transcendent deity. A transitional piece indeed.

But if God is dead in the universe of Mercier and Camier, the narrator is not. He makes his presence felt all along the road. And that is a flaw the fledgling novelist will have to learn how to control. That Beckett will eventually learn how to control his propensity to intrude upon the universe of his novels becomes clear in the collected *Stories*. The evidence is to be found in the examination of their structure(s), to which we now turn.

We shall find traces there of the continued existence of Watt, as Beckett continues to look for an answer to the question implied in that name.

The structure of the *Stories* considered as three in one, is already a miniature trilogy, complicated by the fact of the individual structures of the three; for, apart from the way in which Beckett presents them, either of the three might have been the beginning, the middle, or the end of a triune narrative. Surely an explanation is in order. Let us therefore begin with a description of the events of the three first-person narratives, assuming the narrator to be the same for all three stories.

In *The Expelled* the narrator recalls the past, a life of continual movement from one location to another, without an apparent end. Thrown out of one habitation, he is unable to find another, and rejects the hospitality of a coach-driver for the comfort of his coach, which is in turn rejected for the insecurity of the streets, where the journey continues indefinitely. Symbolism: the expulsion from the womb and projected return thereto; nothing that is only a symbol of it is an adequate substitute for the womb.

The Calmative takes up the narrative following the death of the expelled narrator, his death being a further repetition of his initial expulsion, and his present location a new symbol for the womb, where instead of growing, his body must continue to rot: a frightful sound to hear, as he first imagines, then utters those words. To calm his fears, aroused by the darkness of his new abode, he tells himself a story, just as his father used to tell him stories as a youth to calm the fears of the approaching night and the temporary loss of consciousness that is sleep.

The story told concerns a journey through a city that has been deserted by its inhabitants. To get there, the narrator is forced to pull himself along the vegetation of his path (like Macmann in Malone's story). After meeting a boy with a goat, he enters a cathedral, climbs its tower, descends, and returns to the streets where he meets a traveling peddlar, selling flasks. He asks for the time and direction toward a haven, but receives no answers. Instead, the peddlar tells him the story of his own life, or loves imagined or lost, and inquires about the narrator's own physiological reactions to the presence, imagined or real, of women. Since the narrator has no money (a shilling, it would take) he can buy no flask; and so is asked to embrace the peddlar in lieu of the customary payment. He does, and while looking down and away from his tormentor, the peddlar leaves him to himself.

The narrator tries to recall Pauline, the pedlar's loved one, and can hardly succeed in calling up the image. The story was badly told. So he strives for the image of the boy with the goat, who had given him candy. This too fails. Only the young girl seen on the steps of the cathedral's rising steps comes firmly to mind, and he can now see her as primly coifed in a bonnet and carrying a prayer book. This revery is broken by the sounds overwhelming his consciousness, coming from inside the shuttered homes of the town.

He falls down, first on his knees, then prostrate, to wait for the Good Samaritan who would lift him up and point out a new direction for his journey. But no one comes, and he is forced to continue, without even the light of the stars to indicate a direction, since the night sky has been covered by clouds.

So his story ends. Why, not even he knows.

The End recounts the death of the narrator, by suicide. That, of course, could not be given in the present tense, since it ends with the narrator's achievement of the oceanic consciousness, just prior to death by drowning. The story is reconstituted in recall, and that could only happen to the consciousness of the narrator following the event narrated, hence after his death, by the same narrator expressing himself in *The Calmative*.

But instead of any story which is only symbolic of his past life, this time the story is of the events connecting his expulsion from a charitable institution to the streets; from the streets to a lodging, and from the lodging back into the streets again. The streets lead to the country, and from the country to the seaside, to the cave of a man who befriends him. The sea prevents his sleeping, so he removes himself to the mountains, where he would perish except for the existence of a cow, whose milk he steals. From the mountain he redescends into the city, where he pursues the life of a beggar. He lives there, at the riverside, in a shack in which there is a canoe just big enough to serve as a home within a home. Making his own substitute womb, he fits a cover to the canoe, and drifts along the current back to the sea, where he opens the bottom of the boat with the knife given him by the man who had befriended him, and slowly sinks beneath the waters, womb and all, all in all.

His initial journey of expulsion therefore ends here, in death following the oceanic consciousness which makes the narrator one with the whole world. The outstanding feature of this final communion with reality is the narrator's regret of not being able to tell the story of his life clearly and in proper sequence.

So ends *The End*.

But the events of all three stories are intercut with the consciousness of the narrator, as narrator, reflecting on the way in which his own story is being told. That is the present of the story; the events being recounted as a result of that reflection can only be rendered in the past. The future is a bastard tense having no relevance for the narrator, since he is now present only as reflecting on his past as it leads up to the present, where the story is told. And the same pattern holds for *The Expelled*, *The Calmative* and *The End*, which were published in that order.

This order, however, is nonnaturalistic; for it represents the following sequence: the events having transpired before death, but exclusive of death itself; the events of the past recalled after death; and those past events leading up to, and including, death itself. A before followed by an after, but the after followed by another startling before; whence a return to the beginning. The reason can only be philosophical, if it is true that the structure of a story is determined by a philosophical idea. In the abstract, however, either story could have been a beginning, a middle, or an end to each of the others. Later this very thought of the interchangeability of beginning, middle, and end of his story will occur to the narrator of *How It Is*,[19] and *Molloy* actually begins with a "middle."

Consider the present case. If the author had chosen to begin with *The Expelled*, he might have continued, as he did, with *The Calmative* and finish with *The End*; but that reason still remains obscure. How much more natural to have followed *The Expelled* with *The End*. For then the symbolism of the expulsion from lodging to lodging would be clearly a reference to birth, and birth must end in death; and *The Calmative* would be the means for arriving with the least pain to that last expulsion of the living human being: to death (or rebirth) from birth, by inventing stories that happen to another. The present consciousness would then be a form of escape from the horror of living.

Or, suppose the author had begun his triune tale with *The Calmative*. We should then have achieved the point of view of a narrator already dead, but recalling the events of his past life leading up to and including that of his death: *The Expelled*, followed by *The End*, the two sets of expulsions, birth and death. That would be nice, but too easy. Yet *The Calmative*, in another hypothesis, might have been followed by *The End*, and *The End* by *The Expelled*. In this way, from the point of view of the dead narrator, we should be informed first of all about how the narrator had come to die, and this would be followed by a series of events prior to the foregoing, but whose rational structure would be the same: a series of expulsions from birth to death. That too would be fine, but a bit too analytical.

Or, finally, the author might have begun his miniature trilogy with *The End*. The meaning of a story is always its end, so why not get to the point as quickly as possible? The point of all living is a dying; nothing can be any clearer; for, if one is not first alive, one cannot subsequently die any kind of death. To live is to be, dying.

Let us therefore hypothesize the beginning of the story as *The End*. It could be followed by *The Calmative*, as, naturalistically considered, a pure consciousness can achieve its purity only in death: first a dying, and then the thoughts of a liberated soul. And in its liberation what would be more natural than the soul's recall of the essence of its life; a continual expulsion: from his mother at birth; from lodgings; from community with others; from self? We end therefore with *The Expelled*. Clear enough, but too naturalistic this.

There remains then the reversal of the order of the sequel stories. Beginning with *The End*, we could be presented next with *The Expelled* to show that the expulsion of death (from this world to the next) is nothing more than another in the indefinite series of expulsions already experienced in life; and then be given the story of *The Calmative* to serve precisely as a sedative for the pessimism of a life that can be lived only as a series of expulsions. A bit too moralistic, this.

All orders, therefore, save one, have a readily apparent rationality to their structure; and it is precisely that one chosen by Beckett to show how Watt's continued existence has come to its end. Surely a new explanation is in order.

Why should the critic assume the narrators of all three stories are the same? And even if they are, that this one narrator is the persona of Watt? He had last been found as friend and counsellor of Mercier and Camier.

To answer these questions in reverse order, the narrator of the *Stories* is the persona of Watt because of the repetition of circumstances. Watt was expelled from society into Knott's house, from Knott's house back into society; from society to a mental institution, where he told his story to Sam. Indeed, the narrator of *The Expelled* walks just as Watt walked; he must walk badly because he has to think of the movements to be made:

I set off. What a gait. Stiffness in the lower limbs, as if nature had denied me knees, extraordinary splaying of the feet to right and left of the line of march.[20]

But now older than Watt, the narrator falls after each swing from right to left in his attempts to keep going in a straight line. In this way his stride takes up a whole sidewalk, so he decides to walk in the street until a first policeman informs him that streets are for vehicles (such as bicycles) and not for pedestrians. When he is back on the sidewalk, a second policeman pulls him up short for taking up the whole walk: "If you can't bloody well get about like everyone else, he said, you'd do better to stay at home."[21] And the narrator notes snidely, "It was exactly my feeling."[22]

But the expelled Watt cannot stay home, he has none, except the earth to which, as dust, he must return. Almost miraculously his third stop is caused by a funeral procession that reminds him of his own ultimate end.

So much for the literary context. The philosophical idea necessitates this context and gives it structure. One becomes aware of oneself only by virtue of a relationship to others, but in refusing the others, one must return to oneself, in which condition the only alternative is stoically to reaffirm one's own freedom and then to deny that by engaging oneself in the appearances of the real world. Whence, Watt's skepticism.

That skepticism will force the narrator of *The Expelled* to deny that he wants to go on a visit to the Zoo, although he has given that destination to his driver, to spend the rest of the day looking for a room that cannot be found, even with the help of the coach-driver; then to refuse the hospitality of the driver (because it too cannot be what it seems, a gratuitous gesture of friendship). Instead of accepting a room, therefore, the narrator goes downstairs into the stable, where he spends a part of the night in the empty coach, only to leave it when the presence of the horse awakens his consciousness that this new haven is not his own. He slips out of the cab, headfirst in a rebirth, projected once more onto the dialectical paths of existence.

Thus, there seem two good reasons for the assumption that the narrator of *The Expelled* is the persona of Watt: their physical and mental circumstances are the same; and the events in their lives portray the structure of the same philosophical idea. Both have achieved the only spiritual essence capable of a self-consciousness that rejects and is rejected by the Other. First it becomes resigned; and then, resigned to itself, it becomes skeptical of the world, and ultimately an unhappy synthesis of the positivity of the world and the negativity of its own mind. The idea is as old as *Murphy*, and was merely made more elaborate in *Watt*.

No's knife is Yes's wound; so Beckett has said. If the Absolute is content to see all things revolve in and around itself, the self-conscious mind of a man must be its negative, unhappily searching for the end that will complete its circle. In this way, incidentally, the events of the *Stories* fill in the missing arc of the circle contemplated by Watt in Erskine's room; and, in the future, the Unnamable will find himself at the center of Malone's circular peregrinations around himself.

The second question is easier still. The narrator of the three *Stories* is one because of the order of the events recounted in each story. We have already seen how each story may be the beginning, middle or end of a trilogy on the basis of this assumption. Working out the consequences of the various possibilities is itself an exercise in Wattian logic. There remains only the explanation of the order actually chosen by Beckett.

And the answer to this problem has already been hinted at by the dialectical structure of *The Expelled*: a journey outward from self into the world, and from the world back into the self. "Hegelian," we called it, and Beckett has obliged us by entitling the English translation of the *Stories, No's Knife* – an imaged representation of the Hegelian idea. A final examination of the actual structure of the three stories considered as one should convince us of the truth of our hypothesis.

The Expelled recounts the events of a person, hence of a self as it is in itself, as an object of the narrator's consciousness. That is the thesis. The antithesis is established when the narrator recounts the events of his past consciousness in the present act of narrating. Hence, the self of the thesis is both itself and its own other. As itself, its own other, it needs no others; and so is depicted in death, when it has achieved its final expulsion beyond the world it has rejected. Then, the synthesis: the in itself must become for itself, a process that is to culminate in its own death; or, heaven forfend, in its awareness that its present predicament is nothing more than a symbol for the ultimate development of the Absolute Spirit, which, too, must always return back unto Itself, encompassing all self-contradictions.

The End presents the consciousness of the narrator as having reached the synthesis of its past and present: its past as continually expelled and its present as its ultimate expulsion into some future, nether world. That is what it is in itself. For itself, it is presently recounting the details of its actual final expulsion, its death by drowning, where it achieves an "oceanic consciousness" (pun naturally intended, but unexpressed by Beckett – the circumstances do that): "The sea, the sky, the mountains and the islands closed in and crushed me in a mighty systole, then scattered to the uttermost confines of space."[23]

But that oceanic consciousness must contain all of the actual and possible universes, otherwise the solipsism will not be complete. So, that final awareness contains a last regret for not having told the story as it should have been told: "The memory came faint and cold of the story I might have told, a story in the likeness of my life, I mean without the courage to end or the strength to go on."[24]

In the end, the solipsist dies, not by living and loving as was the case with Belacqua; nor by retiring from the world as was the case with Murphy; but by achieving that one consciousness which contains the whole of the world, as Watt's persona does in the *Stories* when Beckett allows that persona to tell its own story instead of putting it into the mouth of a third person, as he had done in *Watt*.

Hence, the sorcerer's apprentice has overcome his last remaining difficulty. There is no other to recount the solipsist's tale, so there is no need of an omniscient narrator. And yet for each teller there is a told. No need for a reader, either, since the solipsist who is a writer suffices unto himself to serve both these functions, provided he can live with the contradiction of using the words he himself has not created and whose meanings are established by the reactions of others.

Consider the time sequence. The writer writes of his past (his consciousness in itself, as it was) in the present (a reflective consciousness existing for itself, now), which slips unnoticed into the past as he reads what he has written in a new present where he exists both in and for himself: thesis, antithesis, and synthesis, which can only become a new thesis with a new antithesis and a new synthesis until the end of time. There is a future, then, only by implication; just as, before, the future of Watt was only implicit, to be made explicit in the *Stories* of *No's Knife*.

One last image: the oceanic consciousness that precedes the death of the narrator is a pure symbol of the success of a solipsist's project, to reduce the world to a set of his own ideas. It can be experienced only in a dream, in a vision, or in the anguished moments preceding death. It is for this reason that all of Beckett's former solipsists died in the end: Belacqua, Murphy, and now, at last, Watt.

But the oceanic consciousness is at the same time a natural symbol for the Absolute Mind, which likewise contains the ideas of all natural things, precisely in a series of contradictory appearances, beginning with sense certainty, and ending in the self-contemplation of the Spirit. That, too, is Hegelian, as a reading of the *Phenomenology of Mind* makes clear. If it is true that Beckett gave up reading the philosophers in 1930, after his perusal of Schopenhauer, Kant and Geulincx,[25] he has performed an almost miraculous task of deducing the later philosophical thoughts of Hegel from the contents of his own experience. The same will be true, later, for the ideas of the existentialist philosophers.

Coming as it does at the end of the *Stories*, the oceanic consciousness of the narrator creates an image that veritably tops off the literary structure leading to that point.

Whence, the *Stories* are ordered as they are for good literary and philosophical reasons. The order presented shows the structure of a philosophical idea, and the surface confusion is only apparent. The surface style and the depth structure of the tales are mutually determinant of a single fictional world; each "mirrors" the other.

This world has only one serious flaw. The being of the narrator, in itself and simultaneously for itself, is portrayed as an unhappy consciousness experiencing its own death as an object in itself. But the reflective narrator can only recount that event as occurring in the past, where it should be happening in the present as an anticipated future of the narrator's transcendence. Yes, there is a flaw in "Watt's" solipsistic death; his on-going world possessed no future.

But Beckett will remove that flaw in *Malone Dies*, the second novel of his larger trilogy. How he has removed it will figure once again in the ruminations of the narrator of the *Texts for Nothing*, who first discovers the difference between the uneven balance of the dialectical progress and the conscious projection of a human transcendence; but that narrator works with the knowledge gained from the voyages of Molloy and Moran, Malone's dying, and the Unnamable's verbal fixation.

In this way Beckett's oeuvre developed from the English Watt to the French trilogy. Having learned the language of the first person in *First Love* and the *Nouvelles* (*Stories*), and how to express it in his adopted tongue, he could set out to share his gift of tongues by framing a still more accurate account of the life of the schizophrenic animal that is a man.

The gift of tongues? What was it for, after all, but the revelation of the truth of the Spirit?

If these three transitional works display a confusing array of literary techniques, the confusion gives way before the structural analysis of philosophical "conveniences" used by the author. The psychophysical dualism of the early works has been clearly left behind in favor of the dialectical progression of human consciousness. The larger trilogy will go further yet. There life will become transcendence, in the full existential sense of that term. And, moreover, the solution to the human predicament of optimism in the face of pessimism is prefigured in the portrayal of these transcendences as creative writers. For tragedy exists only in art, not in the sufferings of our workaday worlds. We need only keep in mind that to appreciate the aesthetic value of tragedy, we must attend to the artistry of our author!

CHAPTER V

THROUGH STREETS WIDE AND NARROW

The solipsist who before was Belacqua, Murphy and Watt is brought to a third death, by suicide, in the *Stories*, but only to be resuscitated in the *Texts for Nothing*. In the meantime, however, he is made to live and die once more, just as the girl in that old Irish song lives and dies each time her song is sung, selling cockles and mussels, alive, alive, oh – sweet Molly Malone,[1] whose name will be used in the creation of two new Beckettian worlds, those of Molloy and Malone. To pronounce the very names the lips must be joined in that primeval, atavistic sucking reflex that draws the sweetness from a mother's breasts. But that phenomenon is on the surface, and quickly fades into the background of the new worlding worlds. Once Malone dies, projecting the story of his own death, his spirit will go on, in the person of the Unnamable, whose song may be heard by all who have ears to hear the plaint of the unfinishable tale.

Are there three stories? Or only one? The evidence seems clear.

That the three novels form a true trilogy, and not merely a series, is stated bluntly in the language of Molloy, who is writing in the first person, and thus telling his own story – that of a decision taken to return to his mother to settle old scores, and what happens to him on the way – as the next to, next to the last ('antepenultimate' won't wash here) story he has to tell.[2] Although the narrator changes identity (for reasons of the philosophical structure of the trilogy) from Molloy to Malone to the Unnamable, we should be obtuse indeed to have missed this insistence.

For the reader, the problem of the identity of the narrator begins with Molloy's difficulty in remembering his own name, as well as that of his mother and of the town she lives in, and whether he has had a son or not (the narrator of *The End*); it is complicated by a second first-person narrator, Moran, who symbolizes the predicament of the reader, being sent to look for Molloy (their paths crossing only once, but without recognition on either of their parts), and is only momentarily relieved when Molloy learns through his writing who, but not what, he is. Malone, on the other hand, is Molloy alone as everyone may verify merely by putting himself in the shoes of the narrator and pronouncing the words, I am alone.[3] After all, facing death everyone is alone.

Malone dies in order to achieve what Molloy could not discover in his search for himself, the missing whole of life which is given to all in death but which no one may experience in life, not even by a deliberate act of suicide. The body in any case is the first to go. Molloy and Malone are both bedridden; Molloy in his mother's room (womb); Malone in another, if not the same, room, waiting for the rest of his body to catch up with his feet, already in an advanced state of decomposition. Both are writers, and it is Malone who finally gives a list of his possessions, completing one of Molloy's abandoned projects, in a last vain attempt to identify his human reality with the things that surround him, rather than by his unique self-transcendence. But it is difficult to die authentically, just as surely as it is death that makes an individual of us all.[4] In the meantime, one goes on living, authentically or inauthentically, as one responds as one can to the call of conscience.

The writer, in particular, is called to speak; but to speak, he must listen to his own inner voice that always speaks in silence.[5] That is the project of the Unnamable, who survives Malone's death only to confuse himself with his own creations, Mahood and Worm. He is haunted by the presence of Murphy and Malone and Moran and the whole host of earlier Beckettian creations because his own voice at times becomes indistinguishable from that of the "man on duty,"[6] who has had the effrontery to speak for them all.

What happens in the lives of Molloy and Malone and the Unnamable is no longer important, for the problem has shifted from what they might have been and might have done to what they are in the instant, telling us their own stories as the tellers of stories. And in this trilogy the reader is veritably bombarded with the babble of voices striving to get their stories told.

Molloy will begin the new parade, speaking in the present about a time in the past when he still had a future. The temporality of his conscious being, a new philosophical idea (not unrelated to Murphy's concern for his horoscope), will be used to structure his tale, but to unravel that philosophical conundrum now would detract from the significance of its telling. After all, Molloy has to write in order to discover what he was, and with that discovery he also learns that he is no longer what he was; and the careful reader must make these same discoveries there where Molloy does, in the words he uses to fix his own self-identity.

The rest of the trilogy will follow, both logically and artistically, upon that project, which constitutes the beginning premise for the conclusions of the Unnamable.

Molloy is divided into two sections with distinct narrators. In the first Molloy tells his story, and in the second, Moran his. The entire Molloy story is composed in two paragraphs, the first, or short one, recounting the narrator's present circumstances, writing in his mother's room, and the second, what the narrator experiences as he writes what he has to write about. Moran's report on his search for Molloy follows, written in crisp, logical paragraphs as befits a mind (falsely) assured of itself before its tasks. Both stories, then, contain a double narrative thread: first, the banal order of events constituting the actions of the narrator as subject; and second, an account of the difficulties each narrator experiences as teller of his own tale. In each case, therefore, there is an active human being engaged in the performance of a task and a reflective, contemplative individual bent on giving an accurate account of his having succeeded or failed in the fulfillment of that task.

We know that Molloy has at least partially succeeded, for he describes himself as writing in his mother's room. But he fails to remember how he has gotten there, perhaps with the help of another. Every Sunday someone comes to pick up his pages, and pays him if they are good. Unfortunately, however, it is the person who pays that decides whether they are good or not. The pages are returned liberally marked with corrections, which Molloy does not understand; but then he never rereads them anyway. The biggest quarrel between payer and payee is about the proper beginning. Molloy, like a dope, tried to begin at the beginning, but the editors wanted something else. So he begins with his own new beginning, the old one having been the end, or the middle, who knows?

For the writer, however, the question is important, since the end is the meaning of the beginning, and every beginning is motion toward the end. Some writers, of course, never begin before they have an end firmly in mind, but that makes as much sense as to say that no one should end a story before the proper beginning has been found; one might never start or end a story that way, and should one decide to close the circle, by the principle of the eternal return, it would be impossible to distinguish beginning from ending.

The appearance of Moran and his two stories, which are presented in mirror image of Molloy's, severely complicate this discussion since it occurs after Molloy's, but the events recounted have occurred temporally before. Just a bit of reflection will show that both levels of conceptual confusion are aesthetically necessitated whether Beckett himself had them clearly in mind before he started writing or whether he discovered the proper ordering by shifting his parts, following the glimpse of a satisfactory ending for *Molloy* somewhere in its middle. For that is just where this novel ends: in its middle, at the mirror-like interface between the end of Moran's story and the beginning of Molloy's. Now for the reflection, which may "darken a few pages."

Let us begin with Molloy's concern for a proper beginning. There are indeed two: the one with his decision to visit his mother (the beginning of a journey for the active self, this); and the other with his decision to write about his return to his mother's room.

The story of the active self, with respect to the reflections thereon by Molloy writing, are all past, therefore ended before the reflection begins. The reflection begins because Molloy-acting cannot grasp the significance of his decision or of its consequences because that self is entirely bound up in the project of the return. The writer's reflections are going on in the present, and will come to an end when Molloy-acting is finally understood by Molloy-reflecting. Thus in a sense Molloy's first order activities are likewise the end of Molloy's reflecting; a beginning, yes, but an end too.

In closing this circle Beckett has defined an aesthetic universe, with its own space and its own time determined by the limits of Molloy's own spatiality (his region, centered there where his body is) and temporality (the succession of his own conscious states).

Critics, unless they choose like Moran to write about themselves instead of the object of their search, must decide where to start for themselves. Molloy, of course, was in different circumstances. His principle was supremely economical. His very writing constitutes the presence of his reflective self, and what he writes about is his own past, his former acting, non-reflective self who had made a decision and undergone the consequences of having made that decision. And as if sandwiched between these two selves of the first person narrator there appears a third self, the authentic one that appears, logically enough, at the beginning, middle and end of the reflective Molloy's account. At his innermost being, Molloy is a conscience that appears implicitly at the beginning and explicitly at the end of Molloy's written account of his journey, and intermittently when Molloy-acting strays from the directives of his past resolve. All of this, of course, needs proof in the events of Molloy's first order active life. So this critic will start there.

With Molloy-reflecting's premonition that he is engaged in writing his antepenultimate story, he begins the account of his active self's journey. From the cover of surrounding hills, he remembers having seen two individuals, A and C (A and B in the French version), who meet and part in opposite directions. One of the individuals is perfectly familiar with the locale; and the other is looking for a direction, and moves about as if under the impulse of duty. Neither sees Molloy. But, from another point of view, Moran later describes a similar scene between himself and a shepherd. The shepherd lives in the Molloy country, Ballyba, and, in answer to the question concerning his whereabouts, Moran is informed that he has arrived in Ballyba by the shepherd, who indicates the ground with the stem of his pipe. Moran then decides that his mission has been accomplished, except for the writing of his report, and begins his homeward trek. In this way, the beginning of Molloy-acting's journey is marked by the same event as the end of Moran's.

Logically, therefore, Moran's journey has occurred in naturalistic time prior to Molloy's, and had the novel been written from the naturalistic point of view, Moran's account would have preceded that of Molloy, instead of following it, as it does. In this sense, also, Beckett's *Molloy* may be said to end with a beginning and to begin with an end; like Watt, Beckett has preferred to relate this story with the second incident before the first.

At any rate, Molloy decides, upon seeing A and C, to make his journey home to his mother; and Moran, after meeting the shepherd, is told to return home to make his report. Both men are at that time severely handicapped for making any kind of journey: Molloy must use crutches and pump his bicycle with his one good leg; Moran has been separated from his bicycle by the defection of his son, and must walk with two gimpy legs, supported only by his umbrella. But like A and C, they take off in opposite directions, and will never meet again.

It is significant, of course, that each begins a journey home; the self must return to itself for the completion of its identity, or, as Sartre has made sufficiently clear, the self must fulfill its "circuit of ipseity."[7] And though each returns home to different sets of circumstances, Molloy to his mother's room and Moran to his dead animals, they both discover themselves in the process of reflecting on their journeys. Moreover, since Moran is sent by his employer to meet Molloy, it likewise seems logical that Molloy's account of his own circuit of ipseity be given before Moran's; for in this way the reader will know what Moran was sent out to find and why Moran will not find Molloy, but only himself, in his search.

Painfully, Molloy mounts his bicycle, and given his physical condition, must rest often. He does this without dismounting, and is arrested because he cannot move on when told to do so by a policeman. What's worse, he has no papers other than the *Times Literary Supplement*, which offers little proof of identification, but which does allow Molloy to wipe, after a stool, if he is so inclined, and has the added virtue of withstanding the farts he sends into the lining of his coat, where the paper is carried. Who is to know what kind of paper is being requested, when one is asked for papers? And who can know his own name, unless he is given time to reflect? Or that of his mother? Or of the town she lives in? Molloy doesn't. But he is left to continue his journey.

In the next episode, he runs over an old dog, killing it outright. But he is saved from another encounter with the law by the dog's owner, who lies about her own destination and the dog's prior condition, and cajoles Molloy to help her bury the dog in the grounds of her home. Strange, since Molloy cannot dig, and his help is limited to his holding the body of the dog while the woman digs the grave. Molloy stays in her home, where he is bathed, shaved, clothed, lodged and fed for as long as he should like and under the conditions he alone stipulates.

The woman's name is Sophie Lousse. But 'Sophie' is too familiar, and 'Mrs. Lousse' too formal for Molloy's taste; so she becomes simply 'Lousse.' Lousse becomes confused in Molloy's active mind with the image of his only love, an Edith or a Ruth, he forgets which. That one he mounted from behind, dog fashion, in order to learn the secrets of love. She paid him for his service, and that was strange too, since Molloy would gladly have made love to her in any way possible merely to know love for what it is in itself. Ultimately she died in a fire, and left Molloy in ignorance. And as Molloy is not so dead as not to feel the full force of his id, the image of his mother comes to amalgamation with the composite of Ruth and Lousse, and he is reminded of his resolution to return home to her.

Molloy continues his journey without his bicycle, using his crutches. He begins in a town, goes on to the sea, where he resupplies himself with sucking stones, and then to the forest. By now he must use his crutches like grapnels, pulling himself along the ground, or to abandon them and roll his body like a cylinder. His temptation to stay in the forest is great:

> For it was not so bad being in the forest, I could imagine worse, and I could have stayed there till I died, unrepining, yes, without pining for the light and the plain and the other amenities of my region. For I knew them well, the amenities of my region, and I considered that the forest was no worse. And it was not only no worse, to my mind, but it was better, in this sense, that I was there.[8]

But Molloy was no Belacqua, to rest under a rock waiting for a forced change of state; he had a feeling that to stay there would be to go back on an imperative, that call of his conscience, to get on with his task. So he continues to crawl into the light of the plain, where he stops:

> I remembered the corn-crakes. The two travelers came back to my memory. One had a club. I had forgotten them. I saw the sheep again. Or so I say now. I did not fret, other scenes of my life came back to me. There seemed to be rain, then sunshine, turn about.

Real spring weather. I longed to go back into the forest. Oh not a real longing. Molloy could stay, where he happened to be.[9]

So ends Molloy-reflecting's narrative.

This same narrative poses a number of hermeneutical problems. Why at the end of the account should Molloy abandon his original project, since it was motivated by his free decision and continually stimulated by its momentary recall? How did Molloy-acting get to his mother's room to assume the contemplative state of Molloy-reflecting? And how does Molloy relate to Moran? All these questions, and perhaps others, may be answered by a single hermeneutical hypothesis: the function of conscience in the definition of a self. And here the fundamental ontology of Heidegger is a sufficiently clear guide. If human reality is determined by its being in the world, one might well pose the question as to just who is in the world. And in doing so, one must face two other ontological problems: the elucidation of the concepts 'world,' and 'who.'[10]

"The world" is the world of everyone, i.e. of no one in particular. Ordinarily, everyone lives in an external world, but becomes lost among the things of the natural environment it provides. That is why Moran cannot find Molloy, who like everyone else lives inauthentically, defining himself by what he possesses, in terms of the things that surround him. But no one is truly at home there, since, being lost in that condition, everyman (*das Man*) has no feeling other than curiosity, no speech other than idle talk, and no understanding, other than a calculated average, of the commonplace. That is all phenomenal existence of the everyday will yield.

To get out of such a state, one must create his own world by assigning values to things in terms of what one's own authentic reality is. In this way, conscience functions to define the 'who.' It is the call from self to self, as Heidegger puts it, from afar to afar, from the self that is to be to the self that is lost in the phenomenal world of everyman, from the authentic self of the future to the inauthentic self of the present, linked to an overwhelming past. What makes this possible is the concept of transcendence: that of a being already there in the world (by virtue of its past), surrounded by things (in the present), yet already ahead of itself (as a flight into the future).

We of course know nothing of Molloy's past before his decision to seek out his mother, since that is the unfound object of Moran's search; or we know everything since we all share Molloy's past of existential inauthenticity; but then so did Moran. He had only to look at the conditions of his own life to find Molloy. And he does just this, by accident. His report is about Moran and not about Molloy. And he acknowledges his failure to find the authentic Molloy, as indeed any other, by denying in the last sentences of his report the very words used to begin it. For the authentic Molloy, one must discover the truth in Molloy's own words. Moran in this way refers us back to Molloy.

How to make all this clear is his problem, not Moran's. But Molloy's equipment is meager. He has studied in turn astronomy (astrology), anthropology, and philosophy, and was forced in the end to turn to magic (logical sorcery). We know about the astronomy, and have already had hints, from the narrator of *Watt*, about the anthropology, here reinforced:

What I like in anthropology was its inexhaustible faculty of negation, its relentless definition of man, as though he were no better than God, in terms of what he is not.[11]

and philosophy fares no better: " . . . my ideas on this subject were always horribly confused, for my knowledge was scant and the meaning of being beyond me."[12] All the magician knows, in the end, is the power of his words:

All I know is what the words know, and the dead things, and that makes a handsome little sum, with a beginning, a middle and an end as in the well-built phrase and the long sonata of the dead. And truly it little matters what I say, this, this or that, or any other thing. Saying is inventing. Wrong, very rightly wrong. You invent nothing, you think you are inventing, you think you are escaping, and all you do is stammer out your lesson, the remnants of a pensum one day got by heart and long forgotten, life without tears as it is wept. To hell with it anyway. Where was I. [*sic*][13]

Molloy, the reflecting narrator who couldn't even keep his tenses straight, or clearly recognize the difference between statement and question, was telling the story of his journey home. And he will be able to stay where he is, at the edge of the forest, because that journey was only a symbol for what was really at stake: the discovery of his own authenticity. And he cannot say how he got to his mother's room because to return there he had to experience another fall into the inauthentic life, entering the world of everyone and leaving his, in order to do so. The other person by whose aid he was allowed to leave the edge of the plain was none other than the inauthentic Molloy, and his existence is dubious, if not denied, because Molloy-reflecting is in search of the authentic Molloy, who is at that time not yet present.

The authentic Molloy, it has been said, makes his appearance at the beginning, middle and the end of the inauthentic Molloy's journey. As the conscience that called, he was implicit in the creation of an imperative by which Molloy projected his return home; he was explicit in the voice offering contradictory imperatives in the forest. There was a self-imposed duty, yes, but not to return home in the inauthentic mode of everyman; for that, it was already too late. But it is never too late to assume the responsibility for an authentic existence; for that, Molloy could remain where he was. The call of conscience is from self to self, not from place to place. And on Molloy's inauthentic journey, he is called explicitly to fulfill his past resolve. His oedipal complex does it once; the sight of a "young old man" does it again. And the latter is the correction of the former, since the oedipal motivation is inauthentic, and the young old man is a symbol for himself, past and future.

In this way, viewing Molloy as an existential self-transcendence, we have solved the hermeneutical problem posed by *Molloy*, Part I. There remains the problem of Part II, Moran's report.

As has already been pointed out, Moran's report reveals two Morans, just as Molloy's does two Molloys, one engaged in the chores of everyday living and the other writing about the one engaged in a journey during which the wayfarer learns the truth about himself. Moran's report begins in a tone of complete self-assurance. He is employed, and is a nominally devout Catholic. His employer, Youdi, sends him on a mission to find Molloy and to write up a report. The contact is made by Youdi's messenger, Gaber, on a Sunday, disturbing Moran's weekly religious duty of attending Mass. Gaber, of course, is Mammon's Gabriel, who not only gives Moran Youdi's assignment, but also an advance briefing, and at the end, when Moran has reached the Molloy country, also gives Moran the message to return home and make his report.

Moran meticulously describes his job, his family, and his religious life prior to setting out. Since he has missed Mass that Sunday, he makes a special trip to the parish priest to receive the Eucharist, in spite of the beer he had drunk with Gaber. The man who came regularly to pick up Molloy's pages also had an unquenchable thirst, so it is not unseemly to suppose that it was Gaber who came to pick them up once Molloy had started his writing (after Moran's failing mission).

Having satisfied the needs of the spirit by receiving the viaticum from the priest, Father Ambrose, Moran satisfies his bodily hunger by enjoying a last meal with his son before starting out at midnight the same Sunday he has received the order to depart. Moran is thus a dutiful, but inauthentic man, living by the catechism and by the rules of conventional morality; he follows both God and Mammon.

Since he is ordered to take his son along, both prepare for the trip by their meal and a nap, and finally leave along the cemetery road, where everyone in the town will ultimately end his journey, with enough provisions to last for an indefinite future. But Moran's leg grows stiff, and must be supported for the rest of the trip. In this, of course, he already begins to resemble Molloy.

Moran sends his son for a bicycle, who returns in three days, not empty handed. But the boy has lied concerning the flat tire on the bicycle in order not to have to pump for himself and his father. Moran perceives the ruse and reinflates the tire.

In his son's absence, Moran meets a man whom he kills, beating him over the head with his own club; and during that struggle he loses most of the keys from the huge ring he always carries with him. We remember immediately that Molloy too had killed an interloper without an apparent reason, since the reason for both murders was the same: the destruction of the other, who places an obstacle in the way of completing one's own self-identity, but also the destruction of one's own past self, which is the chief obstacle to the same end. And the lost keys likewise symbolize Moran's loss of connection with the things he cherishes most:

Does this mean I shall one day be banished from my house, from my garden, lose my trees, my birds of which the least is known to me and the way all its own it has of singing . . . , lose and be banished from the absurd comforts of my home where all is snug and neat and all those things at hand without which I could not bear being a man, where my enemies cannot reach me, which it was my life's work to build, to adorn, to perfect, to keep?[14]

He muses thus at the outset of his voyage, in response to the voice of a misguided conscience:

And the voice I listen to needs no Gaber to make it heard. For it is within me and exhorts me to continue to the end the faithful servant I have always been, of a cause that is not mine, and patiently fulfill in all its bitterness my calamitous part, as it was my will, when I had a will, that others should. And this with hatred in my heart, and scorn, of my master and his designs.[15]

During Moran's journey he is to learn that all foreign voices are alien to his own conscience, that of God as well as that of Mammon.

That is why, once he has been abandoned by his son, he gives way to his fear and panic. In the dark he begins to see figures that aren't there:

But I confess I attended but absently to these poor figures, in which I suppose my sense of disaster sought to contain itself. And that I did not labour at them more diligently was a further index of the great changes I had suffered and of my growing resignation to being dispossessed of self. And doubtless I should have gone from discovery to discovery, concerning myself, if I had persisted.[16]

But Moran persists in spite of himself. The day he meets the shepherd who confirms his presence in the Molloy country (the scene witnessed by Molloy),

Moran has a violent scene with his son, who absconds with the bicycle and all of Moran's money, save fifteen shillings. Moran is forced to hobble the rest of the way leaning on his umbrella in lieu of a walking stick.

It is then Gaber brings the order to return home. Youdi's only comment of Moran's performance is that life is a thing of beauty and a joy forever. So by now we can be sure that Youdi has returned Moran's scorn. Unable to answer the list of theological questions he poses himself, he is likewise abandoned by God. As a result, he turns to more personal questions, one of which is

> Would we all meet again in heaven one day, I, my mother, my son, his mother, Youdi, Gaber, Molloy, his mother, Yerk, Murphy, Watt, Camier and the rest?[17]

Only Youdi, Gaber, Yerk and Molloy are involved in Moran's journey, but like all other of his narrators, Moran is aware of Beckett's earlier heroes. They all had split personalities, and they all were deeply engaged in seeking their own self-identities. Would they all be saved? Along with all their voices, Moran attends, finally, to his own:

> I have spoken of a voice giving me orders, or rather advice. It was on the way home I heard it for the first time. I paid no attention to it.[18]

The voice of his conscience heard before his journey was here repudiated as false, and it is true Moran has difficulty defining himself spiritually or intellectually. His body kept getting in the way. Following the rigors of his journey, his face, his hands, his belly no longer are the same.

> And yet the gist of the sensation was the same as in the far-off days when I was well-shaven and perfumed and proud of my intellectual's soft white hands. And this belly I did not know remained my belly, my old belly, thanks to I know not what intuition. And to tell the truth I not only knew who I was, but I had a sharper and clearer sense of my identity than ever before, in spite of its deep lesions and the wounds with which it was covered. And from this point of view I was less fortunate than my other acquaintances.[19]

He was indeed less fotunate than Molloy, for Molloy had not yet made his personal commitment, and so had not yet had to suffer the anguish of its writing.

When at last he reaches home, he finds it abandoned and all his animals dead, except the wild birds that never needed his care in the first place. He announces to Father Ambrose that there will be no more religious duties performed, and as if in explanation he muses to himself:

I have spoken of a voice telling me things. I was getting to know it better now, to understand what it wanted. It did not use the words that Moran had been taught when he was little and that he in his turn had taught to his little one. So that at first I did not know what it wanted. But in the end I understood this language. I understood it, I understood it, all wrong perhaps. That is not what matters.[20]

What it was saying now was to write that report. So Moran begins, this time with the truth: "It is midnight. The rain is beating on the windows. It was not midnight. It was not raining."[21]

The first two sentences of that last contradictory set of statements were the opening words of Moran's report. The last two deny their truth. He had not been looking for Molloy at all, for he had succeeded only in finding himself. To discover who Molloy was, the reader is therefore shunted back to Molloy's own pages.

In this way Moran is a symbol for all readers whose critical report should send their own readers back to the original source for any authentic information; and Molloy obliges us with the true account of his own search for self-identity. So he is the symbol for any author. And, to end this critical tale, each is a symbol only for a part of every author's personality who must first write something to find out what he has to say, and then read it to confirm the revelation.

Thus Molloy and Moran, like all previous Beckettian creations, merely symbolize their creator's own split personality: the reader and the writer talking to each other, in silence, and in an effort to produce unity between the I and the me. And this is the fifth, and the last, level of significance amenable to our hypothesis. Count them: the original split of the created character into active and reflective voices; the appearance of the authentic; the writer and the reader in separation; the author of the work itself.

Youdi was perhaps wrong, but if life is not a thing of beauty and a joy forever, *Molloy* surely is.

Malone Dies depicts the continued existence of the principal narrator of Molloy, who had discovered in writing his pages who, but not what he was. As Molloy, the narrator was already aware that, alive, he could not succeed in describing his essence. He had none. Recall the medieval definition of an essence as what the definition of something refers to, so that the essence of Socrates is what the definition of Socrates would refer to if Socrates had a definition. But individuals cannot be defined either in Aristotelian or in other terms until they are dead, when their existence may be summed up in terms of their historical circumstances' having just come to closure. Molloy understood this well:

For the particulars, if you are interested in particulars, there is no need to despair, you may scrabble on the right door, in the right way, in the end. It's for the whole there seems to be no spell. Perhaps there is no whole, before you're dead. An opiate for the life of the dead, that should be easy. What am I waiting for then, to exorcize mine?[22]

To achieve this essential whole, left undiscovered by both Molloy-acting and Molloy-reflecting, Molloy must die, as Malone.

Let us recapitulate the historical events here. Molloy-acting resolved to return to his mother, to establish a relationship between his then present and the beginning of his past, his life in his mother's womb. That project misfires because it is too late for such a reunion, and he must be satisfied with his mother's room. All this is past for Molloy-reflecting, who discovers in that past his own true self, a conscience calling him to authenticity, but which permits him to remain where he is, at the edge of a forest.

There were inauthentic imperatives, hypothetical these, that called him back to his mother; and indeed he gets there with the help of his inauthentic self. Then he makes a second resolve, to tell his story, in whose words he discovers that moving from place to place, even if one must roll there, has no meaning for the authentic conscience, which issues the only true categorical imperative. He could, therefore, stay where he was. But when he tells this story, he is already in his mother's room, now reflecting; and he stays there until the end of his report.

He is still there when conscience calls once more; but he is alone, without his mother. The project to return to her was misguided, looking backward into the past; the new project is to look ahead, into the future in anticipation of his death as Malone. As a transcendence, then, the narrator is pulled first into the past towards his mother's womb, where his life has begun, and then into the futute towards the grave, which is only another womb for the narrator's future resurrection as the Unnamable.

Thus, *Malone Dies* is the necessary philosophical consequence of Molloy's partial self-discovery. But in Heideggerian terms, if the 'who' is defined by a response to the call of conscience, from self to self, from the future to the present that has become lost for its entanglement with an inauthentic past, the whole of the project of human transcendence must be defined in terms of its passing. The new hermeneutical concept is, therefore, (human) being fleeing towards its end as a being-unto-death. The literary problem is to express this being, authentically. For my critical purposes, it will be parsimonious to explicate here the existential concepts of this additional hermeneutical hypothesis. In this way, we shall have a clearer picture of the philosophical structure of Malone's dying.

For, in Heidegger's fundamental ontology there are two ways of dying, as there are two ways of living, depending upon whether one takes one's exit from the real world (that of everyman) or from the inner world of the solipsistic self. As dying is ordinarily understood, for example, everyone dies a real death, whether from heart failure, old age, or malfunctioning kidneys. But in speaking of this kind of death, the inauthentic person always speaks of the death of another, never of his own, unless in jest, or in fear and trembling before the certainty of that real event. One must always wait to die, and that is one way of facing the future – waiting for the inevitable to transpire in the common world of everyman when there will be one less citizen to vote, a place to be filled in the office or factory, or a loved one to be mourned. It is for this reason that death as an event in my life always happens to another: my enemies will be gratified, and my true friends, sorrowful. But as a dead person I shall feel nothing, for my transcendence will have come to an end.

The question then arises as to how to make one's death happen to oneself, how to face the future not as a future now of the real world, but as a now future in one's own world. Given the nature of existence – that condition by which the human individual exists already ahead of himself as a projection towards a freely chosen future – that possibility is an existential choice. Malone at one point chooses to imagine his death, inauthentically, as a reversal of his birth. Presumably he was born, head first, screaming from his mother's sex; and since his legs have already died (like Socrates who grew cold from the hemlock from his feet to his head), Malone sees himself, if to be sure he had not already died in the forest as Molloy,[23] as being expelled, feet first and without complaint, from "the cunt of existence":

I am being given, if I may venture the expression, birth to into death, such is my impression. The feet are clear already, of the great cunt of existence. Favourable presentation I trust. My head will be the last to die. Haul in your hands. I can't. The render rent. My story ended I'll be living yet. Promising lag. That is the end of me. I shall say no more.[24]

In these words, which interrupt the narrative of his last hero, Macmann, Malone's consciousness switches from his task of telling that story to his own condition, in anticipation of that ultimate silence where he shall hear no more voices; not the real event portrayed in that indelicate image, but the mere possibility of his no longer being there, in the world. Up to this point of his existence he is not yet complete, not yet an essence; and he shall need the help of his creator to die in order to achieve that essence. Really, therefore, he can do nothing to die; but as the last possibility his transcendence already is, he can resolutely project his own ultimate silence; that is his authentic death. And he must face it in anguish, since there is nothing there, really, to be feared. For this reason Malone does not fear his death, but he cannot escape its anguish, which is that peculiar state of mind in which he is thrown back upon himself, groundlessly, as a pure possibility of no longer being.[25]

The novel has no structure different from this ontological conception. To understand this structure, we need only remember the difference between death as a real event and death as the last possibility everyone is, of no longer being anything at all. It is death that individualizes and isolates; authentically, it is always one's own, unrelatable to anything beyond, never really overtaken (since it is always a possibility), certain and indefinite as to time and manner.[26]

In the beginning Malone errs with respect to the authenticity of his death, however, because he expects to die within a specified period of time: "I shall soon be quite dead at last in spite of all. Perhaps next month."[27] So his story begins, but immediately he becomes aware of the indefiniteness of an authentic death, and projects the possibility from April or May until June 24 (St. John the Baptist's Day), or . . . , or . . . , etc. He shall have to wait, but while waiting he shall amuse himself by telling stories. If to die is to lose consciousness, one could stand that easily:

> The loss of consciousness for me was never any great loss. But perhaps I was stunned with a blow, on the head, in a forest perhaps, yes, now that I speak of a forest I vaguely remember a forest. All that belongs to the past.[28]

Looking to the past, Malone sees himself as Molloy; his present reflection must be on himself as a future non-being. And his time predictions won't turn the trick, for they are of an inauthentic event in the lives of others. For this reason, when he reflects on his immediate projects – to recount his present situation, three stories, one about a man and a woman, another about an animal, and another about a stone, and finally, the inventory of his last possessions – he becomes aware that his original prediction was inauthentic:

> For this must be going on now for over a week, it must be over a week since I said, I shall soon be quite dead at last, etc. Wrong again, That is not what I said, I could swear to it, that is what I wrote.
>
> . . . .
>
> Yes, I shall soon be, etc., that is what I wrote when I realized I did not know what I

had said, at the beginning of my say, and subsequently, and that consequently the plan I had formed, to live, and cause to live, at last, to play at last and die alive, was going the way of all my other plans. I think the dawn was not so slow in coming as I had feared, I really do. But I feared nothing, I fear nothing any more.[29]

His anguish banishes his fear, as his awareness turns from the world to himself; and if his death is to be authentic, he must actually die in the act of telling his last story. In this way the anguish of getting his story straight will be the same as his anguish felt as a being undergoing his being-unto-death.

As a consequence Malone's story, like those of Molloy and Moran, contain two immediate frames of reference: on the one hand, Malone on Malone's immediate condition, reflecting on his own demise along with the attendant anguish of not being able properly to finish the lives of his characters; and, on the other, the substance of those lives, recounted in the third person. In this way, Malone is both a first- and a third-person narrator; and his progress or lack thereof as the latter continually modifies his state of mind as the former. He calls this living and making live, or playing; but whether it is one or the other, both projects are unified in his attempt to face death authentically:

To show myself now, on the point of vanishing, at the same time as the stranger, and by the same grace, that would be no ordinary last straw. Then live, long enough to feel, behind my closed eyes, other eyes close. What an end.[30]

That is Malone's end, projected as a possibility in his own words, and realized in actuality by those further words lent to him for the occasion (the same act of grace) by his own creator, Beckett.

Such, in a nutshell, is the next to the last story of M . . . , alone, this time, but whose being is still in the world. And we Morans of this world must wait until the narrator reappears as the Unnamable to discover what it means to have achieved the status of an essence once one has achieved one's individuality by dying, alone.

We may begin with Malone's efforts to forestall that total anguish which would incapacitate him for any action whatsoever. He makes definite plans: first, to describe his present condition; then to give his three stories; and, finally, to finish the inventory of his possessions. We remember, of course, that it was Molloy who first projected a description of his possessions; for that is one way the inauthentic man may try to define who he is, and Malone was once Molloy.

If it does no good to look into the past for a definition of one's present duty, it doesn't do any good to deny the past, either; no one lives authentically all the time. A past choice is authentic when it is repeatable in the present, and the only present moment in which it would be meaningful to define oneself at all would be the moment of death. Thus Malone repeats the unfulfilled resolve of Molloy, timing its execution with the moment of his death when he possesses nothing but his voice, slowly but relentlessly giving way to silence. In order, then, the fulfillment of his present tasks.

He is in a room, bedridden, capable of moving only his hands. He will not die of uremic poisoning because his kidneys still function. He eats and drinks and excretes, and the various pots necessary for this activity are emptied and filled, as the situation requires, by a female attendant who seems to come with the room. His chief connection with the world is a hooked stick that enables him to draw objects toward and to push them away from himself. His body is impotent; and as for his mind,

> All my senses are trained full on me, me. Dark and silent and stale, I am no prey for them. I am far from the sounds of blood and breath, immured. I shall not speak of my sufferings. Cowering deep down among them I feel nothing. It is there I die, unbeknown to my stupid flesh.[31]

Malone's spatiality, his physical presence to the world, is reduced to its absolute center, as if in fulfillment of every solipsist's dream, just as his temporality, the consciousness he has of himself in that state, is approaching the point of no return. And the anxiety of that consciousness is reflected in the style of its expression: hurried, tedious, interrupted by thoughts of self, of the felicity or infelicity of the moment's image.

Malone begins the story of his masculine character, Saposcat, an anagram for one about to take leave of both his blood and his wits; Sapo, for short. We read

> Sapo loved nature, took an interest. This is awful.[32]

So Malone begins again to describe Sapo's interest in things natural, especially in the birds, the hawk in particular, and finishes a paragraph with an image that presents on the surface what it refers to in the signified depth:

> He would stand gazing at the long pernings, the quivering poise, the wings lifted for the plummet drop, the wild reascent, fascinated by such extremes of need, of pride, of patience and solitude.[33]

There follows a pause for a new paragraph, and the narrator expresses his satisfaction: "I shall not give up yet,"[34] as his attention is brought back to himself. These alternances between his stories and his own condition continue until they become one in the last trailing words.

Although his parents are hopeful of a career in the liberal professions for him, Sapo is very early dim of wit, and loses himself looking for the meaning of words in his reading texts; he prefers the direct contact with things. Not even the new pen his father bought for him can help him pass his examinations. He visits the daughter of the Lamberts, a poor family nearby, but talks to her rarely; and one day after the death of the Lambert family mule he announces to her that he will return no more. He goes away, as usual, stopping often to contemplate the phenomena of nature. Later he will assume the same attitude under the name MacMann; but it will be raining, and he will end up rolling in the mud.

The second story concerns the animal. It was the Lambert's mule. Old man Lambert, who is a hog butcher, buys his animals when they are being brought by someone else to the slaughterhouse. He had bought the mule there thinking to be able to get another six months' work out of it, and succeeds in getting two more years'. He and his son bury it, bending its legs to fit into the six foot hole. It therefore has to be replaced; but since this story concerns only one animal, it could end with its burial (giving Malone a fine image of the inauthentic death of man, as well): a beast of burden worked beyond its years and finally reunited with the humus from which it sprang. But that image is partially defective: the mule has no choice but to die in that way; and some rabbits, such as the one prepared for the Lambert supper, die from fright before actually being butchered by the cook. That perhaps is a better image for the inauthentic death.

The story of Macmann, 'Saposcat' having been exchanged for the more universal 'Macmann,' continues in order for Malone to throw light on the condition of all sons of men. He takes the job of a streetsweeper, only to leave the streets more precariously dirty than he found them. His catatonic stupor in the rain presages the last episodes of his life, which transpire in an asylum, St. John of God's.

There he will meet Moll, another patient who cares for him in more senses than one. They are both old and impotent, so he must fold his sex to place it into hers; and, as if in a spasm of joy, she ejaculates, "Had we only met sixty years before!" Then, upping the ante, she cries, Seventy years before. Her fetish is the crucifix. She wears two, one on each ear, and they dangle as she moves about. Macmann asks why she wears two crosses, and like a woman she replies with a question, Why two ears? The real reason was that they were the crosses of the thieves (one of which was saved, and that's a good percentage); the true cross was in her mouth, since her last remaining tooth had been ground into that figure. Macmann runs his tongue across it in their embraces. But the tooth, or real cross, is loose, and falls out one night, just as her hair, too, has begun to fall into shreds. In a hurry to get at his own anguish, Malone kills her off, as he himself is shortly to be killed off by Beckett.

Yes, Beckett is there too. His name is Lemuel, always a brute and sadomasochistic bully. He is Moll's replacement as caretaker of Macmann. A masochist, he enjoys dealing himself blows with a hammer, on the leg, on the head, anywhere it will hurt without killing himself; for if that should happen he should no longer feel pain. Besides Macmann there are four other patients in his sadistic care. Each embodies characteristics of Beckett's former heroes, from Belacqua to Molloy. They are taken on a picnic one fine day to visit an island (there's the rock or stone, and hence the third of Malone's stories), where Lemuel kills the two innocent sailors charged with bringing the group to the island. That, by the way, is where the Unnamable's hell is located.

Here, too, Malone finds a reflection of himself. Lemuel wields the hand that kills, even as Malone himself has killed the mule and Moll, with malice aforethought. He is the creator, so the killer; and Malone feels for the moment the fear of the lord.

Returned to inauthenticity in that fear, Malone then thinks of his possessions; but they have all long since been gone, from the momentoes of his life as Molloy to his stick, to his own creation, Macmann. Only his voice is left – but enough pencil? –

> Lemuel is in charge, he raises his hatchet on which the blood will never dry, but not to hit anyone, he will not hit anyone, he will not hit anyone anymore, he will not touch anyone any more, either with it or with it or with it or with or[35]

It is difficult for Malone to concentrate; he must look forward to the end, where nothing is more real than nothing,[36] that's a thought he had in conceiving his own story of Macmann, that last possibility of no more possibility to be in the world.

He continues writing nervously:

> or with it or with his hammer or with his stick
> or with his fist or in thought in dream I mean
> never he will never
>
> or with his pencil or with his stick or
> or light light I mean
>
> never there he will never

never anything

there

any more [37]

And so Malone dies, by the same grace that had created him, authentically, resolutely anticipating that last possibility he had always been. That would be enough to prove his anguish, if that anguish were not apparent in the style of his writing. Indeed, in his anguish, he confuses himself (the pencil and stick) for Lemuel, his murderous creator.

Whether his pencil had finally been reduced to nothing, or the space in his copy book entirely used up, his voice is extinguished along with his breath. No use to check there at his mouth, try the other orifice, to see if Molloy has finally stopped his miraculous farting.[38] Whatever spirit, whatever air there was left in that mortal envelope has already escaped into the surrounding air. Breath, voice and heart, all ceased simultaneously. It is all there in Malone's dying words.

An arc has been missing in something, either in the narrators or their stories, projects both, ever since Watt's perception of Erskine's abstract painting where there was an essence to be seen, but which was missed by Watt in his dazed condition. Incomplete, confused, mad, they all are led, in *Malone Dies*, out of the safety of their asylum to venture forth once more in a mass expedition to an island, where "Lemuel" Beckett is the man in charge. That expedition, that island, that directorship, are given further exposition in *The Unnamable*. Otherwise Malone's story, like his transcendence should he have failed to come to an end, would have remained incomplete.

The Unnamable projects the condition of the narrator who was once Molloy, but who subsequently became Malone to die in order to achieve the status of a completed essence. In this condition there will be no more changes, no more free choices, no more calls to conscience, no more transcendence; for an essence cannot be said to exist as standing out from itself, like a man. At best an essence subsists in a network of relations that are both real and conceptual, in reverse order as an idea and its meaning. As real, these relations subsist, on the one hand, between the Unnamable, a projection from the past, and his predecessors, from Malone to Murphy and beyond; and, on the other between the Unnamable as narrator and his reflections in and on his own creations, Mahood and Worm. As conceptual, he is nothing but the words Beckett has found to create his essence, the better to see therein his own image, his own essence, in reflection. On the surface, then, all is sound and silence, as words send out their rippling images.

It seems obvious, then, that the locus of the narrator's new habitat is the same, from this last point of view, as that of his previous states, when as an existent transcendence Molloy started out in search of himself and discovered a conscience that was at variance with the hypothetical imperatives of his project to reach a certain place (his mother's room) within a specified period of time (before it is too late), and which calls him to a rendezvous with his authentic self that can happen at any time and any place; and when, as Malone, he was torn between the fear of dying a death in the real world and the anguish of facing his only authentic death by assuming the responsibility for projecting his own last possibility, that of no longer being in the world to do anything at all, if not to continue contemplating his essence as an individual still searching out his own self-identity.

As the Unnamable, the narrator is precisely no longer in the world, but out of it; no longer a possibility of any sort, but a predictable necessity, both literary and philosophical: a mere function of the meaning of words; from the past, the words of Molloy and Malone that made him what he has become; and now, in the never ending present, of his own words that point him towards three new distinct sets of relations: those he has with his creatures, with himself, and with his own creator, the man on duty in this entire voyage, Samuel Beckett himself.

Since no one is in a position to understand the least thought that is not expressed clearly in words, not even he who must speak those words, the Unnamable will undergo the agony of finding the proper words; and, already having achieved the status of a subsistent essence, being but no longer existing,[39] that agony cannot come to an end, and the narrator must continue to suffer beyond the end of his story, related in these very words:

> . . . strange pain, strange sin, you must go on, perhaps it's done already, perhaps they have said me already, perhaps they have carried me to the threshold of my story, before the door that opens on my story, that would surprise me, if it opens, it will be I, it will be the silence, where I am, I don't know, you must go on, I can't go on, I'll go on.[40]

If the Unnamable's story ends in this way, obviously it will not end; for the apparent ending is only a new beginning, just as every graduation is properly called a 'commencement.'

One can never discover, in this world or outside of it, what one is. That is for others to decide. Every Molloy needs his Moran; every writer his reader; every author the two sides of his split personality, the one according to which he is writer; and the other, the reader, both seeking to communicate, to become reunited as one, via the vehicle of the created work. Thus instead of ending the Unnamable's tale of woe, Beckett turns it over to our critical review, where, if he and we are lucky, its significance will continue to reverberate until we ourselves lose consciousness.

The Unnamable, in effect, poses a host of hermeneutical problems; but in this case, as in the past, such problems are not unsolvable, provided we find a workable hypothesis; and a theory of the absurd is not a workable one. We begin with a negation – just as the Unnamable begins to define himself by a negation, once he understands that where he is has no earthly meaning – that the novel is about the narrator (it is written in the first person, until the narrator becomes confused and identifies himself with others, with "them," with "him," with other voices in other silences) only on the first level of signification; that his true identity is of an essential nature, and so cannot change; that this unchanging essence is schizoid, a creating creature; that to be in such a state is agony, whether in the performance or in the knowledge that one is likely not to be understood, either before, during, or after that performance unless the reader is attentive to the words that create this creating creature to the same degree that the author has been in writing them.

In sum, then, the novel does not represent anything at all; it presents on its surface (the words) a field of depth significations from the appearance of an essential consciousness, still looking for what it is in its most authentic form, and still, in spite of its solipsistic beginnings, at the mercy of others. That's one kind of hell, as Sartre's Garcin has said;[41] but it is only a concept unfolding itself in depths of spiraling significance. To feel that significance, one must enter this eery universe, where to the eye and ear nothing is fixed because everything fixed there is attached to a word, either by writing or reading the words. Abandon hope, all who wish to enter here; for there is no exit.

Listen to the words of that false beginning: "Where now? Who now? When now? Unquestioning. I, say I. Unbelieving."[42] We have slipped beyond the portal that opened at Malone's passing only to find ourselves in the Unnamable's hell. Where? Who? When? All unimportant: Malone's transcendence, his spatiality, his temporality, his conscience, all left behind. The important thing is the now, which shall never end. Out of habit, for one can become only what one has already been, and therefore unquestioning, the narrator must say 'I,' but it will be without conviction since the question still remains as to just who this narrator is. And as the title of the novel indicates, no one will be able to say exactly, in a word, what proper denomination may be placed upon the essence that has been achieved.

This project is not so easy as that of a Malone or a Molloy; they existed, and existing did everything they could to define their own being; the Unnamable might be they, or Murphy, or Watt, or any of the others. After all, these former characters when they were present at a time that is now the real and closed past have made their claim on the Unnamable's present condition: it was they who projected him out of their own pasts into the future which is the Unnamable's present. That is a claim of some sort, and one that the Unnamable will examine in some detail.

He has not lost his capacity for observing, but what he observes has become fixed as the routine observed in Knott's household, where really there were no changes. This critical observation is not a gratuitous comparison, for the philosophical reasoning is the same: in entering Knott's service Watt had arbitrarily entered into an essential relationship, that of servant to master, where freedom is sacrificed for the security of continual repetition of the same, pre-ordained tasks.

Whom does the Unnamable serve? That is what must be discovered. If he were alive, and free, he might be said to serve himself, but he is not. His old habits, like the one of taking his own existence for granted, are only a second nature that hide him from himself. That was a lesson Beckett had learned by reading Proust, but Marcel was alive trying to resuscitate a strangulated past. The truth lies beneath the habits. What is the 'I'? A grammatical fiction, as someone else has said?[43] If so, the substance of that fiction must still be made clear, and that is precisely the Unnamable's task, still a project in the present that shall never end.

The voices of Malone and Molloy and all the others will force their claim, only to be rejected by the Unnamable's growing lucidity. He cannot serve them; for he is in thrall to another, the one who says 'I' to refer to the Unnamable, meaning 'he,' the Unnamable, in order to hide himself as best he might.

He is the man in charge, the same author who mercilessly killed Malone by running out of breath and playing as if that breath were Malone's own. Shameless he is, still a brute and a bully, badly concealed under the trappings of a fictional Lemuel. They are all there asking to be served, what can the Unnamable do but obey? As he says,

When all goes silent, and comes to an end, it will be because the words have been said, those it behoved to say, no need to know which, no means of knowing which, they'll be somewhere, in the heap, . . . they have to be ratified by the proper authority, that takes time, he's far from here, they bring him the verbatim report of the proceedings, once in a way, he knows the words that count, it's he who chose them. . . .[44]

And, of a sudden, the musing recalls Molloy's situations,

in the meantime the voice continues, while the messenger goes towards the master, and while the master examines the report, and while the messenger comes back with the verdict, the words continue, the wrong words, until the order arrives, to stop everything or to continue everything, no, superfluous, everything will continue automatically, until the order arrives to stop everything.[45]

But that is in a moment of reflection, when the Unnamable has gone through his confusion and has begun to see the light at the other side of the tunnel with the help of his own remembrance of things past, his experience as Molloy.

In the meantime, he has sought, like all inauthentic men, to define himself by his place. Where is he? In hell, perhaps. He looks for the lower depths, but cannot find them. He notices that he is either at the center of a circle or on its circumference, either traveling around the shade of Malone or it around him. His task has only begun; no order has been received to stop. He must continue. Was it really Malone, or Molloy wearing Malone's hat. But why suppose that? It's far safer to assume, when one must continue, to be at the center of things, fixed there forever. Molloy, Malone, Mercier and the others, including Watt, seem to be there. But what could they have been guilty of to deserve such a fate? A conscience doesn't say what one is guilty of, only that one is both guilty and responsible. That is what the others had taught the Unnamable, along with a bit of knowledge about love and destiny and death, even about God and hearing voices of silence, but that was in the past, when the Unnamable was Moran. He may, then, assume his own responsibility and create his own subjects. That's the only way for a writer to go on.

The first of these creations is Mahood. That's manhood, without the 'n.' Where is he? Deep inside the unnamable, as yet unexpressed because he was first thought of as Basil, whose name gives no clue to his identity. Humanity always lacks one of its essential parts. But what is essential? Certainly not the legs. Mahood has lost one of his legs and misplaces the other in his travels around the circumference of the earth. He is forced, like Molloy, to use crutches for the peregrinations that, having started out from his family and moving in a circular pattern, must return him home. He sees them all there waiting for him: his father, mother, wife and children, all in the presence of his grandparents, urging him on. But alas, he has lost an arm too; so how is the crutch to be used? Better to sink into the ground there where he is, in sight of his goal, but still at a tantalizing distance from it.

The Unnamable here enters the universe of Mahood. Who is speaking? No one knows, for places and persons have been transposed. Either the family is now at the center looking out at the circumambulating Mahood, or he is at the center enclosed within his own windowless monad:

> I found myself in a kind of vast yard or campus, surrounded by high walls, its surface an amalgam of dirt and ashes, and this seemed sweet to me after the vast heaving wastes I had traversed, if my information was correct. I almost felt out of danger! At the centre of this enclosure stood a small rotunda, windowless, but well furnished with loopholes . . .[46]

These loopholes, of course, mirror the structure of the entire universe, if Leibnitz was right,[47] however confusedly. At least they were there to permit the impression of the expressions of others.

But now minus his legs and both his arms, since he is reunited to the primal slime, he achieves his final shape, an egg that might roll wherever it lists. In this shape, trussed up in a jar and tended by his faithful feminine keeper, who finds his offal useful for her garden, Mahood becomes Worm.

This should be no surprise, since nothing but the desire to move is left his physical reality and since Watt's own narrator has forewarned us that to speak of God one must speak of men and that to speak of men one must, of course, speak of termites. But Worm is a termite who thinks; the narrator's Cartesianism returns, long enough to suggest that the Unnamable's penis is still intact. His hands, or a horse's rump might produce an erection. Vain hope; the noble tissue only stirs for a moment and then recedes into what it always was, an insignificant blight upon its owner's conscience.

Where was Mahood's family now? Forgotten. There remained only the thoughts of two females:

> But the bouquet was this story of Mahood's in which I appear as upset at having been delivered so economically of a pack of blood relations, not to mention the two cunts into the bargain, the one for ever accursed that ejected me into this world and the other, infundibuliform, in which, pumping my likes I tried to take my revenge.[48]

Sursum corda! That was Belacqua, who should have staid under his rock in Dante's antepurgatory. Besides, the whole thought was but a physiological commotion,[49] unrelated to the business at hand. That, it should be recalled, is to define humanity; a lobster has no mind because it has no temporality and cannot project an authentic death prior to being boiled alive. Dying quick, it never dies quickly.[50]

The question here is of the mind, where

It all boils down to a question of words, I must not forget this, I have not forgotten it. But I must have said this before, since I say it now. I have to speak in a certain way, with warmth perhaps, all is possible, first of the creature I am not, as if I were he, and then, as if I were he, of the creature I am.[51]

The dialectic is still of the other. Even so, memories cannot be forgotten without ceasing to be memories, and the memories of a forgotten voice will resound once more: "Faint calls, at long intervals. Here me! Be yourself again!"[52]

Who is the authentic Unnamable? A voice to be sure, but not the voice of his past, Malone, Molloy, Watt or Belacqua; they are now what he is not. Nor can he continue to be the dupe of Mahood, or even Worm. He has created them, in the vain attempt to give himself a name:

Worm, I nearly said Watt, Worm, what can I say of Worm, who hasn't the wit to make himself plain . . . ? Perhaps it's by trying to be Worm that I'll finally succeed in being Mahood, I hadn't thought of that.[53]

But the narrator of *Watt*, Sam, had. Worm is in thrall to the Unnamable as he himself is in thrall to his own creator. It is for this reason that the narrator now can say, "Worm, Worm, it's between the three of us now, and the devil take the hindmost."[54] As the word is pronounced, the lips must part to exhale that warm breath.

Lucifer, we shall recall, is half buried in ice, surrounded by the flames of hell. In his mouth are Brutus, Cassius and Judas Iscariot, traitors to a trust, all. The Unnamable would certainly be the hindmost should he be faithless to his own trust, to discover what he is. Worm's jar, now fixed with a collar so that all movements are prevented, is on an island. That's where Mahood had finally understood himself to be. But where is the island? At times, when his keeper is Marguerite, the island is off the continent; but at others, when she is Madeleine, it is not far from the Rue Briançon. What matters where he is? Knowing only that, anywhere in the world, he would still be in ignorance:

Worm, to say he does not know what he is, where he is, what is happening, is to underestimate him. What he does not know is that there is anything to know. . . . Feeling nothing, knowing nothing, he exists nonetheless, but not for himself, for others, others conceive him and say, Worm is, since we conceive him, as if there could be no being but being conceived, if only by the beer. Others. One alone, then others.[55]

To be because one is conceived, that is the ontological proof; but it works, if it does, only for the Perfect Being. How for a man to try? " . . . by the absurd prove to me that I am, the absurd of not being able."[56] But there is the other mode of conception. Here, too a failure; for the maid has not yet begun to menstruate that would give birth to Worm, or the testis decended that would put life into that descending egg. Perhaps Worm is the spermatozoon ejaculated into cold sheets, where he must wiggle his roundabout way until he too is as cold as they.[57] That's a law of thermodynamics.

This play on the difficulties of rationalism and empiricism to solve the mysteries of life is not new to our narrator. He must have recalled Malone's friendship with Jackson, the intellectual Israelite who tried to teach his parrot the Latin phrases of the dispute between Locke and Leibnitz. The parrot could get as far as Locke's *Nihil in intellectu quod non primus in sensu erat*, but the celebrated qualification, by Leibnitz, *nisi intellectus ipse*, drove the poor parrot into fits of fluttering wings and meaningless croakings.[58] No critical irrelevancy this: the Unnamable was Malone, and Malone had a series of friends with 'son' in their names: Johnson, Wilson, Nicholson, and Watson; but being sons of man, like Macmann, they were all "whore-sons" anyway.[59] Now Macmann was brought to the island by Lemuel, and Mahood, another whoreson pure and simple, was left there to become Worm in the Unnamable's reminiscences.

Worm is a voice, with the impudence to suggest that he is the Unnamable himself. But that certainly cannot be, for Worm exists only in the Unnamable's words, just as the Unnamable himself exists only in Beckett's words:

> ... Worm says to me, the subject matters little, that my purveyors are more than one, four or five. But it's more likely the same foul brute all the time, amusing himself pretending to be a many, varying his register, his tone, his accent and his drivel.[60]

So why bicker about pronouns?[61]

The other words used by those voices may be instructive still. Consider Worm's plight, devoid now, as the Unnamable, of all transcendence:

> ... but suppose, instead of suffering less than the first day, or no less, he suffers more and more, as time flies, and the metamorphosis is accomplished, of unchanging future into unchangeable past. Eh?[62]

That's what happened to Malone when he breathed his last; that's what will happen to Worm, who as a human being must be self-defining in working out his trajectory, like everyone else transcending himself into the future, unless it is true he doesn't exist. And the Unnamable insists upon his existential discovery:

> No one asks him to go to those lengths (to be burnt alive or to throw himself out a

window). But simply to discover, without further assistance from without, the alleviations of flight from self, that's all, he won't go far, he needn't go far.[63]

Just on, without respect to the lying words of others, to find those words to make all this clear.

Some critics claim that a story begins with the idea of an emotion. So the Unnamable satirizes this falsification of authentic human speech by inventing a story full of emotion – in the discursive idea, but not in the words. I shall give its resumé here.

A man meets a woman, and they fall in love. They marry the better to enjoy their love, when he is called to war and reported killed in action. So his lady love meets another, falls in love and marries this second suitor the more to enjoy her second love. But her first husband returns, and she goes to the station to meet him, where he dies, from the emotion of meeting her again. So she must go home all sad for having lost her first love once more. And upon reaching home she finds her second husband dead from the emotion of the thought of having lost her to her first. Her mother-in-law is likewise sad because it was she that cut the poor bastard, her son, down. Aim at happiness, and come up with sorrow: that was the theme of *More Pricks than Kicks*, was it not? So the Unnamable was already projected at that time. The emotion of a story is not in the facts of its representation, but in the structure of those events, as presented on the surface of the narrator's prose. By now the Unnamable has understood his Heidegger, for whom a state of mind is "equiprimordial" with understanding and its expression. Neither of these three "openings of the world" can be said to precede the other.

Besides, the narrator is engaged with his own emotional struggle, precisely the coming to an understanding by the expression of a state of mind. And the narrator exists only in this process: "The mistake they make of course is to speak of him as if he really existed, in a specific place, whereas the whole thing is no more than a project for the moment."[64]

What then is the Unnamable? Nothing more than a projected meaning in the words of his master, where he will be as if the interface between two worlds, changing positions from sound to silence, from word to reverberating meaning, the one internal which speaks and hears as the other hears and speaks. Indeed, Murphy's solipsistic world has achieved a new meaning: there are no longer two worlds, the one larger and the other smaller, but only sounds and silences held together by the gestalt of a vibrating tympanum.[65]

So the solipsist will either hear or speak; and to speak conscientiously, so said Martin Heidegger, he must learn to listen to the silence.[66] That is the only way he can go on, and go on he will – to tell us How It Is.

The trilogy is therefore a continued effort to plumb the depths of self-consciousness, and will continue until the final *Ping*. In the meantime the song of the dead and dying will continue to fill the air. – Cockles and mussels, alive, alive, oh!

Since the triune story of the trilogy brings Beckett's earlier treatment of time (*Whoroscope, Murphy, Watt, Stories*) up to the level of conscious temporality as it is lived by a human being projecting a future that may flow significantly through a vibrant present to a vaguely remembered past, he has already succeeded in showing, both on the surface and in the deeper structures of his prose, How It Is. But since the philosophy was comparatively new and difficult, perhaps if the story were told once more, it could be better understood.

Beckett's pessimism turns from the theme of time and death (eternity) to how an author must communicate his understanding of the world in terms of these themes.

CHAPTER VI

"I WAS MY FATHER AND I WAS MY SON"

The speaker is the narrator of the first *Text for Nothing*, who recalls the story of Joe Breem(n) first told by the father of *The Calmative's* narrator. The new narrator, then, has a past stretching beyond the trilogy back into the *Stories. The Texts* were written in 1950, after the trilogy and after *Godot*, but were published in 1958, along with the *Stories* that were written in 1945. Published together in English as *No's Knife*, the *Stories* and *Texts* (likewise published together in French as *Nouvelles et Textes pour rien*) might be interpreted as forming a continued narrative in which the teller of the former stories has become the one to whom they were told. Nothing would be less surprising. Every writer tells the story of his life in one way or another, not that he speaks of himself necessarily; but in creating his story he has discovered who he is, raising his feelings to the level of an understanding expressed in the words, and as writer, he is the first to react to those words. Thus, whether he speaks autobiographically or not, the writer is both the teller and the told, and the told in a dual sense: first, and most obviously, he who first reads what has been written; and second, he whose consciousness is created in the act of writing the story.

In the light of the trilogy this identification of the teller and the told is as clear as the Heideggerian ontology that has lent its structure to Beckett's greatest achievement. To read the texts as if they were unconnected with the trilogy is to look on the references therein to the situations of Molloy, Moran, Malone, the Unnamable, even Pozzo of *Godot*, as threads for the development of future stories rather than for what they are, facets of a life already lived and recalled by the current narrator, who finds himself with no place to go, with nothing to be named, and with no words to name it. Is there a way out? Yes, the search for self-consciousness must go on, and perhaps only it is capable of generating its own subject matter. No's knife is still Yes's wound; the writer will find himself by reflecting on his present situation, beginning at the beginning, when, as a boy, the dead narrator of *The Calmative* first heard the story of the brave Joe Breen(m) that provided the motive for his own suicide, as the narrator of *The End*.

For this reason, the structure of the *Texts* will reflect both the Hegelian dialectic of the *Stories* and the concrete fulfillment of time, along existential, ontological lines, in the temporality of the larger trilogy's principal characters. Reflecting as he does, the new narrator experiences the anguish of the writer worried about his having said all he has to say about the art of writing; and the same anguish is perceptible on the surface of the works, as the collection of the *Texts* presents the plight of the poet at its most unhappy stage: having to go on with nothing new to say, and knowing that even if one were to say it, that would be less than all that might have been said. As long as there is breath, alas, that much of one's soul must come out; and what it is to be only the words can tell.

Just as the implicit future of Watt's universe is made explicit in the *Stories*, their own implicit future is made explicit in the *Texts for Nothing* as the initial step beyond the trilogy. The in itself of the past consciousness (*The Expelled*) and the for itself of the then present reflecting consciousness (*The Calmative*) were rejoined in the consciousness existing in and for itself in *The End*. If the dialectic is to continue, this new consciousness must itself become a new beginning, a renewed "immediacy," as Hegel would say, for further mediated development. And the new words are to establish the "means" necessary for the mediation.

Simply put, the first synthesis was achieved in abstraction from what the narrator really was. Already bent in upon himself, as the expelled, the narrator continued with his negations, which, to the extent they made their cuts into the phenomena of the real world, left the nature of this world unexpressed. The negation that is the power of the solipsistic mind must be reexpressed in full concreteness. As the first step beyond Beckett's trilogy, the new context will therefore be an examination of the literary frame of reference itself, this time to show the narrator examining his own conscience, forced in spite of himself to go on:

> Suddenly, no, at last, long last, I couldn't any more, I couldn't go on. Someone said, You can't stay here. I couldn't stay there and I couldn't go on. I'll describe the place, that's unimportant.[1]

Note the tense, which refers us to a previous existence of this narrator; it will come as no surprise that he recalls the same story that the father of *The Calmative's* narrator told him as a boy, the story of Joe Breem (or Breen) who jumped into the waves with a knife between his teeth, to do what was necessary and return.

The narrator could not stay there, wherever he was, since his future was unexpressed (and so his search for self-identity was left unclear); but he could not go on either, for his story had already ended in the suicide that produced his oceanic consciousness. He had died as he had lived, unhappily; and the stories of Molloy, Moran, Malone, and the Unnamable confirm this hypothesis. How to begin again? By describing the place; that surely is not the point of the new story even though Molloy's story has shown how a person may be identified with a region. Such a lead may go somewhere to define this new negative entity that is the self of the creative writer.

In a passage reminiscent of *The Expelled* and of *Molloy*, the present narrator, thinking of himself as a past project, hears his controlling voices say, "All you had to do was stay at home. Home. They wanted me to go home."[2] Indeed, why undertake any voyage when it is easier to stay home? Because there is no home where life is nothing but a series of expulsions, or a flight from here to there and back again; to live is to transcend oneself toward some future: authentic if self-determined, and inauthentic if determined by others.

Writing itself is a voyage that pictures that transcendental flight of man; and Beckett is himself in the same rôle as his narrator, as symbolically he has always been there, having just finished one story and worried about whether it has said anything significant. Who was the 'I' of the *Stories*? The persona of Watt, to be sure; but also the teller of those stories, and the four stories of the trilogy – all depicting voyages having as end the return to self. The narrator now reflects on his immediately past achievement and on the possibility of going on.

Death will not stop this procession of narrators because the creative consciousness, in and for itself, is above all imaginative, both imagined and imagining. It can fit itself with a body, a head, and circumstances to meet the needs of the moment. No longer committed to a real world, its place is truly unimportant; and time, a timeless moment of imaginative reflection:

> How long have I been here, what a question, I've often wondered. And often I could answer, an hour, a month, a year, a century, depending on what I meant by here, and me and being, and there I never went looking for extravagant meanings. . . .[3]

Here the narrator is lying, unless it is Beckett talking; and in that case the lies are his. The narrator's question was simple enough; it called into dispute the past and the several answers proposed in the past: nothing for the present but the question itself, and the recall of the past, unsatisfactory answers. *The Texts* were written to expose the truth of this nothing.

The setting, too, recalls the past. In *The End* the narrator had seen sea, mountain, plain, city, and forest, and all these at once in his oceanic consciousness. Like Joe Breem (or Breen) in his father's story, he threw himself into the waters. And remembering this, he experiences the first epiphany of the *Texts*: "Yes, I was my father and I was my son . . . ,"[4] he who tells the story and he who is told. Up to now he has lived the schizophrenic existence of a man talking to himself, with no external point of reference to check the sense of what has been said. That's the force of the calmative; it takes our cares away from the trivia of the real world, only to replace them with the anguish of self.

As it happens, that is where the narrator is now, wrapped up in his own ego-centric feelings:

And this evening again it seems to be working, I'm in my arms, without much tenderness, but faithfully, faithfully. Sleep now, as under that ancient lamp, all twined together, tired out with so much talking, so much listening, so much toil and play.[5]

At this point in non-time, therefore, the narrator is his feeling of fatigue, of being worn out by the experience of talking to himself: a pure creative imagination, in full possession of its past words, partially failures and partially successes, but now engaged in a new project in which the future is to be made present – in anticipation.

The question is no longer "Who?" but "What?" Hence, a return to the badly disguised question of Watt's name, which will receive the same answer, the return to the question itself that provides no answers, if not the inevitable denial of Knott. The living narrator has transcended himself into an essence, from a living existence he has been reduced to a non-living thing – a series of meanings embedded for better or for worse in his own creator's words.

What is this consciousness existing in and for itself, if not the end and the means of a new expression? The Unnamable's predicament is reestablished in the *Texts for Nothing*, where the 'nothing' refers not only to the absence of an abstract idea to be formulated in words, but to the character of the one who must discover a concrete idea in its expression. The twelfth *Text* is explicit on the nothingness of the writer's consciousness. No consciousness is a thing, but it may be a function.

Caught between that function of his personality that speaks and that which listens, there is a third:

And this other now, obviously, what's to be said of this latest other, with his babble of homeless mes and untenanted hims, this other without number or person whose abandoned being we haunt, nothing. There's a pretty three in one, and what a one, what a no one.[6]

And Geulincx's ethics is still there, left unstated: where you are worth nothing, there you shall wish for nothing, since that book was his as the narrator of *The End*. But Geulincx does not make clear the essence of these two nothings, as Murphy was already well aware.

Between the first and the twelfth texts the creative imagination must express itself to find out what it is: as the power of negation, a pure nothingness of being with respect to the positivity of the phenomenal world. Hegel had already made this clear, well before Sartre's *Psychology of the Imagination.*[7] Being in itself, as absolutely undifferentiated, a negation of all determinate being (a this, a that, or another), it is nothingness itself. Each thesis must pass over into its own antithesis.

The *Texts for Nothing* are therefore thirteen texts for describing Being; not texts for something to be sure, but for a being that must define itself by denying what it is, a being in the world, one that must create its own world by projecting a future (as illustrated in *Molloy, Malone Dies, The Unnamable*) which does not yet exist. It is not yet this future, no longer its past, except in memory, and presently not of this world. There's another pretty three in one: a human transcendence laid out in its three temporal "ekstases"; its being is to stand out from itself, existing forever at a distance.

The philosophy here, of course, is Heidegger's, but it grows naturally out of the Hegelian scheme; for, if being in itself must in any present moment become its own other (a non-being at any other moment), only to return to itself enriched and fulfilled by that negation, there lacks only the concrete beginning and ending of that process. Hegel's scheme is only ideal; Heidegger's restatement of it changes the abstract dialectic into the concrete stuff of human consciousness, temporality itself. The suffering must be made real, not only in the representation, but on the presentational surface of the writer's prose as well. It is moments such as these when every critic must realize that no running account of the poetic prose of his texts is likely to do justice to their achievement, alas, not even this one.

The *Texts* shall remain abstract in their representation, because they represent the culmination of the self-returning consciousness. Their prose is full of the anguish and the doubt of the power of the word to express its thought, and that is the only level of concreteness open to the human self-consciousness existing in and for itself. It is, as such, a stage the creative imagination must go through if it is to fulfill its heritage; to be for itself it must express itself by becoming aware of what it is in itself. For the future, then, a new diremption in which it shall express itself, in and for itself, as a writer of significant prose.

In this way, the *Texts* prepare for still another expression of the same creative imagination as previously presented in *Molloy, Malone Dies* and *The Unnamable*. Indeed, Molloy and Malone are referred to by name in one of the texts; and *How It Is* is the expected fulfillment of the current project.

As a synthesis of the dialectical antithesis of the *Stories*, the *Texts for Nothing* project the existential reality of a new negation, a new consciousness, another narrator, somewhere in the future. Constantly they bring up the past, and constantly foreshadow this nebulous future, in which the narrator will relate how it is one learns to tell a story. That will be another return to the beginning – as *How It Is* translates only one connotation of the word-sounds generated by pronouncing *Comment C'est* (*commencer*).[8] There is to be then another return to a new beginning.

The first of the thirteen *Texts for Nothing* opens with the reminiscence that "someone" has said the narrator must continue, to get on with his story; but with one (not immediately stated which) just ended, he feels he cannot go on. The recall of the story of Joe Breem, however, informs the reader that the narrator is the same as that of *The Calmative* and *The End*.

Since when has he been in his present location? Forever; as a timeless essence, like the Unnamable, he can experience nothing more than the continual recall of his past, where he was both father and son to himself, telling and listening to the story of the brave Joe. Thus, a teller and a told: a voice, a silence, and a tense gestalt (grave now, to become acute later on) of their union, just as every writer may be symbolized in this way, a consciousness attending to its own inner workings as a pure imagination.

At one time he must have possessed both body and "head," that thinking part of the body; but the questions this head puts at this time and in this place are indeed unanswerable. All it can know for sure is that it had a past (by virtue of its reminiscences), and that in recalling that past it is still conscious. The in itself has become the for itself; and having returned to itself in death, it is now in itself for itself, and for itself alone. For this reason it must be nudged into telling still another tale; for it had already achieved its essence in death, the dual essence of teller and told, which could not of itself go on. He needs that awesome third voice, in this case of the author, his creator, to go on.

The second *Text* poses the present predicament in a question: Where can one go, now that one is fixed as an essence for all eternity? Up there? Back to the real world? The narrator recalls acquaintances from the past: a Mrs. Calvet, and a Mr. Joly with only one leg and a half. To go back to the upper world, one would have to know where one is, otherwise one risks too quick a return by throwing oneself from a cliff into the mud of the sea. Where is he?

Ce n'est pas la mort, ce n'est pas la tombe, loin de là, ça ne peut pas être la tombe, ce serait trop fort.[9]

No, he's lost in his imagination, looking for and commenting upon the words that refuse to come to him. He's lost in those lost words, but even should they come he could only be there: "So long as the words keep coming nothing will have changed, there are the old words out again."[10] These words situate him at a distance, as he can only recall the past, the farm of the brothers Grave, for example, whose image throws off light, giving him a new sight: "There, it's done, it ends there, I end there. A far memory, far from the last, it's possible, the legs seem to be still working."[11] The old trick still works: the words create the image that gives the light by which he can see; and the same breath will provide the words to create the body to feel the vigor of those old legs for going on to complete his return.

The voice does go on in the third *Text*, but fighting, now, with its words:

. . . it's understood, there is nothing, no more phrases, let us be dupes, dupes of every time and tense, until it's done, all past and done, and the voices cease, it's only voices, only lies.[12]

One must believe that the imagination creates the real, knowing full well that what it creates is false; its freedom is in its very denial of the reality of the world: its freedom to reject the strictures of that larger, confining, outside world. One must will oneself a body – any kind; he chooses the body of a man, who will have had a governess, to assume the responsibility for his being, and that man will have a friend. Still struggling with his words, he jacks himself out of the past into the future, as that grammatical tense indicates a change in direction of his view; he could go on remembering only the past, but that would get him nowhere:

That's so that I'll never stir again, dribble on here till time is done, murmuring every ten centuries, It's not me, it's not true, it's not me, I'm far. No no, I'll speak now in the future, I'll speak in the future. . . .[13]

The tense of his sentence startles; he is still placed in the words, but the words now indicate a future; he is resolutely determined by their projection to some other place equally far away which enables their user to transcend his self-enclosure in the past.

These words imploring the future embody a personal decision, an existential commitment; but like all such decisions taken in the real world, they forebode nothing but anguish. Where can it lead, but to the death of the friend and of self? So the imagination falters, and falls back into its old inauthentic ways, separated from reality by its own internal negations: "Departures, stories, they are not for tomorrow. And the voices, wherever they come from, have no life in them."[14] It is difficult to start even when one knows how, but the future has been won.

After all, where would one go from where one is when one is a voice that stutters before its own responsibilities? That's the predicament of the narrator in the fourth *Text*. He is still the same unknown: "Where would I go, if I could go, who would I be, if I could be, what would I say, if I had a voice, who says this, saying it's me?"[15] Unknown to himself and unknown to the other, vacillating between an 'I' that is neither a body nor a "head," and that Other, who continually demands that he speak. Who is that Other, the author? Yes, assuredly, but not only he; for it is the conscience as well, calling him to continued resoluteness. The narrator objects, the Other is most assuredly his own creator, the author himself:

He has me say things saying it's not me, there's profundity for you, he has me who say nothing say its not me. All that is truly crass. If at least he would dignify me with the third person, like his other figments, not he, he'll be satisfied with nothing less than me, for his me.[16]

Even authors need an other to achieve their self-identity. But having learned the lesson of those past figments of his imagination, this author does not like to speak of himself, nor to let himself be seen.

If such a person, this author, wants a life of peace, let him lose his life like everyone else. Hiding himself in others! That is good enough for Molloy and Malone; they were "happy mortals,"[17] who had a life in front of them. But the narrator cannot escape being like his creator; both are creating creatures, and both can truthfully say, as does the narrator: "What am I doing, talking, having my figments talk, it can only be me."[18] He proceeds by speaking and hearing his own inner voice – the conscience.

Two moments are needed in this conversation: an action to be reflected upon – some kind of life; and a dogged determination to continue the reflection. The life guarantees a presence to the world, as we begin here to hear Heidegger's voice, and permits a project beyond it, from self to self:

. . . there are moments, like this moment, when I seem almost restored to the feasible. Then it goes, all goes, and I'm far again, with a far story again, *I wait for me afar* for my story to begin. . . .[19]

And that is precisely where the narrator would go, precisely what he would be, if he could.[20] His going and his being would then be the same, a flight from here to there, from past self to a future self in this languishing, anguished present.

At this stage of the developing awareness, however, the narrator speaks in both Hegelian and Heideggerian terms, as he has moved from the dialectic of the Other, between himself and his author, to the self-determined projection of his own future. If the author limits the narrator, the narrator is free to return the favor. He has only to proceed with the words that become more and more Heideggerian. And when he does, the return to the past will no longer be to achieve the completion of the in itself for itself, but the conscientious assumption of one's own central anguish.

That story must be written, but with the materials supplied by the facticity of one's own haunting past. Moreover, the dialectic of the author and character will be left behind, to be supplanted by the inner dialogue of the narrator with himself. In the schizophrenic world of self-consciousness, there is no such thing as an interior monologue. Schizophrenia? Yes, but schizophrenia with a purpose. No longer the suffering of inner self-contradictions, but the pure anguish of facing the task before one, writing that story.

Text V takes on from here, with the narrator holding the pen. Like Moran in search of Molloy, he tries to be what he is searching. He still hears that voice, now more of conscience than of his creator, telling him it will be difficult to go on. Why change? Why does anyone change? Why did Pozzo leave his chateau and servants? He had an excuse, at least, living in another play,[21] and being, precisely, *pozzo*, i.e. crazy, or better, deluded into thinking his mastery could not be mastered by his slave.

Holding his pen, however, the creating narrator finds it difficult to face anything but his past. He recalls previous deaths: ". . . I have even gone to the bottom more than once, under various assumed names. . . ."[22] After all, he had done it twice in *The End*; first in anticipation, and then in the actual deed. The future holds open only a judgment on his soul. There's theology to the rescue again; perhaps someone will have pity on him, even have the goodness to explain what he has been guilty of, to be suffering so much. "But sufficient unto the day, those are other minutes."[23] The pen falls in the fatigue of the moment.

The sixth *Text* ruminates on the equivalence of life, words, and unnamable essences. No secret this: the life of the writer is in his words; before their discovery they have no name and after it they need none; and the narrating character is entirely confined to the writer's words. Those guardians of the past, Watt's and Macmann's male nurses and the self-hiding author, are a thing of the past, at last now at rest and sleeping unless another Malone wakes them up in still another story of Macmann. The narrator must go on alone into that new world.

What has happened to those bodily tissues? Fixed, like all the rest in the essential words. The new world will open on the Unnamable's past horizon:

> . . .what is it, this unnamable thing that I name and name and never wear out, and I call that words . . . with what words shall I name my unnamable words? And yet I have high hopes, I give you my word. . . . [24]

The only way out into that new world is to open one's mouth, first creating the body with the head that will tell the story of how it is.

Text VII is a regression, the statement of a recurring doubt. Has everything already been said? Why should one speak when one can only speak of oneself? Surely one person, call him 'X', who has lived a real life is much more interesting. Why all this introversion? Because Mr. X, lost in the Southeastern Railroad Station, was expelled out into a train compartment, where his apparent motion could lead nowhere. Watt's past experience leads the present narrator to conclude that train stations are to be stricken from the list of places to be visited; one meets the weirdest travelers that way. He must look for himself somewhere else, but he is already on the way. A return, then, to his own self-project.

The new beginning is formulated in *Text* VIII. There are longer pauses as the narrator searches for his inspiration, that breath that can now come only from himself. When the words do come they are without end, hence without significance, since the end marks the significance of the story's beginning.[25] Not able to find an end is the cause of the pauses, and the occasion of more reflection on the past:

> Well I'm going to tell myself something (if I'm able) pregnant I hope with promise for the future, namely that I begin to have no very clear recollection of how things were before (I was!), and by before I mean elsewhere, time has turned into space, and there will be no more time, till I get out of here.[26]

His lost past throws him into a projection of the future.

Nothing remains now but the imagination to salvage a time significant in itself: like that future in which the narrator (as Molloy) traveled back to his mother's womb: " . . . nothing but fantasies and hope of a story for me somehow, of being able to go back, or on, somehow, some day, without hope."[27]

What is this new direction? Telling a story, having some place to go; suffering, to make another suffer. However, all this, too, is false, told by another, by the ventriloquist with his dialectical tricks; but the narrator is no dupe: ". . . I never saw the light of day, any more than he, ah if no were content to cut yes's throat and never cut its own."[28] The unhappiness of the dually affirmed contradiction cannot be avoided. Hegel, then, was right?

The author gives names to his characters, as humans do to their pets, to be able to call them; and distinctive characteristics, so they can be recognized. But he cannot deny that he lives in the other, and so the other in him. So, the character, this narrator, must await the assistance of his creator to achieve a life, ". . . begging in another dark, another silence, for another alm, that of being and ceasing, better still, before having been."[29] The reminiscence is of his life as a beggar in *The End*, but this new death will be as vain as any other.

What is the way out? What the "issue" to be expected. The French of *Text* IX plays on the equivocation of the term, but in the English translation Beckett is satisfied to assume that a "way out" constitutes the essence of his narrator's predicament. The narrator is looking for a way out as for something of importance, the probable ending of his story. But he can find only the cemetery, a dead symbol of an issueless story. The importance, however, is to begin:

> . . . Ah, says I, punctually, if only I could say, there's a way out there, there's a way out somewhere, then all would be said, it would be the first step on the long travelable road, destination tomb, to be trod without a word, tramp tramp, little heavy irrevocable steps. . . .[30]

The end can only be the beginning when the imagination has found its target.

No, stories aren't written that way, the issue is made apparent in the writing itself, where beginning, middle and end may suffer permutation, as both Watt and Beckett had demonstrated once before. The real beginning is with the start: the narrator imagines he has had a mother, whose warm, humid womb will haunt him until he has returned to its only earthly simulacrum, the grave. Molloy's project is hard to forget; it's those lip-tightening 'm's.'

Between the womb of birth and that of death there are others, such as the prisons of the State, and the institutions of the charitable, where the air is stifling. The issue is death, and so is the way out – of life; the writer has only to state the point flatly: the way out passes through the cemetery.

> There's a way out there, there's a way out somewhere, to know exactly where would be a mere matter of time, and patience, and sequency of thought, and felicity of expression.[31]

The very terms 'sequency' and 'felicity' are both sequential and felicitous. For the felicity of expression there is no need of the body. Like Moran the narrator projects a return home, past the cemetery road, and abandons himself into his own being-unto-death, his own ultimate silence.

It was Malone, after all, who had shown the way, telling stories to create time by projecting his own temporality towards its end, just as he had shown how to produce the connection of ideas and the felicity in their expression. The philosophical idea must give its structure to the work, and Heidegger's previous statement of the same idea was anything but felicitous in the expression. And if Malone's death is structured on the ontological description of human abandonment unto death, and the Unnamable's story, on the nature of essential being, there still remains the account of a writer's abandonment to his own essential disclosedness.

The abandonment of the narrator takes up his thoughts in *Text* X. Indeed, abandoned to the sole condition of a voice that speaks, he need fear nothing, neither the loss of body, nor mind; nor his birth, his life, or his death; for one is abandoned to the project itself: always a passing, never completed, never surpassed. Condemned in this way continually to speak, to break the silence, he cannot give up, must continue to the end, wherever it may be found. Someone must be found to express the silence itself, to discover the words that would express the nothingness of a consciousness that is silence between sounds and dark between flashes of light. That will come in the new beginning – in the panting of the narrator crawling through the mud in *How It Is.*

Texts XI and XII sketch out in retrospect Molloy's existence as narrator of his own story. "Knowing no one, and by no one known,"[32] that's how he will wind up; but this narrator must begin by setting his mind to the task, gathering himself together, before he becomes "too scattered by the words"[33] for the attempt. A moment's reflection, and the narrator is on the road once more:

> And it's still the same old road I'm trudging up yes and down no, towards one yet to be named, so that he may leave me in peace, be in peace, be no more, have never been. Name, no, nothing is namable, tell, no, nothing can be told, what then, I don't know. . . .[34]

The one to be named, however, after Molloy, after Moran, after Malone and the Unnamable will call himself 'Bom,' an 'm' somewhere so the lips will be forced to close in pronouncing the name.

The narrator will pass through the Molloy country on one leg; or better yet, like the Unnamable, without legs, hands, arms, amputated all over; without tears either, for his sighs will suffice to create the mucus of his body, those sighs coming from the heart with reasons of its own. With the heart, any imagined being is complete, or at least human enough to get along. Its beats are sufficient to measure the solipsist's time, which will be laid out in the stretch of a new temporality, expressed in three voices, like those of Molloy seeking to respond to the conscience that calls him to authenticity. With all this world, the narrator may mount his bicycle.[35]

Here, now, at the beginning of *Text* XIII, the narrator is concerned with the authenticity of his own voice. Other voices assail him with their sounds, coming as if from nowhere. From Molloy, perhaps, but that was in the past; this one calls him to the future; here, now, only the babel:

> ... it wonders, that voice which is silence or it's me, there's no telling, it's all the same dream, the same silence, it and me, it and him, him and me, and all our train, and all theirs, but whose, whose dream, whose silence....[36]

If only there were only one, the question would be simplified. But there are always more than one, so the question is natural:

> ... what more is it waiting for now, when there's no doubt left, no choice left, to stick a sock in its death-rattle, yet another locution. To have rounded off its cock-and-bullshit in a coda worthy of the rest?[37]

A moment's intemperance, a cry of despair and disgust.

When at last the grand orchestration will have been made, silence will be the beginning and silence the end; the middle will be made of words that must lie, pretending to be what they are not: "... so many lies, so many times the same lie lyingly denied, whose the screaming silence of no's knife in yes's wound, it wonders."[38] Over the din of conflicting voices we thus once again hear the voice of Hegel, explaining the unhappiness of complete self-consciousness:

> It's true, yes, it's true, it's true and it's not true, there is silence and there is not silence, there is no one and there is someone, nothing prevents anything.[39]

Silences before, silences after, and silences in between as the sounds of the words project their silent, reverberating meanings.

The voice of the narrator continues its murmuring, holding out an end of peaceful silence and promising a future when all this is to be made clear; not only as an idea, not even one concretely expressed in the structure of one character's world, but presented on the very surface of the writer's own world. That will be the symbolic result, as Beckett proceeds to show how one learns to communicate – authentically – with one's fellow man.

The voices will be heard as the characters crawl in the mud, but that is what man has always been – *homo*, from *humus*, moist, molded earth into which someone has breathed the gift of soul, and who in turn exhales that breath into new creative worlds.

Down, then, into the mud, back to the earth, the poet's vision returns to the sources of its own expiration and the center of its own spatiality, the better to understand its projection into a new temporal existence. The motion is spiraling, inwards and downwards, like Dante's Hell; we need only enter the vortex – with or without abandoning our hope of a clear understanding. If Heidegger is right, that understanding will be the fruit of our response to the feeling of the words.

There is a way out, but it leads to an end few authors have had the courage to project – the extortion of their meanings from the agonizing cries of their fellows. What more can the Self expect from the Other, or the Other from the Self, but a chiasmatic crossing of the paths,[40] and a final abandonment: a dual project of self-creation moving from the alone to the alone, from Molloy through Moran to Malone. The solipsist in each of us, both writer and reader, cannot be denied.

But listen to the voices of silence. They will explain how he who was his father and his son came to be both by being himself. If these are truly texts for nothing, at least they are for the nothing a man can be – a fading mediator between father and son, both of whom are brought into existence by the narrator's creative consciousness. How does one celebrate one's own nothingness in the positive scheme of the world's phenomenal appearance? By showing How It Is.

CHAPTER VII

VOICES IN THE MUD

Before listening to them, a brief review of the paths leading to their conjunction may focus the reader's attention.

In *The Unnamable*, the narrator of the larger trilogy was reduced to a voice attached to an egg shaped mass trussed up in a jar: Worm; and before Worm, he was Ma(n)hood, circling around a lost island home or it around him. Since all motion is relative, who is to say? Not Mahood. He is a visitor to the narrator's consciousness, an image, just as the narrator himself, a naught like him of the *Texts for Nothing*, is only a visitor to his author's consciousness. Worm is Mahood completely amputated, except for the useless penis, immobile, lost in a never-ending present. They or he speak, who can tell? Not Beckett, for he is in the same situation, really, as his creatures are imaginatively. Nothing, no one can be named, even in this purely literary, imaginative universe where, once the end has been found, everything has become fixed – someplace between the words and their significations.

The larger trilogy retells an old story, that of the *Stories*, with a different philosophical structure. When the consciousness of the *Stories* has achieved the summit of self-consciousness, first in itself, then for itself, and ultimately in and for itself, it has ended a dialectical journey exhibiting the eternal return – to self; and as the *Texts* explain, all for nought. Although the order of the *Stories* is rational, it is not natural.

A new order, still nonnatural, develops in the trilogy, as the solipsistic narrator sets out on a new journey. Molloy sets out to find his mother, looking to the past for his significance and not finding it there unless by ricochet as the voice of conscience that calls him forth. The call directs him to the future, as Malone, in whom he dies, thereby achieving completion, and essential determination, but at the cost of losing his transcendence. Cut off from all transcendence, fixed in relation to the voices he hears, The Unnamable is condemned to the unchanging present, surrounded by his voices and an equally unbearable silence. Where now? Who now? When now? Three questions directed to this now that cannot end, since a human transcendence has there become fixed, essentially, for all eternity.

But the "when," the "who," and the "where" pose the eternal question of human reality: time and temporality, and so consciousness; facticity, existentiality and fallenness, and so transcendence; space and spatiality, and so the body and its natural environment. Indeed, in this new order, human reality is a projection from a factitious past that exists no longer to a transcendent future that is not yet in a present that continually evanesces into nothing, a mere flight unto death where there is no more fleeing.

We know how that story ends, with all its questions answered. The "now" is forever; the "who," the voice that asks the questions; the "where," some place beyond the confines of the natural world and within the context of the literary configuration. The dialectical journey of the *Stories* has been replaced by the existential condition of human reality, *Dasein*, whose being is to be there in some world. Hegel has been replaced by Heidegger.

How It Is is told by a new narrator, who expresses the desire to recount his story in a natural order, a simple before, during and after (past, present and future) to replace the dialectical swings and the existential abandonment to a flight hither and thither. Pretending, like the Unnamable, to hear a voice not his, the new narrator explains how it is, how it has come about; but he must do so as he hears it, in bits and pieces, when the panting of his breath has stopped to allow the hearing:

> my life last state last version ill-said ill-heard ill-recaptured ill-murmured in the mud brief movements of the lower face losses everywhere [1]

The syntax startles, but it successfully presents on the linguistic surface an aural image of the narrator's condition.

A translation into a more natural syntax would tell us as much about the world being signified, but would fail to communicate the essential, a world that exists between the verbal surface and its signified depth. The prose would go limp, lose its tension, and with it, its creativity; consider,

> My life in the last state, given in the last version, was ill-said, ill-heard, and consequently ill-recaptured. It was ill-murmured in the mud, by brief movements of the lower face, with losses everywhere.

The bits and pieces would have been restored to a larger unity only to lose their significance, as the poetic tension is replaced by prosaic statement. The "natural" order of *How It Is*, therefore, is not to be found on its surface. That order is pure invention.

Yet it communicates an important message. Just as the Unnamable had found the ventriloquism of their author too apparent in *Murphy, Watt*, and all the others, so the new unnamed (neither Christian nor, surely, otherwise) narrator deplores how ill *The Unnamable* had been murmured in the mud. Pause. The image is psychophysical: the voice, a symbol for the consciousness; the mud, of the body and of its earthly abode. But the Unnamable was a disembodied consciousness, lost in a constant present, suffering its unchanging state as it was created, itself creating a literary context of unique significance. True, this body continues to wend its less than merry way, ashes and dust to dust, from bloody cunt to sodden earth, all humus, *homo*, a man, indeed; but full of care, and taking care less his nullity lose even that significance.

Such is the story that had to be retold in a clearer image.

In the new context, the hero will be alive, not dead; on a voyage, not yet arrived; nowhere, to be sure; but all when. Time will be measured by stretches, by the snatches of silence interrupting the bits and pieces of conscious awakening when the voices are heard. Then the hero may quote the words he has been given by his creator, just as he hears them, without adding, subtracting or hypothesizing.

His name will depend upon his function at a given moment of his transformation (he recalls Haeckel, less famous, perhaps, than Lamarck and Darwin; and we shall recall Worm, the fulfillment of a previous narrator's despairing attempt to treat of a man in any terms other than those fit for a termite); as executioner, he who makes the other speak, he is Bom; as victim, he who is prodded into speaking, Pim. Before and after he is nameless. But there are a witness and his aide: Kram and Krim; the one notes and the other writes, if indeed they are not both one. Another pause for another image.

The Unnamable's situation has been simplified. There is one symbol for the created creature, another for the creating creature (Pim and Bom respectively); and another, Kram/Krim for the creator, that witness who needs no witness, the author of all. Pim speaks to Bom, then abandons Bom to become Bom to another Pim; the original Bom then abandoned becomes Pim to another Bom, to whom he is forced to speak of how it was and of how it is. Kram speaks only to Bom, in a ducklike *quaqua* that must be given sense.

There is no world other than that supported by a consciousness. In the mud all is dark and humid, with tactility the only clue to reality other than the sounds and silences. But every Pim is forced to recall how it was "up there," in the light. And besides those of his memory, his consciousness can fantasize images of all sorts. Sorting them indeed is the problem, selecting the true from the false; for all is not clear in the kingdom of the light, nor all unclear in the kingdom of the muddy dark. When the blue, i.e. bodily, eyes of Bom close, he is returned to the dark; and when the others, those behind the blue, go blank, he enters the black – sleep, unconsciousness, a momentary respite; but often only to dream, when he hears the voices again, or sees the events of the lighted world as if his blue eyes were open once again to its stimulus.

Each Bom, continually created by generations of Krams, will have retained his "humanities," a smattering of anatomy, geography, physics, and mathematics; and two philosophers are implored to supply a missing connection. Malebranche, the French occasionalist, is the authority for a dog's having the same idea as its mistress and her young lover: on the same occasion God (Kram) willed that they act in the same way, to continue their walk; and Heraclitus the Obscure grounds the flux of experience being undergone in a river of time, into which no man, it will be recalled, can ever step twice.

This literary world begins with the before and ends with the after, the natural order of a world left obscure enough by the naturalness of the order. Heraclitus' obscurities had no other source. And the beginning and the ending have a true middle, a "during" after the "before" and before the "after" when Bom_2 meets Bom_1 changing him into Pim_1 of the current revolution. In the cycle, there will be another Pim for the old Bom_1 and Bom_2 will become Pim_2 to another, new Bom, formerly some other Bom's Pim; but those events go beyond the frame of this particular story.

How It Is is the story of all Boms, (Bem) and of all Pims, both groups at the mercy of all Krams. And, once again, we are given it in tripartite form: that enormous stretch of time before Pim, where Bom suffers his own solitude; an equally enormous stretch of time with Pim in which Bom teaches Pim a new language, ending with Bom's abandonment; and finally a new enormous stretch of time during which Bom's new solitude is spent awaiting his own Bom, some one else's former Pim, to whom he shall be forced to tell his story as a new Pim, still under the watchful eye of the recording Krim, aide to Kram. The cycle thus plays itself out from solitude to solitude, passing through the fateful coupling of each Bom and Pim.

The when and the where, space and time, and hence causality, all become unimportant in this new universe. There is a natural order, but no natural world, which has been reduced to a memory or a fantasy image of the solipsist who must continue to endure his solitude in total abandonment; the mediating succeeds only in establishing the differences between an abandoned Bom (Pim) and the new Bom (Bom) who shall be abandoned by his Pim, telling us all the while how it is. Still another pause: another image.

This nonnatural world is the universe of pure literature in which creature (character) is abandoned by creator (author), who must suffer the indignity of becoming creature of another creator, only a creature himself. In the process God is abandoned too, for the cycle is self-productive and self-reproductive unless the whirling of this world is itself a symbol of God's abandonment of his own creation: if so, there would be a natural movement in levels of symbolic significance from the literal, to the allegorical, to the anagogical. The moral is always the same: in the words of Geulincx, not directly spoken in *How It Is, Ubi nihil vales, ibi nihil velis*. Two nothings surrounded by a where and a there indicating the hypothetical absence of a value and a prescribed absence of will.

All this is true, but misleading. This literary world is structured, like all the others of Beckett by a philosophical idea. Where the (real) world is of no value, there it would be folly to wish its presence. This imaginative world has the structure, as if in the abstract, of all Beckett's former literary worlds, the essence of how it is when to be is to be a creator; and to be a creator, to be the discoverer of the significant word. What happened to Pim "up there" is banal; to Bom, beyond recall. The "content" of any story is only a means to present a coherent structure, and that structure is to be found in the relations of the creative words, in the tension between signs and signifieds.

In the words of Bom, as he heard them from the mouth of (quaqua) Kram, he must hear the words to see the images, and see the images to understand the world. Feelings have no other source, since sensations only link us closer to the mud, to our bodies and to their ties with the earth: existential fallenness, it's called, the inauthentic life of lostness in the present. To leave that condition behind us, we too must hear the call of conscience.

Whether, like Molloy, we listen to the call of conscience, from afar to afar, to establish our authenticity, and, like Malone, resolutely anticipate our ownmost final possibility, in death, in order to achieve our essence, or, like the Unnamable, we trade off our lostness in the real world for that of the imaginative, there remains only one last resource open to us: the living of that imaginative present, now; for only in living it can we come to know how it is. By a turn of the imagination the structure of *How It Is* has become epistemological, i.e. patterned discourse on human knowledge, as it is revealed in a literary work of art.

Porfessor Hoffman had an intimation of this fact, but found it in all of Beckett's works.[2] He was a victim of his own knowledge of historical tendencies; for him, Beckett became understandable as one of many twentieth century novelists. Only Ruby Cohn, of Beckett's former commentators seems to have grasped the essence of Beckett's plight:

> Beckett's motif – if it can be reduced and abstracted from the comic concreteness of the fiction – is that words are thoughts are emotions, that fiction is our only knowledge, and all knowledge a fiction written in a foreign tongue: 'that was to teach me the nature of emotion ... it was to teach me how to reason,' says the Unnamable. All Beckett's fictions merge, finally, into the unique, suffering artist-human, ironically probing the meaning of fiction. This is Beckett's unique achievement, uniquely achieved.[3]

But this conclusion was reached in the discussion of the larger trilogy. It catches the reason that trilogy had to be rewritten in *How It Is*, but misses the clear existential structure of the trilogy and fails to provide an explanation for the identity of feeling, thought and expression, which is indeed the epistemological structure of *How It Is*. Once more, then, into the philosopher's bag of tricks.

The trilogical frame of the *Stories* was presented in nonnatural order presenting the dialectical structure of self-consciousness. It was replaced by the trilogical frame of *Molloy*, *Malone Dies*, and *The Unnamable*, still nonnatural, but reflecting the temporal ekstases of a human transcendence. All this, says Bom, was "ill-said," "ill-heard," "ill-murmured."[4]

The trilogical frame is retained in *How It Is*, but it is in natural order, reflecting nothing at all. Its frame has nothing to do with its structure. Timeless in its infinite repetition of cycles (solitude, coupling, solitude), its significance must be found in an enlarged moment in which before, during, and after reveal the timeless truth of the creative artist's condition, where sign and signified come together in sudden epiphany. Slowly, painfully, but suddenly the light will shine through the darkness of the mud. For the nonce only the idea is lacking, but that is already there, in the tragi-comic structure of the human reality depicted in the novel. What, indeed, are the structures of knowledge that permit our author to equate feeling, thought, and expression? A question once again to which Martin Heidegger has offered a hermeneutical clue.

Ironically, perhaps, this clue must be turned back in upon itself. To make it clear, we must elucidate the concept of hermeneutics itself.

According to Heidegger,[5] the hermeneutical process is circular; the interpreting consciousness goes out from itself into the situation to be interpreted, and returns in its successful attempts, to itself, enriched for the experience. It can be applied to any primary human experience in which the subject is looking for a sense to that experience: in the first, or creative order, it applies to an author consulting the facts of his life; and in the second, or recreative order, it applies to anyone who reads the written report of the former. The author of a literary work symbolizes the first, and its reader, the second. Moreover, since the author is his first reader, both functions may be embodied in the same person. For Molloy and Moran the first and second orders of hermeneutics were personified in distinct subjects; in Malone and the Unnamable both characters embodied both functions. In *How It Is*, there is a triple layer of hermeneutical expansion: Kram gives his words to Bom; Pim speaks to Bom; and Bom returns the words to Kram via Krim. But in all three cases, the process is the same.

The circularity of the process is evident in that no one can make an interpretation of anything, be it an experience or its written record, without having been acquainted with the experience or the record. Implicit, then, is the knowledge of the context to be interpreted. This knowledge is slight, but not nothing; it is what lays before one; and for this reason the process for interpreting its sense is called a clarification of the fore-structure (*Vor-Struktur*).

The fore-structure is implicit in any situation where a sense has been felt or understood. The purpose of analysis is to clarify in what sense the situation (experience or record) constitutes a meaningful experience. It has three components: first of all, a *Vorhabe* (a fore-having), i.e. the project itself of making or finding sense that creates a concrete context of a manifold of real and possible impressions; secondly, a *Vorsicht* (a fore-sight), i.e. the cautious, intentional circumspection of the initial context, with particular attention to its closure; and lastly, a *Vorgriff* (a fore-conception), i.e. the anticipation of the sense of the situation in the formation of an hypothesis that relates the context with its principle of closure.

Readers of this book will recognize the process used in interpretation of Beckett's works to this point. In each case they are expected to verify the hypotheses by reading the original texts to determine whether those texts do indeed come to closure by the principle alleged. Beckett had already used this technique in his criticism of Joyce and Proust. And since what a literary artist is engaged in doing in the first place is to create a significant world by the use of words, philosophy is in a privileged position for suggesting hermeneutical hypotheses: cosmology is a philosophical, metaphysical, discipline. But what would happen, if, instead of cosmology, the author has chosen to write about his having achieved sense. He would still be creating a unique universe of significance, but that universe would come to closure by virtue of the epistemological ideas implicit in the author's knowledge. And that is how it is, in *How It Is.*

Although epistemology may be thought independent of cosmology, it was the purpose of Heidegger's earlier philosophical investigations to show that it could not be independent of ontology; for the being who knows, or comes to know, experiences his being as a form of knowing. Indeed, knowing is a paradigm case for *Dasein's* being as being-in-the-world. What is it to know?

There are as many ways of knowing a world as there are of being acquainted with oneself. And this, it shall be recalled, is as a transcendence, a projection to a future that will become a present and ultimately a new past. Bom continually projects the second and third phases of his formulation in *How It Is*. And this projection prefigures a transcendence that is embodied in the three temporal ekstases: past, present, and future. To concretize these dimensions of time, however, the narrator must live the past as a now past, the present as a now present, and the future as a now future. Bom easily slips, prompted by Kram, from "how it was" to "how it is." The literary tour de force is to present all three temporal ekstases in the mode of the immediate present. For this reason, we read Bom's despair of treating his relationship with Pim in the past: " . . . this won't work in the past either I'll never have a past never had."[6] The narrator is locked into the present.

Yet he remains open upon the world and his being therein. This openness to the world is called by Heidegger *Erschlossenheit*, disclosedness, that quality of human reality that makes solipsism an impossible solution to life's individual problems. Open to the world, it must feel how it is to be in that world. *Befindlichkeit* is the German for this state of consciousness; marvelously ambiguous, it means affectivity as well as a finding. *Wie befinden Sie sich*? means both How are you? and How do you find yourself? And to give an answer to these questions, one must consult one's feelings at the moment of the question. That is a kind of knowledge, eminently associable with the fore-having of the circular hermeneutical situation. We know ourselves in relation to a context by virtue of the feeling that relationship provides. It is a feeling of not being at home (*Unheimlichkeit*) in the real world, and of anxiety in the face of one's own: uncanniness on the one hand, and personal insecurity on the other.

But feelings unexpressed are lost to the atmosphere, if not to the darkness of the mud. And to be expressed they must be understood: enter the foresight of the situation, the anticipation of the context's closure. With the achievement of closure, the particular intentionality of the consciousness has become individualized, concretized into a sense: something is there to be understood.

This understanding needs articulation, and the articulation, expression in the creative word: 'creative,' for it is not sufficient to express one's own individual experiences in the terms of everyman. That's the essence of inauthenticity, *Gerede* instead of *Rede*, idle talk instead of speech. The creative writer, whose task is to make explicit what he has already known implicitly through his feelings, must discover the words that create his universe. Whence the anxiety of the authentic author. He will know that he has found the words when his context assumes the structure of the sense implicit in his understanding. That is his philosophy, the idea that structures his created universe.

With a completion of the process he will have returned to his being-in-the-world now enriched in a double sense: his feeling will have been confirmed by the expression of the understanding he has gained thereof; and his expression will have made it possible for him to communicate to another. His work is creative of a new society in which his own contribution is retained as it is, if not forever, at least as long as there are readers intelligent enough to enter the process of communication with him. Unto now we have been trying to do just this by employing the same (phenomenological) hermeneutic that Beckett has been using to create his works.

But a caveat is in order. Affectivity, understanding, and the word are, in Heidegger's terms, *gleichursprunglich*, i.e. equiprimordial; neither can be said absolutely to precede the other in time. To begin with a feeling to be expressed is an exercise in translation, not literary creation. That is the fallacy of romanticism, and Beckett is no romantic. His solitude is the solitude of the writer anxiously seeking to know his own feeling in the words he must use to express his understanding. That's how it is with creative writers; their suffering is to be alone, to consult only their own inner resources for a discovery of their meaning. Their words are thoughts, are emotions, as Ms. Cohn had perceived in her reading of *The Unnamable*; but neither precedes the other in the process of creating.

For this reason, *How It Is*, which embodies the epistemological ideal of the hermeneutical expansion of human experience, must be written in the all engulfing present; the novel is a search for meaning in the literary artist's search for meaning. The first search is that of Kram, symbolic of Beckett himself; the second, that of Bom and Pim, symbolic of the teller and the told, the two (perhaps schizophrenic) functions of the personality undergoing the hermeneutical expansion.

Such is the hermeneutical hypothesis for the interpretation of the structure of *How It Is*. To test it, we need only apply it to the details of the context and gauge the extent to which it shows a single significance developing in the closure of that context. If the philosophy is clearly understood, the novel's structure should be equally clear. That is, now as before, the challenge of criticism.

The narrator begins his journey nameless, projecting a story of his life, in three parts, constructed according to a natural order. Such is his fore-having, the abstract intention to construct a context where he and others may discern a significance. The three parts shall be: first and present formulation, before Pim; second and next formulation, with Pim; third and last formulation, after Pim. Whatever sense these formulations are to have must be gauged by the specific intent brought to closure in the incidents of these three formulations. Such is his foresight: a journey from solitude to solitude, via that enormous stretch of time spent communicating with Pim, torturing Pim in order to make him speak. His fore-conception is an expansion of his self-awareness, to be acquired by the voyage from solitude to solitude.

The natural order he projects is therefore only a ruse, an empty frame for the telling of his story. Neither *Watt*, nor the *Stories*, nor the trilogy had been recounted in natural order, but in one closer to the structure of the philosophical idea embodied in those formulations, but they were ill-said and ill-heard. Perhaps a separation of frame from structure will finally make the point that a structure is no empty form, but the very concreteness of the novelistic universe. That's the hazard of formalism, to mistake the form for the significance, and the error of all classicists, as critics. Neither Kram nor Bom is a classicist, and Bom very early lets it be known that he must say what he hears as the voice (quaqua) murmurs to him in the mud.

But the hermeneutical circle may be displayed in a natural frame. It is a question of openness to the world, so the as yet unnamed Bom will begin his account of how it feels to be found where he is affected with the uncanniness of his situation and open to his own anxiety. He receives the words of Kram in bits and pieces; they create the images; and these images lend their sense to the developing whole. But not by virtue of their meanings alone: their signification will be found throughout mirrored on the surface of those verbal bits and pieces, just as they are heard. His "walk," for example,

> ten yards fifteen yards semi-side left right leg right arm push pull flat on face imprecations no sound semi-side right left leg left arm push pull flat on face imprecations no sound not an iota to be changed in this description.[7]

In the mud, digging in the mud, he pushes and pulls his way forward, not to any place in particular, beyond a rendezvous with Pim, but to that other rendezvous with himself when he is to be abandoned by Pim. From solitude to solitude, forever, with no hope for change since justice is what it is in this world – but that is a point for development in the third formulation.

The "not an iota to be changed in this description" indicates two things: first, that the words are as he has been given them, and secondly, that they jibe with the experience. It may be contrasted to all the occurrences of "something wrong there," indicating either an error in his hearing or a dislocation from the experience. Both the right and the wrong have their place in the correct showing of how it is to achieve significance in writing. The writer still needs his reader, perhaps only himself backed off from the attempts at finding the word, to judge the accuracy of the statement. Here, indeed, they must be the same since the narrator is alone, without even a symbolic Moran. There will be no extraneous visitors in this world, in any of its formulations.

First formulation, before Pim. In this section, the narrator establishes his relationship to himself, a project of knowing, to be told in the tripartite form of a story:

> how it was I quote before Pim with Pim after Pim
> how it is three parts I say it as I hear it.[8]

No time is lost in laying out the context, in laying out the first arc of the hermeneutical circle. The state of mind that goes along with it, implicating a relationship between self as knower and self as known is anxiety; authentic this, there is no fear. Fear could be related in the words; but like the Unnamable's last story such words would be about a feeling, not a presentation of one. The narrator's anxiety is expressed in the very syntax of his words. He has lost all connections with the conventions of speech; they yield only idle talk.

He is surrounded by voices, the quaqua of his creator and his own murmured responses to them. His past is there but only as a series of unconnected images, separated by the panting rhythm of his own exhaled breath as he murmurs inaudibly in the mud. Time is indefinite, measured only by the rhythm of this variation in the sounds and silences. The place is nowhere, i.e. anywhere on the earth, in the furrows of its mud. His body is mud lost in the mud. His eyes, open, see only the dark; when they close, the black. But his consciousness is intact. His other eyes, those of his reason, of his imagination and his memory, may function even in the blackness of the black, as he is given the words to express the images that occur in the state of free association; absurdly, for there is no reason for these occurrences outside of the words he has been given, is being given. The giver of the words will later be identified as Kram, the external witness to his own inner anxiety to get it all straight, just as it happens. Should he fail once again it shall not be for a lack of attention to the detail of his consciousness.

His first image is of some creature or other he had watched through a spyglass at night through windows: his keeper or his parent; for he sees that creature now as less ugly, less stupid, less dirty, and less cruel. That's the effect of distance; he continues

> . . . and you saying to myself and you bad to
> worse bad to worse steadily
>
> something wrong there
>
> or no worse saying to myself no worse you're
> no worse and was worse
>
> I pissed and shat another image in my crib
> never so clean since
>
> I scissored into slender strips the wings of
> butterflies . . . [9]

And the light goes out on his infancy and childhood.

His second image is of his mother, her loving care expressed tenderly, but interrupted by his own care for his sack, in which he carries his provisions, tins of food and a can opener; by his thoughts of there being no visitors (they would only speak of themselves anyway; no loss there); by his thoughts of himself as a monster of the solitudes, whose skin others would like to bring home as a trophy. Then a complete break to recall the conditions of his present as a re-sounding board for an external voice:

> Suddenly afar the step the voice nothing then suddenly something something then suddenly nothing suddenly afar the silence [10]

After the silence, a realization of his abandonment:

> life then without callers present formulation no callers this time no stories but mine no silence but the silence I must break when I can bear it no more it's with that I have to last [11]

There are to be no others in this formulation. Except of course in a dream, when the panting stops and the bodily eyes close to open those of his subconscious. He dreams of an impossible love, closing his free hand over his face without touching it; he wills to but he can't, and that inability of adjusting bodily activity to the act of will reminds him of his forgotten mortality.

In the third image his mother returns, this time praying and teaching him to pray, but even then he had difficulty in repeating the words. He should have remembered, for now he knows:

these words of those for whom and under whom and all about the earth turns and all turns these words here again days nights years seasons that family [12]

In this project there will be only the words. The family, lost loves, the very light of day are things of the past. Now it is how it is before Pim. No one exists to pray for his salvation; in sleep, when the dark gives way to the black, he falls back into the foetal position like Belacqua in antepurgatory: "Belacqua fallen over on his side tired of waiting forgotten of the hearts where grace abides asleep."[13] And no one to pray to, if not his witness, who speaks first and then has his aide note down all.

The last image is of himself as a lad of sixteen in the flowering season of April or May, with his girl friend walking her dog. A scene of love on a picnic, which vanishes like the others, leaving him there alone, occupied with his sack, the mud and his direction toward Pim.

That is enough for him to imagine himself as a redirected transformation, from higher to lower species; his "humanities" help him give it expression:

. . . made or worse transformed à la Haeckel born in Potsdam where Klopstock too among others lived a space and laboured though buried in Altona the shadow he casts [14]

For, to talk of a man, even if that man be himself, he must still speak of him as a termite. And no gloss will suffice for an understanding of the words. The anxiety they express is there to be read by him who has learned to read, all bits and pieces, recalled in flashes and lost in the all enclosing dark.

With his sack, his hunger, his thirst, his body, and his groping consciousness the narrator has laid out the general context in which he is to achieve his self-fulfillment as the knower known. His fore-sight is his direction toward Pim, and ultimate abandonment.

Second formulation, with Pim. Having met with Pim, if bumping into the exposed buttocks of another can be called meeting him, the narrator proceeds to baptize his companion; it's more commodious to be able to refer to him by a name:

no more than I by his own account or my imagination he had no name any more than

I so I gave him one the name Pim for more commodity more convenience it's off again in the past [15]

Pim is a creature like himself, dragging a sack through the mud with the same kind of contents. He also has a watch, wondrous to behold; for he has need of none. It can only be a reminder of life "up there" in the light where the passage of light unto dark measures the duration of a day and where time is of the essence. Here in the total dark one can hear nothing but its insignificant tick.

The narrator explores his discovery. That one is an old man, like himself: a man, because his exploring hands came up with a pair of testicles, and he has not forgotten his anatomy. The old man's head is covered with hair; and given his age, it must be white to those who can see. To communicate with him, the narrator must scratch the skin of his companion; his own murmurs cannot be heard. On the exposed buttocks only two letters are necessary for this, since the anus itself suffices for the O:

BOM scored by finger-nail athwart the arse the vowel in the hole I would say in a scene from my life he would oblige me to have had a life the Boms sir you don't know the Boms sir you can shit on a Bom sir you can't humiliate him a Bom sir the Boms sir [16]

That is the way the narrator imagines his introduction to Pim, his own name being a well known reference to the human hindquarters. A possible confusion there. To carry out this intention to communicate his companion must be taught the language; and so the lessons in tactile communion begin.

First, to sing: dig the finger nails into the armpits. Pim screams in pain, and receives a blow on the head that drives it into the mud: end of screams, with nothing yet learned. Repeat the process: same results. One communicates, however, by taking the point of view of the other. Bom does first, thinking that Pim must be thinking that he (Bom) is asking for something; as indeed Pim does. At the next blow Pim sings, repeating the sounds Bom had heard him make before, pleasant to the ear though inaccessible to the mind. Pim stops singing, and is given another scratch in the armpit. He goes on; he has learned. To make him stop: another blow on the head. For "bravo," a blow across the buttocks.

Next, to talk: prod him in the buttocks with the sharp end of the can opener. Same initial failure, as above; and same final success. For "louder," the dull end of the opener into the kidney; for "softer," the finger in the anus; for "stop," again the blow on the head.

In sum,

table of basic stimuli one sing nails in armpit two speak blade in arse three stop thump on skull four louder pestle on kidney

five softer index in anus six bravo clap athwart arse seven lousy same as three eight encore same as one or two as may be [17]

But if Pim is left to his own devices he can sing or say what he wishes. Bom must devise a way of indicating specific questions: the nails again, to carve his directions in Roman capitals "from left to right and from top to bottom" on the back of his victim.

In this way Pim learns his new name, and that of his tormentor. The question causing most confusion is that concerning Pim's life in the light above. He remembers little of his father, except that perhaps he had been in the building trades; nothing of his mother, except her Bible; and most of his wife, Prim or Pam. It might have been that Pam was too prim, shaving her pubis and all. Their love ran the usual course: daily conjugations at first, then every three days, then on Saturdays, and after that only on sparse occasions, just to get rid of the desire. As a result their love was spent. But Pim tried to revive it with anal coition, and Prim Pam threw herself out the window, broke her back and died; but not before forgiving Pim his bestiality. His parents he had forgotten, and his wife he would like to forget. What does stand out is his constant fleeing, his uneasy transcendence:

never anyone never knew anyone always ran fled elsewhere some other place my life above places paths nothing else brief places long paths the quickest way or a thousand detours the safest way always at night less light a little less A to B B to C home at last safe at last drop sleep [18]

And the story told is Molloy-acting's life in a nutshell.

Other questions are of Pim's life before Bom down here in the mud, of his life here now with Bom, whether he loves Bom, whether he shall die one day, whether he thinks of or believes in God. Concerning the latter and that belief's relation to his present situation we hear:

if he talks to himself no thinks no believes in God yes every day no wishes to die yes but doesn't expect to no he expects to stay where he is yes flat as a cowclap on his belly yes in the mud yes without motion yes without thought yes eternally yes [19]

Significant in this discourse is the jump over Malone's resolute anticipation of death into the essential condition of the Unnamable. There will be no dying, because Pim is already an essence (the character who must be made to talk), fixed in the words of his creator, who is not the God of all above in the light, but Kram, who may be believed somedays and on others not, because his messages arrive only intermittently.

As for the love between creator and created, the author and his character, it can only be sadomasochistic, like that of Lemuel for the wards of St. John of God's. In this way, Beckett has substituted Sartre's model of the concrete relations with others for the Heideggerian model of concern (*Fürsorge*). Our mode of being with the other may be concernful, but it necessitates alternating moments of inflicting pain and of receiving pain in return. In any life there is sure to be more pricks than kicks. And both Pim and Bom, following the advice of Geulincx, resign themselves to their fate: the arbitrary decisions of Kram.

And of a sudden Pim is gone, leaving his half empty, half full sack behind.

Third formulation, after Pim. Here Bom is returned to his unnamable self, since there is no longer a Pim by which to differentiate himself. The structure of the hermeneutical circle has sketched its second arc: from the initial project out of solitude to communion, and from the communion to abandonment: fore-having and fore-sight accomplished, there remains only the achievement of the fore-conception.

The hypothesis that would close the context must be stated, indirectly, in the images created by the words. We shall not say, Kram will not say, but the unnamable Bom will be forced to say what the hypothesis is. Justice, he thinks; each tormentor must become the tormented. Boms and Pims change roles, and nothing is lost, neither the dialectic of *Watt* and the *Stories*, nor the existential ontology of the trilogy. Pim's consciousness for-itself comes to exist in-itself for Bom, and Bom's consciousness becomes in and for itself in *How It Is*. To be a character is to experience how it is; to be an author to express how it is: a life and its account in a story, but always the same, a voyage outward and a return home to self, existing alone and anguished to know how and why.

The narrator once more becomes unnamable, since he is either Pim or Bom depending upon the moment we catch his words in the flight. Suppose there to be a million characters; the millionth comes back to the one. Select any three, no, four; something is always unknown and therefore only conjectured, according to a principle:

two there were two of us his hand on my arse someone had come Bom Bem one syllable m at the end all that matters Bem had come to cleave to me see later Pim and me I had come to cleave to Pim the same thing except that me Pim Bem me Bem left me south[20]

This new visitation, Bem, is that part of Bom's personality that turns Bom into a new Pim. It is he, the reflective self, that must make the old, active self give its report on how it is.

Before Pim, first formulation, Bom was Bem, an old Pim having abandoned his then Bom, to become Bom to another Pim. The sacks are the clue. When leaving a Bom a Pim leaves his sack; if there were no Kram to provide the missing sack the whole parade would go awry. Therefore, it would seem to follow that there have been, are, and will be generations of Krams (Bom with Pim had imagined the thirteenth generation: so many novels of the same author) to provide the means for physically continuing the journeys; otherwise on the journey one could not climb over the pile of abandoned sacks on the circular path.

The voice of Kram is responsible for all these voyages:

the same voice the same things nothing changing but the names and hardly they two are enough nameless each awaits his Bom nameless goes toward his Pim

Bom to the abandoned not me Bom you Bom we Bom but me Bom you Pim I to the abandoned not me Pim you Pim we Pim but me Bom you Pim something very wrong there [21]

The circle spins too quickly for an exact register of the names.

The important thing is the order; and since naturally solitude precedes coupling and coupling the second solitude, it matters little where one has begun the account. The way to the other is always a return to the one, but with an amazing absence. Consider the case of only four characters:

victim of number 4 at A en route along AB tormentor of number 2 at B abandoned again but this time at B victim again of number 4 but this time at B en route again but this time along BA tormentor of number 2 again at A and finally abandoned again at A and all set to begin again [22]

Thus, if Bom is number 1 (what else?), he knows 4 as victim and 2 as tormentor; and number 2 (Bom's Pim) becomes Bom to 3's Pim, but he is abandoned by 3 (Pim_2) to become Pim_3 to 4's Bom. Hence, number three in the chain is known only indirectly by number 1. No clearer explanation seems possible for the continued knowledge of the successive Beckettian characters from Belacqua to Bom. They have all been put into this circle of justice (personal torments following the tormenting of others), by Kram, (as the Unnamable had said) the man in charge.

There is no need then to describe Bom's new existence as Pim. There can be no changes from the second formulation. As each character approaches the absolute center, in Watt's time the house of Knott, he will spend his moment in service and then depart. This, I take it, is no gratuitious comparison, since Bom's reasoning is characteristically Wattian.

The panting becomes more furious as the image of the whole becomes clearer for the speculation:

that yes a panting in the mud to that it all comes in the end the journey the couple the abandon when the whole tale is told the tormentor you are said to have had then lost the journey you are said to have made the victim you are said to have had then lost the images the sack the little fables of above little sciences a little blue infernal homes [23]

Such in a nutshell is the "activity" of *How It Is*.

There above, all along, has been the anonymous voice of Kram; only he has been responsible:

and this anonymous voice self-styled quaqua the voice of us all that was without on all sides then in us when the panting stops bits and scraps barely audible certainly distorted there it is at last the voice of him who before listening to us murmur what we are tells us what we are as best he can [24]

But even to do this, the author must continue to torment his creature; and the story ends with Bom playing Pim to Kram's Bom:

all these calculations yes explanations yes the whole story from beginning to end yes completely false yes

that wasn't how it was no not at all no how then answer how was it then no answer HOW WAS IT screams good [25]

As justice, the will of the creator, would have it, Bom is forced to respond as Pim to a new Bom's will. Loving each other in the only way possible, each inflicts and suffers pain; and here there is no window out of which to jump. The knowledge one has gained of one's self is bought at the expense of the other, but only if the other lends us his words.

Affectivity, understanding, speech, such is Dasein's disclosedness; but like anything else to be disclosed, it needs a hermeneutical expansion: the projection of a context, the anticipation of that context's closure, and the understanding of the principle by which the context actually closes. We can grasp all this only by attending to the speech of the narrator for an understanding of his situation that presents his feeling on the very surface of his words.

How It Is is the sparsest, densest, most intense expression of the solipsistic predicament (consciousness centered upon itself) Beckett has devised. No more than a few hundred images, incessantly repeated, burn their funding significance on the reader's consciousness. From bits and scraps the whole is made, but only as they lend themselves to hermeneutical expansion. The only meaning in the artist's search for meaning is that it must be created by his words, in solitude, speaking only to himself. There will follow a moment of communication with others, as the reader is forced to respond with what he knows of life "up there" in the same manner as the author's character. But the communication is only momentary, and will pass away for another moment of anxious solitude, the reader to be lost in his own, as the author is freed to project his into still another story.

When the author finishes his word, the reader has the record of an experience, just as before, at the moment of its projection, the author possesses the experience itself that discloses the essence of his being. The one hermeneutical process is the mirror image of the other. Whence the frame of the Molloy tale, with Moran's account structured in mirror image of Molloy's, the symbolic reader's of the symbolic author's. *How It Is* is Beckett's clearest picture yet of what it means to create a literary work of art: two moments of anxious solitude framing an uneasy communication with his creative fictions, two moments of anxiety relieved only by the sado-masochism of the creative mind. Each final abandonment has continually re-occurred between the creative movements: between *More Pricks* and *Murphy*, between *Murphy* and *Watt*, between *Watt* and the *Stories*; between the *Stories* and the trilogy, and between the trilogy and the *Texts*, and between them and *How It Is*.

That's how it has been, and we shall continue to expect more of the same until Beckett himself loses contact with the earth, with no more disclosedness and no more breath with which to disclose even that.

According to one critic,[26] by the way, *How It Is*, succeeded in "stupefying" professional readers. They had not yet learned the possibilities of abstraction in art, by which the linguistic surface contributes as much to the expressiveness of the context as any amount of levels of expressed depth signification. As readers, they too must return to the sources.

Three shorter texts, published as *Residua*, contain one vignette and all that Beckett has been willing to publish of two larger novels. They are useful to project an end to his novelistic odyssey, and the titles he has given the tracts predict such an end – at least in the idea. They are *Enough, Imagination Dead Imagine*, and *Ping*.

The first two of these titles would seem to need no explanation, and when the third is interpreted as the last sound of an extinguished consciousness, it too will fall into sense. If we turn now to a consideration of these texts, it is with the knowledge that Beckett had already discovered what he had set out to learn, that the pain of writing clearly is an unsure calmative for the pain, in the first instance, of living. He has earned his title to the unhappy consciousness, and will go on murmuring in the mud.

How It Is was published in French in 1961, and the author's English translation of it, in 1964. The three opuscula of the present concern were published in 1965 and 1966. If the demonstrated continuity of Beckett's works is to be sustained, his *Residua*, comprising a bare sixteen pages in the English translation,[27] must exhibit a structure that depends for its essence upon some presumably missing facet of the earlier works.

Since *How It Is* abstracts the significance of solipsism into three successive moments – solitude, coupling, solitude – and presents these moments as real events of the imaginative consciousness condemned forever to its unhappy freedom, one might be tempted to conclude that Beckett's inquiry into the creative process had achieved its end. No doubt that novel stands as it is: a minimum of images, perhaps a few hundred at most, incessantly repeated and incessantly cut by their surrounding silences into a temporality now known only in the privacy of the reader's own consciousness as it attends to this last painful voyage of the peripatetic narrator.

That would be a mistake. For *Enough* (1966) repeats the relationship of Bom and Pim in the reminiscent consciousness of a new narrator, another young old man such as the one who called Molloy back to his initial resolve to find his place in nature. This narrator has circled the earth under the protection of an old man. The two are one: spiritually by the union of their desires, and physically by that of their bodies, attuned to the demands of nature. The image is frank, homosexual and unashamed:

> Attitude at rest. Wedged together bent in three. Second right angle at the knees. I on the inside. We turn over as one man when he manifests the desire. I can feel him at night pressed against me with all his twisted strength. . . . With his upper hand he held and touched me where he wished. Up to a certain point. The other was twined in my hair. He murmured of things that for him were no more and for me could not have been.[28]

That image cannot but recall the coupling of Bom and Pim. The old Pim's murmurs could only relate what this new Bom could not understand. For, in spite of their mutual desire for union, when the old man says "Leave," the young one does. That is the scene of his "disgrace," the refusal of the other, which cannot be disclaimed as an actual past event, but which is now itself disavowed by the narrator's commemoration of his loss.

The old man, bent over double so that his upper body is dragged parallel to the earth, symbolizes the source of all knowledge and its record in speech. The two had been on the road to language,[29] united by the forces of nature, eating, smelling, and trampling the flowers in their path, until the old one has nothing to give. Another abandonment, another expression of it, with the same doubts as before:

> Other main examples suggest themselves to the mind. Immediate continuous communication with immediate departure. Same thing with delayed redeparture. Delayed continuous communication with immediate redeparture. Same thing with delayed redeparture. Immediate discontinuous communication with immediate. . . . [30]

However combined, however permutated, the communications and the departures will alternate successively until the end of time. And that is precisely the point of view of this new unhappy consciousness, abandoned to his nature as the recorder of life's significance in a single sustained image.

How can this process end? Imagine that the imagination itself were to die, that one could no longer see, no longer invent a story to tell. *Imagination Dead Imagine* (1955) pictures this event. It's the same idea as one of Descartes, who had asked his readers to suppose that one's imagined existence is false; it would still remain true that one had imagined the truth of the falsity. The quintessential stroke of *No's Knife*: by denying, one affirms one's existence as a consciousness. And consciousness is awakened by its perceptions, the least of which is, can only be, a contrast. This new narrator, therefore, perceives his environment as a pullulating, pulsating sequence of changes: from white and hot to black and cold in alternating systems. The consciousness of it is the image itself, the bracketing of a yes and a no.

Within this environment the narrator sees a couple, bodies circumscribed in semi-circles united in a common chord. Only their eyes move, to open and to close. They open longer than is humanly possible, and can see only the other; but each can see the other only in those passing moments as their respective eyes open or close, out of phase. The essence of the one consciousness remains hidden from the other, except for that fleeting second. Only the reader's intuition of the contrast between the convulsive light and the stillness of the pair can grasp the full significance of the human struggle for unity. Should the imagination imagine itself dead, only the imagination would be alive. And that would be enough to come to an understanding of human abandonment.

Ping (1966) explains why. Here the visual image is reduced to white on white, and the narrator is alone contemplating himself. Legs as if sewn together, hands as well, his action is to contemplate the pulsating changes of his consciousness. Eyes light blue almost white, nails pail rose almost white, slightly at a distance; once again, consciousness and its object, presented as the consciousness occurs:

Head haught eyes holes light blue almost white fixed front silence within. Ping elsewhere always there but that known not.[31]

Each "ping" announces a sound that produces an image of something not the consciousness itself, the formation of a word, the introduction of a "meaning," beginning with the breaking of the silence (first contrast) and then the appearance of the visual image on the background of the white world (second contrast). A silence, another ping: the process is repeated.

One of these pings introduces a ray of hope:

Ping perhaps not alone one second with image same time a little less dim eye black and white half closed long lashes imploring that much memory almost never.[32]

The dead imagination has imagined the existence of another; or seen her or him, him or her (all combinations and permutations possible here), in that one second of crossing visual paths. Not a memory, not a clear perception; but the confirmation of the eternal hope for an understanding fellow:

Head haught eyes white fixed front old ping last murmur one second perhaps not alone eye unlustrous black and white half closed long lashes imploring ping silence ping over.[33]

Within the context of this fragment the imploring lashes have received their saving grace. The narrator has understood, and has relayed the experience in his own act of beseeching the reader's understanding of that desire for a union that we all know must leave both parties, in the end, alone, all fixed where we are (inside our own consciousness); and without that creative ping that announces a new image we should all equally be lost, perceiving nothing but the glare of a white hope on a white ground.

In this way, Beckett has brought consciousness to its barest: a feeling, an understanding, and the evocative word in a single intuition. Nothing more need be, can be, said.

Enough; the imagination lives even in death; the ping announces its appearance. It would be harder for any critic to imagine, for any writer to write a more succinct account of the essence of the human predicament. On the surface of this prose, in the rhythm of the recurring images, is the beating of the human heart: the craft of fiction at its cutting best. No's knife in Yes's wound. The Unnamable's imponderable questions have been answered:

Where now? Who now? When now? Nowhere now; no one now; forever now: there is no particular time; consciousness is temporality itself. Every conscious image affirms itself by denying the world that responds by denying it. Such is the poetic plight of the author who would both tell and show us how it is.

Beckett has and will produce other stories. I have traced in the foregoing his own existential projection as a writer of creative fiction, using as a principle of criticism his own suggestion, that philosophy may serve as a convenience or inconvenience to the writer for structuring the universe of that creative fiction. He has borrowed from many philosophers to give substance to an essentially pessimistic view of both life and anyone's attempts to disclose exactly what life is.

In the earlier stages the basic pessimism of the human project is tied to a notion of the Self as a psychophysical dualism, the separation of mind from body and through the body as necessarily related to an outer world. The early heroes sought solace in the separation of their personal worlds from the imperious demands of the outer: each solipsist seeks his salvation by retreating into the safety and repose of his own consciousness. But that is an empty freedom, hallowed in Stoic philosophy, and always ending in death. Outside of itself there is a world, haunted by others, who in spite of everything conceived to the contrary determine who and what the solipsist was.

The dualism gave way to the dialectic of human experience, but the dialectic is an empty, rationalistic form. It needs the stuff of history. Naturalistic philosophers look upon time as the ultimate form of historical development, but that too is a pure conceptual form. To supply a matter for his form anyone who would tell us how it is must discover the concreteness of human temporality, which likewise comes to an end in death.

Throughout the changes in philosophical conveniences, then, Beckett's message comes to a single statement: the meaning of life is death. The only hint of optimism shining through this dark vision is the possibility of transcending man's fate by the creation of an aesthetic vision made more powerful for its inclusion of the truth of our own condition. But to do this a writer must find the words, and must continue until all the words have been said. 'Beauty' is too pale a word to describe the effects of Beckett's fiction; his works considered as a whole partake of the sublime.

In the second part of this treatise, I shall attempt to evaluate Beckett's dramatic writings in order to describe, eventually, the phenomenological character of his contemporary tragic vision. Our model will not be Aristotle; for he was only describing what the Greek tragedians had provided for phenomenological analysis. Contemporary plays will yield a contemporary understanding of our experience of tragedy, which is not foreign to admixture with elements of comedy. Who dares speak here of comic relief? Beckett's tragedy is based upon an inversion of value by the human consciousness – a thesis now to be explored.

PART II

CHAPTER VIII

CREATING A SCENE

Written in 1948, *Waiting for Godot* was copyrighted by *Les Editions de Minuit* in 1952, and translated into German in 1953. Beckett's own English translation was published by Faber and Faber in 1955. Following its French production, where the audience reacted by creating their own scene of violence, the play was performed in London, and then came to America, where, in Miami, its producer sold out a two-week performance, billed in advance as follows:

BERT LAHR, THE STAR OF 'BURLESQUE,' AND TOM EWELL, THE STAR OF 'THE SEVEN YEAR ITCH,' IN THE LAUGH SENSATION OF TWO CONTINENTS – SAMUEL BECKETT'S 'WAITING FOR GODOT.'[1]

The results of this fiasco are well known: two thirds of the first-night audience, foregoing the fisticuffs of that first French audition, left the theater after the first act, and the management was plagued by requests for refunds from a good portion of the presubscribed audience upset by the initial adverse publicity. Good drama may become successful theater only with the intelligent cooperation of producer, director, and actors; and good symbolic drama must come to closure in the understanding responses of an emotionally moved audience.

But the conditions were far from ideal. Infected with the uncanniness of a world in which space has been constricted to a single scene on a country road leading from nowhere to nowhere, on which two "bums" spend a seemingly unending amount of time doing whatever flits into their heads in order to kill time and thereby bring an end to their sorrows, members of that audience, themselves seeking an escape from the realities of life, could hardly understand they were expected to laugh at their unwitting and unwilling immersion in them – or that there existed another rationale for their failure to do so.

The play is still the thing to capture the conscience or consciousness of an audience, but these unsuspecting consciousnesses were set up to expect something other than the play that was to tell them it is vanity to expect anything at all in life except death – and that only at an unpredictable hour. And if the written accounts of that first night may be believed, the actors themselves panicked, fought with each other over the other's laughs, thereby compounding the overall confusion.

Critics, too, were of little help. 'Absurdist' they called it, confusing content for significance and totally ignoring the philosophical structure of the drama. Perhaps the best approach to reading the drama of any author is to take up the role of the director, examining the literary context for the structure of the universe it creates. In that way phenomenological criticism will be found an integral part of the play's communicative system, with a statement of the hypothesis giving the criterion for an understanding of the whole play, precisely that which enables the director to give his directions to the participating artists.

And here as elsewhere in Beckett's own artistry, the structure of the drama concretizes a philosophical idea – one in which, per se, there is no humor, but which is expressed in a style of high comic irony, whose tone is one of playful seriousness. Beckett's flare for the oxymoron may make one smile, but on second take the absurdity of its expression will strike deeply, leaving its impression on the mind that has just been revealed the tragic essence of life.

For this purpose Didi (Vladimir) and Gogo (Estragon) also served, as the curtain rises on their long-standing wait, talking as if at cross purposes, yet communicating what each has to contribute, within this new pseudo-couple relationship, to the unique determination of the human condition. Gogo, the earthy one, stinks of the feet; he is weary, hungry, easily excitable, sleepy, a prey to his dreams, and eager to experience that last erection to occur should the pair succeed in hanging themselves. Didi, the airy one, stinks of the breath; he is lost, and resigned to his waiting, necessarily united to Gogo lest they both achieve their goal too soon, but full of the defenses of the intellect: a knowledge of causes and consequences; patience and diffidence; and at a distance from his suffering; yet no less weary, hungry, excitable and sleepy, no less conversant with his dreams than Gogo, but, unlike his fellow, full of the awareness that marks the presence of mind. They suffer as one, wait as one; as one symbolize the schizoid unity of the human personality.

As the scene opens Gogo is struggling to remove a shoe that is too tight. Giving up, he says "Nothing to be done."[2] To which Didi sententiously replies:

I'm beginning to come around to that opinion. All my life I've tried to put it from me, saying, Vladimir, be reasonable, you haven't yet tried everything. And I resumed the struggle.[3]

Whence, in the opening dialogue the reader is clued in to the dual personification of the characters. Gogo speaks of his specific predicament of extricating himself from his shoes; Didi, of the general course of life's responsibilities. The relation is as stimulus to response. The one character is tied to the accomplishment of an immediate task and an equally immediate response to the failure, 'There is nothing to be done,' that becomes a projection of the significance to be developed throughout the whole play's "action"; and the other, to a conviction concerning human destiny, likewise a projection of the play's inaction, this time at the level of full human disclosedness. It takes only a moment's reflection to realize that waiting is an action that is no action.

Both characters live the uncanniness of their situations, uneasy with themselves as with each other and anxious to defend themselves against continual failure. Alive and united, they both face the task of awaiting their end; and each holds up a mirror for the reflection of the other. Conscious of self as conscious of the other, neither may keep his secrets to himself; for when the body is wounded the mind must register the pain, and no's knife is still the source of yes's wound. On two different occasions the intellectual Didi will refuse to listen to Gogo's dreams in order not to experience his own horror in their report.

Philosophically considered, the play is both Hegelian and Heideggerian, as were Beckett's *Texts for Nothing*. Curiously enough, however, there is a missing third for the complete definition of self, that triune nothing; and he or it hangs over the unhappy wayfarers with a sense of impending immediacy: the personality of Godot, he who would end their suffering, and thus possibly an analogue of death, God, the author, whose voice in this context must remain unheard, the end itself, the promise of a way out, or all of these at once. Godot, like the author himself, leaves the pair to their own existence, to exercise their freedom by whatever choice is left open to them in their bare essential condition.

The Hegelian structure is evident in the composition of the couple and its reflection in the couple of Pozzo and Lucky, the master and his slave. Consider. Conscious of themselves in their consciousness of the other, they have achieved the level of self-consciousness that transcends simple sense certainty, perception, and the force of understanding: each has passed the first stages of the dialectician's account of experience alone; and when they come together in the opening scene Vladimir exclaims, "Together again at last! We'll have to celebrate this. But how? . . . Get up till I embrace you."[4]

Estragon rejects this advance for a querulous discussion of their separation, when each was left to what he was in himself: Gogo, the consciousness in itself, tied to its immediate pains or merely lack of pleasures, who nonetheless exists for Didi, the consciousness for itself, full of reasons, excuses, and the superiority of distanced reflection. When they return together to form the unique pseudo-couple, they are the consciousness existing both in and for itself; and at the opening of the play have so existed for fifty years, momentarily separating and momentarily reuniting to achieve their symbolic unity. Like Bom and Pim, they come together only to part again.

Their unity is reinforced by the appearance of Pozzo and Lucky, the new couple directly personifying the dialectic of the master and the slave. But like Gogo and Didi this second apparition is a pseudo-duality, since internal to every full human personality there are both dominant and subdominant conscious moments. Ironically, too, readers of this reflected relationship may perceive the left-right transference of any mirror image. For it is Lucky, the slave with the ironic name, who thinks and works, while Pozzo, the self-deluded master, gorges himself and indulges his physical self-presence. Chiasmus, the figure is called, that criss-cross of competing functions at the heart of which stands a single man, like the Savior nailed to his gibbet. Waiting for Godot, Didi and Gogo can get no further up the scale of human consciousness as the experience of that phenomenon is treated in Hegel's *Phenomenology of the Mind*, but they will return to the figure on the cross.

Meanwhile, they find themselves there – on the road, but not knowing where. *Dasein*, being there, in the world; uneasy within it and anxious before the possibility of getting out of it, their inauthentic decision is to stand and wait. Vladimir's first words to Estragon were: "So there you are."[5] They have pulled him out of his reflection on that initial "nothing to be done" and state his recognition of a presence to his presence. On the stage both characters are reduced in essence to a presence.

Estragon's reply is more typical of the man not sure of where he is than of a mere foil for an author's propensity for the absurd, and communicates the essence of the human predicament: "Am I?"[6] – a response that hardly ever fails to elicit a laugh. Any response at all, including the one given, would answer that question affirmatively; but, given in the form of a question it immediately casts doubt on the reality of the situation. And in a dialogue the here and the there of the speakers is purely relative: each person's here is the other's there. Gogo's reply therefore is not nonsense, but suggestive at once of the dialectical relationship of self-consciousness and the exact location of the action of the play: in the functional relationships of conscious awareness.

Each is there and each is here, another chiasmus resulting in an ambiguity that must be worked out in the "space" of the shared consciousness, as the two "bums" care for each other in the structures of a common temporality. Their being there now is a being with one another, and will not cease until the one or the other dies or otherwise comes to an end.

The German language of Heidegger's philosophy gives us all the clues we need: "da" may mean both here and there; "Dasein," (human) existence, ambiguously located as a flight between a here and a there; "Mitdasein," being in the world with another; "Sein," Being, the meaning of which remains as elusive as the person of Godot, and necessarily so, since the meaning of Being can be revealed only through the living of a life dedicated to its search. It can only be frittered away in an inauthentic wait. That is the "religious" motif of the play: the suffering couple's search for a transcendent significance that cannot be fulfilled within the immanence of time, but which one cannot forego, either, without losing the significance of self. There is truly nothing to be done, for however hard one tries, consciousness can only turn inward upon itself; the transcendent will remain its very denial. Each passing outward toward the other and returning to self, the couple is on the cross! At the chiasmatic crossroads of the *Mitdasein*.

In *Waiting for Godot* the unhappy consciousness is split conveniently into four, two pseudo-couples reflecting in mirror image the struggle to exist, each couple personifying a particular facet of the peculiarly human conscious experience. Each individual in both couples is abandoned to his fate, the experience of self. And how that is done depends upon how time passes. Didi and Gogo must continue to wait; Pozzo and Lucky to exploit and be exploited. Only death can unify them all.

So much for the "laugh sensation of two continents." The starkness of the tragedy, however, is relieved by the humanity of its expression. Distanced from the action, we may respond with a smile at the "absurdity" of the dialogue; but pulled into the center of a struggle for a meaningful existence by the second take of our understanding that affirms the simultaneous truth of contradictory theses, we must, in the end, suffer the fate of the unhappy consciousness, at least in sympathy. Gogo is a likeable "clown" and Didi is his tormentor, timelessly; and the reader who has grasped their essential union, who can find traces of both in his own personality, will successively torment and be tormented, sado-masochistically, by himself. All our individual times will pass away; but when they do, so will we. Our dignity is to know this to be a fact. To wish for surcease of sorrow is vain, an empty way of passing one's time. Not even Jesus was spared his crucifixion.

The tragedy of Didi and Gogo is not to have learned the difference between a real, but inauthentic, death and the authentic, anticipatory projection of a final possibility, really to be there no more. Heidegger has a word for that, too: *Nichtmehrdasein*, no longer being there, no longer existing, which obviously cannot happen, really, as long as we exist. So the human comedy must go on, with all its tragic consequences. The experience of it is of the sublime.

The literary problem of the dramatist is to present this philosophical structure in gestures, both vocal and corporeal. Beckett uses two acts, the second of which is virtually the repetition of the first: Didi and Gogo whiling away the time, an appearance of Pozzo and Lucky, and a return to their abandonment.

In Act I, Vladimir suggests that repentance may relieve the load of their suffering. But for what? Gogo suggests that their only crime (sin) has been to have been born, at which response Didi gives way to a loud belly laugh; for the crime of the conscious human being is to be conscious, to possess a conscience that calls each Dasein to accept its own being responsible (*Schuldigsein*), being guilty for whatever happens to it. To make his point, Didi asks Gogo to recall the story of the two thieves crucified with Christ, only one of whom was saved. From what? Hell. But why then did the other curse the Lord for not being saved from death? Decidedly salvation is a curious notion. From their own decision of waiting for Godot there can be no salvation.

Telling stories may help pass the time. Gogo recalls the story of an Englishman in a brothel, but Didi won't hear of it. Hanging themselves is a more direct solution to their problem, but there is a fifty-fifty chance of failing; so they go on waiting. To placate his bodily needs, now that the erection of the hanging has been foregone, Gogo sucks his carrot. They are united to each other, as they both are to Godot, both on the cross, supplicating their surcease.

Enter Pozzo and Lucky. Pozzo, the master, as his name suggests is crazy, at least deluded into thinking that his slave exists only for his service. His brutality occasions the sympathy of the earthy Gogo, who tries to wipe away the unlucky Lucky's tears; for his trouble, he receives a kick on the shin. Lucky stops crying as Gogo begins. That, says Pozzo, is a law of nature: the amount of tears shed the world over doesn't vary; when one person stops, another starts. In his sympathy, Gogo has replaced Lucky, and the former continues to cry that he cannot take any more of this.

So sympathies switch, from the overburdened servant to the mistreated, unappreciated master. For the "correctness" of this gesture, Pozzo promises a good deed in return. He had already allowed Gogo to eat the chicken bones that would normally have fallen to Lucky, and now makes Lucky perform a dance: the representation of a man caught in a net, as indeed are all four of these characters. Then, in reply to Didi's request, Lucky is made to think. The essence of that thought is a biting satire on the academic mind, beginning with an expression of theological providence, man's suffering therein, the earth's changes from heat to cold.

The tone is comic, the substance supportive of the play's significance; first, concerning God:

Given the existence as uttered forth in the public works of Puncher and Wattman of a personal God quaquaquaqua with white beard quaquaquaqua outside time without extension Who from the heights of divine apathia divine athambia divine aphasia loves us dearly with some exceptions for reasons unknown but time will tell and suffers like the divine Miranda with those who for reasons unknown but time will tell are plunged in torment. . . .[7]

then, man:

. . . it is established what many deny that man in Possy of Testew and Cunard that man in Essy that man in short that man in brief in spite of the progress of alimentation and defecation wastes and pines wastes and pines and. . . .[8]

finally, man's earthly abode:

. . . in the plains in the mountains by the seas by the rivers running fire the air is the same and then the earth namely the air and then the earth in the great cold the great dark the air and the earth abode of stones in the great cold alas alas in the year of their Lord. . . .[9]

To stop the tirade Pozzo has Didi remove Lucky's hat, without which he cannot think. Gogo expresses his relief with a laconic "avenged!"[10]

He had been avenged, for it was he who preferred the dance, a bodily expression of a consciousness in itself, instead of the thinking, the intellectual expression of a consciousness existing for itself. But Lucky's thought, gathered from a myriad of unfinished works of others in academic fashion, was inauthentic, a pathetic babbling of idle talk (*Gerede*), rather than authentic speech. He is not only slave to his master, but to his own lostness in the world and words of everyman (*das Man*). But Pozzo is no more authentic than his slave, for in his dependence upon Lucky's service, indeed in his very brutality, he is enslaved by his servant, who accepts his brutal treatment with stoic renunciation.

Vladimir is amused: Lucky's performance helped pass the time: a remark on the mark for its psychological fitness, but one which is countered by Estragon's physical sense of time: "It would have passed in any case."[11] Once again speech at cross purposes; strange, but far from absurd given the personifications of the characters. Inauthentically one waits to let time pass; authentically one controls the sense of time's passing.

Godot's messenger appears with the news that Godot will not arrive this night.

Returned to themselves, Gogo removes his shoes, to walk barefoot, like Jesus. The comparison is irksome to Didi, who suggests that surely Gogo does not take himself for Christ. The exchange is touching:

Vl. Christ! What's Christ got to do with it? You're not going to compare yourself to Christ?
Es. All my life I've compared myself to him.
Vl. But where he was it was warm, it was dry!
Es. Yes. And they crucified quick.
Silence.[12]

Usually broken by another laugh! Serious readers of Beckett will recognize the old story of Dante and the Lobster.

Preparing themselves to spend the night asleep, Gogo muses that it would have been better had they lived apart (to suffer and not to know that one is suffering), that they were not made for the same road. But that is not certain, since to the skeptic nothing can be certain, and the end has not yet come. Didi suggests, however, that it is not impossible for them to part; and Gogo states that it is too late for that now, no longer worth the trouble. But they do, as they sleep between acts I and II, where their skepticism achieves momentary relief.

In Miami, two thirds of the audience left the matter there.

Act II opens on a song by Didi, a round about a beaten dog. Gogo makes his appearance and reports having been beaten by ten assailants, angry with Didi for letting him go away. Gogo allows himself to be cajoled into saying he is happy to be back with Didi; but now that they are happy, what is there to do? Still nothing; their very words deny their true state of mind.

At this stage of its development the play almost goes dead (from the naturalistic point of view) as the dramatist draws out the psychological action into a list of eight disconnected events, allowing time to make its appearances within the embarrassed silences. Two of the events, the first and the last concern the pair's awareness of the tree, which has put out its leaves since Act I.

Notice of the leaves by Didi have no effect on Gogo, since he can hardly recall what happened "yesterday." That, after all, is in keeping with his nature as a consciousness in itself. Second, an embarrassing failure to start a conversation; given Gogo's state there is nothing to talk about. Third, they acknowledge their inability to think at all, out-of-tune, as all egocentric consciousnesses are with nature that would give them something to talk about. The uncanniness of their world begins to strike home. That is interrupted by Gogo's recollection of Lucky's kick on the shin; the wound has begun to fester. The new shoes he finds fit, or would do so, if he had a pair of socks: the fifth event, followed by more embarrassed silence, and Gogo's request that they get going. Didi reminds him that they cannot; they are waiting for Godot. Finding Lucky's hat, the sixth event, allows them to play the game of switching hats, thereby symbolically changing roles. But Lucky's hat on Gogo is no better than his own: both are awful.

They continue the game of Lucky and Pozzo, insulting each other with epithets varying from "Moron" to "critic," aptly referring to each other's function as a consciousness in itself but not for itself and for itself but not in itself. Beckett had changed his original French "architecte" to the English "critic" to reinforce the denigration of the reflective consciousness that fails to express itself creatively in the first instance; but unless the philosophical personifications are clearly understood, the repartee is heard in English as a kind of theatrical in-joke. The seventh event is their reconciliation. To celebrate it, Didi does an imitation of a tree, and Gogo imitates him. But the last imitation bears little fruit since Gogo is haunted by the idea that God sees him. He asks for pity, and removal from the tree, as the Bible tells us did Christ, three times, when he has had the vision of His own coming crucifixion. The eighth event is the projection of their own ultimate failure at suicide.

Pozzo and Lucky reappear on the scene, Pozzo now completely blind, and at the mercy of the troubled couple. The pity they asked for themselves is now requested of them. At first they refuse, being only human, and then bargain for granting their favor. Pozzo's erstwhile haughtiness is inverted into abject supplication; and when Gogo tries to take out his vengeance on Lucky, he hurts his own foot. Lucky has gone dumb, and cannot sing a song in payment for their help to Pozzo. Exeunt Pozzo and Lucky, with Pozzo now completely under the control of the dumb Lucky. The dialectic of the master and slave has achieved its full expression, leaving Vladimir and Gogo once more to themselves.

The next stage of the dialectic is stoicism: acceptance of the inevitable, resignation; and the next, skepticism: Didi wonders if it was all a dream, as Gogo indeed has already fallen asleep.

Godot's messenger arrives with the same news as the night before: there shall be no transcendent aid. This time the boy did not even see Lucky and Pozzo, while the night before he was frightened by them and had delayed his arrival until they had departed.

As Didi explains the message to Gogo they decide to attempt suicide by hanging themselves on the tree, but Gogo's belt is too weak, and breaks in the test. Having thus lost his belt, Gogo likewise loses his pants, as they drop to the stage, and he is told to pick them up.

The closing scene is the same as for the first act. "Shall we go on?" "Yes, let's go." But they remain rigid as they had the night before, when the curtain falls, promising a continual renewal of the previous scenes until they can scare up the courage to project a future of their own. Until they do, their time will be perennially spent waiting.

And in this play, the last level of signification to be associated with the words and actions is the compulsive awareness of the meaningless passage of time: at the heart of the matter, the temporality of the human consciousness itself. The consciousness here is that of the reader, or those of the audience.

He who would laugh at that would be laughing only at the meaning of his own being, which if one has followed the structure of this play, has been revealed to the reader's consciousness as a swing from the uncanniness of this world depicted within the play to the acceptance of one's own responsibility to be what one has to be, as guilty as either thief, and with a fifty-fifty chance of salvation; for one will either decide or not decide to face one's own responsibility. And in either case a choice will have to be made.

Endgame (1954–6), as has been humorously but judiciously remarked,[13] might have been called "Ending the Waiting Game." *Waiting for Godot* established the interminable wait of the middle game, and anticipated a conclusion that would bring final surcease to the sorrow of Didi's and Gogo's existence; but they were lost in inauthenticity, and so could not end the game. Didi explained it all as he mused over the sleeping Gogo:

> Astride a grave and a difficult birth. Down in the hole, lingeringly, the grave-digger puts on the forceps. We have time to grow old. The air is full of our cries. . . . But habit is a great deadener.[14]

That was presumably one of the laughs projected over the footlights in Miami: as compact an image of the pessimistic view of life as might be found, but one that recurs in Beckett's literature of life's beginning and ending, here compressed into a single action to express the identity of the two processes – living and dying. Our habits only make us forget this fact.

Yet nature would tell us, if our language does not, that living and dying are the same process, that obstetrician and gravedigger perform similar functions, at either end of life seen as a series of expulsions. The obstetrician is usually in a hurry to elicit the first cry that announces the first free exhalation of human breath; the gravedigger, on the other hand, takes his time, and the air is already full of human cries of suffering. We hear them so much we end up not hearing them at all. All his masterly life Hamm has refused to respond to anyone's cry for pity.

Hamm and Clov, the heroes of *Endgame* are the victims of the habit of forgetfulness. In the process they have lost touch with the source of their being. Inured to human suffering, united inseparably as master and slave, each of whom would die without the other, as useless apart as hammer without nail, nail without hammer, they will take the step that ends the game of human suffering once and for all: final separation, with its accompanying loss of self.

The puzzles of hermeneutics are less complex than in *Godot*, since solutions are always simpler than an understanding of the problem to be solved. Once it is understood that all in life is for nought, one need only accept the ultimate denial, anticipating it with resolution. Man's victory over pessimism is the acceptance of his temporal finitude. The game must come to an end.

Philosophically considered, from an existentialistic point of view, the source of man's unhappiness and its solution are the same thing; both reside in the will. If man did not insist upon individualizing his drives, of separating himself from nature (if not from Schopenhauer's primeval Will), he would have no cause for unhappiness; nature would always take its course.

Hamm and Clov both know that something is taking its course, whether Clov projects his telescope out the window overlooking the land or that overlooking the sea before making his report; but each is slow to accept that within, too, something is taking its course. Linked as master and slave in self-consciousness, they have become resigned to their separation from nature. As they have passed through the moment of stoicism on their phenomenological way (as described in Hegel), they can accept what they cannot change, even to the last "reasoned dismissal" (the Stoic term for suicide) for which Didi and Gogo had lacked the means.

Clov is quicker to react than Hamm, because he personifies the sensitive slave with full possession of his sight, and Hamm is still a victim of his old master habits. Blind, he cannot see; but he can still command and expect to be obeyed. As long as Clov is there he is likewise self-conscious, and still possesses knowledge of the past course of natural events. His task is to make sure there are no surprising novelties, no gratuitous event to redirect nature's tendency to run down thermodynamically to zero inertia, where the whole universe will come to the still.

A brief scene epitomizes the entire play:

Clov: Why this farce day after day?
Hamm: Routine. One never knows. . . . Last night I saw inside my breast. There was a big sore.
Clov: Pah! You saw your heart.
Hamm: No, it was living. . . . Clov!
Clov: Yes.
Hamm: What's happening?
Clov: Something is taking its course. . . .
Hamm: Clov!
Clov: What is it?
Hamm: We're not beginning . . . to . . . mean something?
Clov: Mean something! You and I, mean something! Ah that's a good one!
Hamm: I wonder. . . . Imagine if a rational being came back to earth, wouldn't he be liable to get ideas into his head if he observed us long enough? . . . Ah, good, now I see what it is, yes, now I understand what they're at! . . . And without going as far as that, we ourselves . . . we ourselves . . . at certain moments. . . . To think perhaps it won't all have been for nothing![15]

Hamm did see inside his breast. What he saw was both his heart and a sore, with reasons Reason itself cannot comprehend.

That bleeding heart was not living in the usual sense, for it was no longer the seat of courage and longing, only the empty and "cracked beater" that guaranteed a continual supply of blood. Being blind, but seeing in his dream, he saw through the walls of the muscle and interpreted the blood as a sign of a festering wound. In its unhappiness, that is what it was. But what was happening? Inside as out, something was taking its course. That turn of the phrase was dangerous; the dialogue came too close to making conventional sense on two levels of signification, and Clov rejected the idea that these two, left to themselves, could ever make sense. No wonder; they are dialectical opposites, and could only go on contradicting one another. That, as always in the phenomenological development of self-consciousness, is the source of their unhappiness.

But Hamm forsees the reactions of thousands of playgoers and readers of the play; within that entire group there may be one being intelligent enough to look upon the scene and get into his head the correct idea. In this universe, however, there are only Hamm and Clov, and Hamm's parents, Nagg and Nell, so the intelligent being would have to redescend to earth in order to make the interpretation. As Hamm muses, he sees further into the future (a conventional prerogative of the blind); they themselves may be condemned never to make sense, but the point of it all does not surpass the understanding. Only the sense-making context is absent. And so in the end it may all be understood: "That's what they are at," the playwright and his audience, using Hamm and Clov as examples, putting them into a context that finally makes a sense. Although in this way the play will come to closure within the interpreting consciousness, Hamm and Clov will be doubly damned. They will continue to torment each other until their creator chooses to end their sufferings, only to have it start all over again when the audience takes up from there.

Endgame is no more absurd than *Waiting for Godot*. One enters the universe of the former by walking through the mirror of the latter. Foregrounds and backgrounds switch placement, along with left and right. The two couples experience only a change of names, and one person a change of sex. Where Pozzo and Lucky were found in the background of Didi and Gogo, the better to mirror their situation, they emerge behind that mirror in the foreground as Hamm and Clov. Didi and Gogo recede into ashbins in the new background, now with the names Nagg and Nell, where their concernful living together goes on essentially unchanged, until Nell dies a natural death, deprived of light.

The surrounding scene has changed from a country road, leading from nowhere to nowhere, to the end of the road: a house surrounded on the one side by the sea and on the other by the earth. For lack of combatants, the struggle of human survival is coming to an end. Nagg and Nell are our last parents: Clov, an orphan brought into Hamm's service by his father who somehow has escaped earth's zero inertia. Even during the day all is gray; the waves are leaden; on earth there is nothing to be seen.

In a glaring example of the "franglais" Frenchmen hate, Hamm and Clov are hammer and nail (*clou*), the tormentor and the tormented. Why then does Clov not leave, as is suggested to him by Nell and even requested by Hamm himself. Clov too is a victim of his habits, and he enjoys his freedom, which is first to suffer, but then to make suffer in reversing rôles with Hamm. In the opening scene, going about his usually servile tasks, he continually smiles, self-consciously laughing to himself. Should he go away, they would both die anyway, like the boy Clov thought he saw through his telescope. So

Clov stays shuttling between his kitchen, and Hamm's living room, between watching the walls and tormenting himself with his thoughts and caring for his tormentor. To prevent the accidental rekindling of earth's inertia, they make sure that every living thing is dead; there is flea powder for Clov's flea, and poison for the kitchen's rat.

Even Clov's whimsical expression, "If I don't kill that rat, he'll die"[16] makes sense in this context; for if the rat is killed, they will know he is dead, whereas if he is left to die, they run the risk of the damn thing's starting up the whole universe again. But it escapes, leaving the issue momentarily in doubt.

All that, however, is unimportant, since the play is situated within the conscious relationships of the two major characters, in their shared temporality (consciousness) that will come to an end in their separation. Ultimately Clov will make his move, packing and leaving; and as the last pawn is thus removed from the board, the king is left unprotected. He cannot turn himself over, since he is still in his wheeled chair, so he covers his face with his handkerchief. At that moment the waiting game comes to an end.

In this play, as in *Waiting for Godot*, nothing happens. We are presented with words and gestures that in themselves signify nothing, but which together succeed in creating a context of inner dramatic significance. Once again the characters are reduced to dramatic presences, and as usual it falls unto the interpreting critic to discern the significance of their presence. We may begin by training our view on the tragic couple.

The vulnerable one is Hamm who cannot see, and who must count on Clov for every bit of sustaining information; for food, clothing, drugs, and an occasional prosthetic device. It is Clov who wakes him up and puts him to sleep; supplies the blankets for warmth and the drug for relief of his pain; brings in the three legged play dog so that Hamm might imagine its companionship, and the catheter so that he might urinate. That too is something that must take its course.

Yet Hamm is the sadistic son of a miserable, neglected, father. Nagg, the father, begs for food and is given crumbs, a biscuit that he shares with Nell. He is forced to listen to Hamm's same story of the arrival of that other destitute father asking for food and sustenance. It was given only when the man accepted entering into Hamm's service. And Hamm reneges on his promises, making Nagg reveal how dependent Hamm was once upon himself. That the gratitude Nagg expected was an error in judgment even he ought to have understood. Nell, it seems, loved the story of the tailor and his English client, who had complained that three months was too bloody long to take in finishing a suit when God had created the whole world in only six days. The tailor retorts that one could surely understand the difference by first looking at the world and then at his magnificent suit. Ungrateful offspring are a part of that botched world, as natural an event as any other that must take its course.

Nagg and Nell are housed in ashbins filled with sand, there being no more sawdust to make their existence more comfortable. Nell takes her full course and dies, at least she has no more pulse; and Nagg cries at the news. But that only proves that he is still alive, as Hamm must interpret the events around him as symptoms and signs. For a moment that presents a problem, given his dependence upon Clov; for should Clov not make his daily appearances, it would be impossible to tell whether he had finally left or only died in his kitchen. The result, however, would be the same: Hamm would whistle, and Clov would not come.

In the meantime, they must continue to play their little game. Hamm prays that the world end with a bang (and not the famous whimper), but that too is a vain wish. Clov refuses to hit Hamm with the gaff or the axe; to place him in a coffin, for there are no more; but does gratify him with a few words from the heart:

> I say to myself – sometimes, Clov, you must learn to suffer better than that, if you want them to weary of punishing you – one day. I say to myself – sometimes, Clov, you must be better than that, if you want them to let you go – one day. But I feel too old, and too far to form new habits. Good, it'll never end, I'll never end, I'll never go.[17]

That's the old, inauthentic Clov; the new authentic one sees himself walking out:

> I open the door of the cell and go. I am so bowed I only see my feet, if I open my eyes, and between my legs a little trail of black dust. I say to myself that the earth is extinguished, though I never saw it lit. . . . It's easy going. . . . When I fall, I'll weep for happiness.[18]

As he actually leaves, following two calls from Hamm, he says ironically for the commenting playwright: "This is what we call making an exit,"[19] but the second entendre is usually lost on those too wary of a playwright's tricks. One may exit the stage, or exit from life in the same act; and if the character is conscious of his role, which he only plays in relation to that of another, he may actually intend both these meanings at the same stroke.

Before Clov makes his exit, Hamm expresses gratitude for Clov's past services; but Clov insists that it is he who should express the gratitude. Once more there is a denial from Hamm: "It's we are obliged to each other."[20] All three statements ring true to context: the slave is obliged to his master, just as the master is obliged to his slave; as that consciousness lending self-consciousness to the other, each is obliged to the other for the very fact of his essential existence.

When Clov at last departs, there is one last move to Hamm; and he obliges the audience by going through his whimpering end, discarding one by one his now useless appurtenances, and hiding his face with his handkerchief.

Thus, where the first level signification places the action of the play in Hamm's abode, situated between land and sea, the inactivity of the play centers that action in the functional, conscious relationships of the characters themselves. But they are personifications of the positive and the negative, the master and the slave, in perpetual mutual negation; symbols of consciousness's having become self-conscious for having lost itself in the other.

One last level of signification, however, places both conscious functions in a single locus: the consciousness of the individual member of the audience, who may be led to understand that original split of his personality into the sado-masochistic entity it has always been, suffering at its own hand, and knowing that the process will not end anywhere this side of the tomb.

To deliver and to inter this additional pseudocouple, the gravedigger works lingeringly, savoring the very words that supply the missing coffin-crib. Death finally individuates us all, the master and the slave, the masterly slave and the slavish master, or the two in one, in peace at last.

If the source of human suffering is the individuation of a will in abstraction from the primal will of Nature, the solipsistic consciousness must suffer until it receives its surcease in the authentic individuation of death. That indeed is strange, but not absurd. The source of the problem of living and its solution can be the same only where there is a difference between authentic and inauthentic lives, authentic and inauthentic deaths. We live inauthentically in separation from others; we individuate ourselves inauthentically by withdrawing from nature, as our consciousness in the normal course of its existence denies it is the object it knows, affirming itself as nought with respect to the world's positivity: such is the problem's source. On the other hand, we die inauthentically by merely waiting for the specific event to occur within the context of nature where all things must take their course; but we die authentically by resolutely anticipating our no longer being there; by no longer defining ourselves by what we affirm, neither as this or a that in the world of everyman. To solve the problem, once it has been made clear, the slave has only to leave, to refuse his rôle. Clov does walk out, and Hamm throws away his baubles, covering himself in symbolic acceptance of his end. Alone he cannot exist.

Like *Waiting for Godot, Endgame* is a dialectical, existential play, full of meaning and not the least absurd.

Although *Cascando* (1963) bears an Italian title, by reason of its structure and symbolism it might very well have borne the German title *Verfallenheit*, for that would have given the clue for the Heideggerian interpretation which follows. As it stands, the Italian gerundive of "to fall," without stipulation of the tense of the auxiliary (*stare*), indicates a progressive, continual action; the tense of the action would ordinarily be supplied by the missing auxiliary. The non-specification of the time may be of some importance, but a more meaningful explanation for the Italian word will be found in its music. It cannot be pronounced without the tongue's tripping between the 's' and the 'c,' a brief hissing of the breath forward caught up and left to flow in a descending liquid slide. Not by accident, to be sure, such a fall will be represented in the "Opener's" story; it's not all in his head after all: part of the story rolls off the tip of his tongue.

Written for radiophonic transmission, the play is short, with only one character, the Opener, who controls two gadgets. When opened, the one produces music; and the other, the voice of a narrator. The music of the original production was prepared by Marcel Mihalovici, and copyrighted by the *Editions Heugel et Cie*, Paris. The temporal flow of the music is united, or is to be united, with the temporal flux of the narrator's words in contrapuntal rhythm.

As the opener manipulates his gadgets, he debates with himself the issue of linguistic creation, attributing the negative thoughts to an unnamed "they" (*das Man*). "They" say, i.e. everyman in his ontologically fallen condition (given to idle talk, curiosity, and ambiguity) says, that the opener does nothing of his own will, since his words are those of another; but he proves the contrary by opening and closing his sound boxes, as he says, time and time again, 'at will.'

Why should they not be wrong here too, as they are when they say the play is in the head of the playwright?

Opener: They say, It's in his head.
It's not. I open.[21]

Having done so all his life, he knows that it is he who discloses what is there to be heard: sounds and their meanings, a transcendence and its world. Since they cannot see his gadgets, they claim all this sound is only his own voice: "They said, It's his, it's his voice, it's in his head."[22] And they are indeed wrong again, for the significance revealed by the recorded voice is in the words themselves, just as they are heard by the radio audience, without the mediation of the written word.

The "Opener," then, is a symbol for the creative playwright himself, who dis-closes the mysterious world of linguistic creation; and the play is to be found in the words and music he controls. The everyday commonplace that exists first in the head of its creator and then undergoes transmogrification into words and actions of actors bespeaks the ambiguity of the inauthentic understanding. The opener knows better, but can only prove his point by letting the words flow from an external source, as the three syllables of the title of this play.

The words concern the development of a story, of a man on a journey outward and a return. The human transcendence theme is clear, as it has been in Beckett, at least since *Watt*. Any real journey is but a symbol for the ontological understanding of man's being as a projection toward some future – and this understanding allows the further interpretation of the missing auxiliary in the play's Italian title. In his everyday constitution man is constantly falling into inauthenticity, where he may forget the commitment to his own being as a mere possibility of being there, and give himself away to the cares of his ontic placement on the earth, surrounded by the things that are present at, and ready to his hand, i.e. by the "objects" of his perceptions and the tools displaced for his manipulation.

In telegraphic style, the information is transcribed in bits and pieces, to be heard in the same way that the narrator of *How It Is* heard his story from above. The words concern the voyage of a Woburn (*Maunu*, in French; and translated as *Missler*, by Elmar Tophoven, in German). The 'M' or inverted 'M' that is a 'W' is pure Beckett. The narrator-playwright struggles to control his story by catching glimpses of his character in a present situation, and by attempting to follow him to his "end."

If we are allowed to invert a "u" into an "n," "Maunu" would become "Man-nu," a convenient "franglais" denomination for man stark naked, deprived of his conventional cover and stripped of the myriad inauthentic and inessential detailed characteristics by which we usually (in our own inauthenticity) recognize a man for what he is. But in so doing, we merely take appearances for the reality of the situation.

Beckett's English translation harbors a Germanic indication of placement, and the superheated commitment of the narrator's character to the accomplishment of his task. That the character will fail to attain his authentic goal is forecast in the German translation "Missler," which may be taken for the name of anyone who bungles a task.

Maunu (Missler, Woburn) is on a journey; he falls face down in the mud, in the sand, on the rocks, in the bilge of a boat which approaches its haven, an island, and then by-passes it, as the narrator watches the scene passing before his imagination. The images are recorded in his anxious words. We need only hear them to understand the anxiety:

Voice: . . . .
come on . . . Woburn . . . it's him . . . see him . . . say him . . . to the end . . . don't let go

Opener: (with Voice and Music) Good.

Voice: nearly . . . just a few more . . . a few more . . . I'm there . . . nearly . . . Woburn . . . it's him . . . it was him . . . I've had him . . . nearly

. . . .

Voice: this time . . . it's right . . . finish . . . no more stories . . . sleep . . . we're there . . . nearly . . . just a few more . . . don't let go . . . Woburn . . . he clings on . . . come on . . . come on.[23]

Such is the voice of the authentic playwright, always aware of his own possibility of failure, looking for the end of the story where both narrator and character will have achieved their ends in the climaxing word that never comes. For if Woburn could achieve his goal, both the narrator and the playwright could find their own inner peace, and fall into the unconsciousness that is sleep. Silence and sound form that minimum gestalt of any self-consciousness, reduced in Beckett's prose to a punctuating "ping."

Cascando is a dramaturgical play, not a simply dramatic one. The story is always the same: the projected existence of a man *mis à nu*, where the projector is the consciousness of the creative artist himself, so close to that of his character that they may be mistaken as one. To say that a play is dramaturgical rather than dramatic is to say that the dramatic involvement is located within the narrator's consciousness, not in the world represented by his words. In *Cascando* this is evident from the comments of the opener. "They say it's in his head"; but how could that be, when the words he hears bely the truth of that "truism?"

To consider the matter correctly, however, "they" might have said that the play is in the struggle of finding the words to communicate the image that is reflected off the character whose existence is imagined by the reflecting playwright. The words and the music of the opener's gadgets are themselves symbolic at a higher level. Both communicate a state of mind, shared by creator and his audience; at times they run harmoniously when the struggle of the narrator grows in intensity; at times, in dissonant counterpoint, when the character eludes the narrator, thereby causing an anxious moment in his struggle for expression. That is the obvious significance of the noises emitted from the opener's gadgets, but each is but a function of a greater context.

The music symbolizes the sensuous surface of the linguistic universe, and the words, its depth. The playwright's task, like that of the novelist, is to bring them together into a significant expression. When the character is anxious the music should be anxious, and when the character loses himself by falling, the music should lose itself in a *quaqua* dissonance. To accomplish this feat authentically, when character, narrator and playwright himself exist every day in the state of complacent acceptance of the commonplace is an arduous task indeed. To do so, each must come to terms with his real being, as disclosed by his being there: an anxious understanding of his own being as projected into the creative words. The play then exists not in anyone's head, but in the projected significance of the playwright's words. Beyond the world of everyman each authentic personality projects a union with his Self. All other unions and reunions, indeed all other worlds, are temptations to be avoided, since they are only so many occasions of the falling consciousness's forgetfulness of being, which a moment's reflection will reveal as a *Nicht-es-selbst-sein*, a not being oneself.

That, finally, is why "they" continue to say it's all in the opener's head, and why he must continue to dispute their commonplace; why, in the end, any ontic situation may serve only an ontological purpose: i.e., phenomenologically, to allow the self to declare itself as it is in itself; the dramaturgical play thus allows the phenomenal disclosure of self to self in thought. The playwright is not the narrator, is not the character, but they are bound to each other in the enactment of linguistic communication. The medium is the message, is the words we hear.

There is missing only the consciousness of the director, who must read or hear this linguistic context with his own anxieties, his own anxious understanding of its significance. And, as before, he need only work out the hermeneutical fore-structure, to test whether the hypothesis above fits the context of the play. Should its enigmas be removed, he would have discovered the essence of the play.

If it is not easy to create, why should it be any easier to recreate? The saying and the hearing are reciprocal processes. Idle talk is no replacement for the creative discourse, whose words and music must reveal the self of the speaker and allow his hearers to understand what has thus been revealed. That is what *Cascando* is all about: the difficulty of making a creative play, given the fact that all plays are about man's everyday, inauthentic existence.

Woburn makes his difficult journey, falling all the while; but those falls and that journey are mere symbols of a meaning to be expressed in the playwright's own world, where only the words count. The words of the recorded narrator and the music are real events in that universe, but they too are only symbols of the relata of a functional equation: the surface and depth constructs of the literary medium.

How is it to live in the playwright's universe? Startling, baffling, uncomfortable; thinking and reflecting on what one has thought, one cannot escape the anxiety of one's own existential predicament. To think anything through to its conclusion, one must find oneself, a mere nothing with respect to what everyone considers the real world. Whether one thinks, acts or suffers, 'alienation' is the word to describe man's inner state.

Let us then live with the impression that the "opener" creates a play by willingly turning on and off his mechanical gadgets. They symbolize the means at man's disposal for externalizing his inner states. As they give off their respective sounds we may come to understand the tranquilizing temptation of the narrator's fallen state,[24] only one degree removed from the false "truism" he combats; for if the play is not in his head, it is not in his gadgets either. They too must be transcended, pointing as they do to the stylistic tension uniting linguistic surface and depth, that conscious feeling expressed within the linguistic medium, where authentic understanding occurs.

Dramaturgy, however, may replace drama only when one has understood dramatic creation. *Cascando*, as Beckett's dramaturgical play par excellence, marks another spiral inward toward the author's self-understanding, and poses another answer to the Unnamable's ontological questions: Where now? Who now? When now? The dramatic narrator, like the novelistic one, is a projection into significance, and their author is the one who must produce the words that create this projection. As such, he needs an alter ego, a reader or hearer to confirm the fact of the creation. The self is no thing with observable properties, but a process of communication: between him or that which opens up a universe of significance and him or that which has perceived its closure.

Where *Godot* and *Endgame* dramatize the inauthentic and authentic aspects of the schizophrenic consciousness, splitting into two sets of divergent twosomes, *Cascando* lets fall the anxiety of its ontological condition, wherein the split is revealed as the two variables of externalized communication, unified in the spoken word. No clearer a representation seems possible for Beckett's solution to the problem of pessimism: it is transcended in art.

As in the beginning, in the end is the word – that by virtue of which speaker and hearer become one. On the stage, however, the word may be replaced by the deed. That's how scenes are created, uniting the word to the deed, to establish the presence of the playwright's characters. Their presence in turn reveals the nature of their world.

CHAPTER IX

VOICES, IN ENGLISH, ON THE AIR

Cascando, Beckett's French radio play with the Italian title and German philosophical symbolism, was not his first creation for radiophonic transmission. Three English plays were either prior to or contemporaneous with that exercise in the direct use of the human voice in service of the creative consciousness. *All That Fall* (1956) and *Embers* (1959) are more dramatic and less dramaturgical than *Cascando*; but *Words and Music,* copyrighted in the same year, 1963, marks a return to the dramatization of the problem of dramatic creation.

If a dramaturgical play is drama raised to a second power, the medium of its expression is still words, directly spoken by a performer who remains unseen by his radio audience. And, as before, the voices are symbolic of the human consciousness in its presence to a world. In a first order drama, the voices are of the characters and in a drama of the second power, of a narrator or of "characters" symbolic of the various functions of the author's own consciousness. In the one case, the characters are open to the influences of the fictive world created by the author, and in the other they symbolize the very openness of the author who is creating the fictional world, a function that may be delegated to a fictional "Opener."

By now the relation between a consciousness and its world is already well known. Beckett has employed the dialectic of the master and the slave consciousness as the source of the central duality of a single self-consciousness in both *Godot* and *Endgame.* In Hegel's account of dialectical experience, beyond the phase of self-consciousness, the phenomenological development of the mind enters the moment of the spirit, actually a return by the self-conscious mind to its concerns in the world. For these concerns, Beckett shifts his attention to the inner structures of the human awareness. The philosophical foundation of the plays as a consequence shifts to Heidegger's structural account of human disclosedness, an existential-ontological description of the way in which human reality is "in" the world. *Erschlossenheit* is the term, referring to the three "equiprimordial" moments of human consciousness coming to the fullest expression of its openness to experience. These constituent moments are: state of mind or affectivity, understanding, and discourse.

For the inauthentic personality, the habitual state of mind is curiosity, the rapid change of interest from one thing to another that settles nowhere and nothing. When this state of mind is revealed in discourse, as it is in the dialogue of the characters of these radio plays, the impression of the hearer may be that the speaker has lost his moorings in the world and has in consequence begun to pursue a private, absurd dream of his own.

The reason is obvious: such talk is idle talk or gossip, the flat expression of the commonplace that communicates nothing to the hearer because the hearer already is fully familiar with the essence of the message. As a result the understanding corresponding to that state of mind is fraught with ambiguity of a non-functional sort–not the proliferating meanings of the creative consciousness as it interprets its experience in wider ranges of enveloping significance, but a constant shift in reference between subjective and objective intent of the curious consciousness.

For the authentic disclosedness of human consciousness, on the other hand, the state of mind is controlled by the affective condition of the perception produced in creative discourse. As the highly personal and authentic words open up a vista on the world they create, they produce an understanding that is a feeling having come round to complete expression.

To interpret such a context of human expressiveness, one must procced with the application of the hermeneutical forestructure: attention to the context as to what one has before one; the specific intent developed within that context (the foresight); and, finally, the manner in which the context comes to closure under the application of the chosen hypothesis (the fore-conception).

That, of course, is the method of analysis used for the interpretation of the novels and plays of Beckett up to this point, and there is no reason to change our procedure for the electronic media. And, if Heidegger is right on this methodological point, the method is circular; that is in making the hermeneutical expansion we already understand implicitly once we have formed our hypothesis, and after the analysis our hypothesis has merely been confirmed. Our knowledge, in every case of successful analysis, has only been made explicit; and where the analysis has failed, we must continue to look for another specific intent, for another hypothesis concerning the linguistic structure of the creative context.

The clue for *All That Fall* should be obvious from the title. Here once again the world is of human *Verfallenheit*, that ontological condition of escape from authenticity by which the inauthentic personality falls away from its true ultimate concern, as a possibility to be determined by one's own act, into the proximate world of everyman, where one's being is groundless. Ontically, so the master has said,[1] a human being may fall only because its being, as Being-in-the-World understandingly and with an accompanying state-of-mind, is an issue for it. To avoid this ontological issue, the human personality afflicted with this condition finds the proximate world of everyone tempting, tranquilizing, alienating, self-entangling, and turbulent.

Fallenness is, however, not a theological conception of a state of original corruption; for there is in man no pristine integral condition from which one might fall into corruption, nor an intermediate state of grace by which one might be lifted from it to achieve one's salvation.[2] The theological and symbolic language merely points to a given ontic fact: as it exists, human reality is ontologically characterized as fallen because fallenness constitutes all Dasein's days as they are spent in their "everydayness," i.e., for the most part, each one like the other. Such, indeed, is the world of the characters in *All That Fall*.

For, on the day in question a dislocated, disjointed couple, Mr. and Mrs. Rooney, are on a trip from their home to a small hamlet of Bogghill (the low and the high); he by train on business and she walking, in an attempt to surprise him at the station. The train is late by fifteen minutes on a half-hour's run, so something extraordinary has happened to distinguish this day, which also happens to be Mr. Rooney's birthday, from another.

The action mounts until the arrival of the train, when the couple is reunited and they begin their return home, walking together. Thus, we have two symbolic voyages converging into a single return, two human transcendences out of kilter, disjoined from their pasts and unresigned to the indeterminacy of their future. What the future will be like is left to the audience's projection, by way of an implicit understanding, of the couple's end, which will come on another day not significantly different from this one.

The play begins with Mrs. Rooney's hike. As she passes the house of a neighbor, she hears a recording of Schubert's musical piece, *Death and the Maiden*. It is perhaps for this reason that Mrs. Rooney walks along, dragging her two hundred pounds as an excruciating burden, sometimes musing over the loss of her own child, Minnie; and at others, over the lack of a true love in her life, past and present, but never forgetful of her seventy-odd years and all the afflictions her aged flesh is heir to. In reflecting upon herself, she can think of nothing but her sufferings in an ungrateful world; given to self-pity, she is lost in the inauthentic world of everyman (*das Man*), where each day is followed by another of equal emptiness.

Six times she is interrupted in her walk. The first interruption is perpetrated by a passing carter, himself walking beside his cart loaded with pig manure and drawn by a hinny. Typical of the inauthentic consciousness, she recognizes the carter by his hinny, and the hinny by its whinny; their subsequent conversation bears out the inauthenticity of their mutual (non-)recognition as their curiosity moves them from point to unrelated point:

Mrs. R: Is that you, Christy?
Christy: It is, Ma'am.
Mrs. R: I thought the hinny was familiar. How is your poor wife?
Christy: No better, Ma'am.
Mrs. R: Your poor daughter then?
Christy: No worse, Ma'am.
Silence.
Mrs. R: Why do you halt? *Pause.* Why do I halt?
Silence.
Christy: Nice day for the races, Ma'am.
Mrs. R: No doubt it is. *Pause.* But will it hold up? *Pause. With emotion.* Will it hold up?
Silence.
Christy: I suppose you wouldn't –
Mrs. R: Hist! *Pause.* Surely to goodness that cannot be the up mail I hear already?

. . . .

Christy: Damn the mail.[3]

And so the "conversation" continues, with each participant hardly doing more than acknowledging the other's bare presence.

When the hinny whinnies, Mrs. Rooney is moved to reflect on that phenomenon, which upon mature second thought, is not strange at all since hinnies are hybrid offspring of a stallion and a she-ass. But that reflection is fatal to Mrs. Rooney, since in her unfulfilled sexuality she has always been obsessed by images of a horse's ass; indeed, at this time of her life a horse's buttocks is not so different from her own.

The hinny resists being put in motion again, and Mrs. Rooney suggests that the carter "Give her a good welt on the rump."[4] As she sees herself in the hybrid's place she exclaims, "Well! If someone were to do that for me I should not dally."[5] But she would, in the second sense of the term. Her sexually ambivalent obsession comes further to light later on, as she muses with her husband on the barrenness of hinnies.

Aware of the strangeness of her language, she is not unperceptive of the image she projects, as she continues to express the bad faith of her existence:

Mrs. R: Oh I am just a hysterical old hag I know, destroyed with sorrow and pining

> and gentility and church-going and fat and rheumatism and childlessness. *Pause. Brokenly*: Minnie! Little Minnie![6]

If the old woman were aware of the etymology of 'hysteria' she would know what we know. She is only a woman frustrated in love and deprived of the fulfillment she had achieved in the past as a substitute for the love she sought but could not find in her husband. The missing child might have made up for her husband's indifference, now as in that rueful past.

Her second encounter is with a Mr. Tyler, retired bill-broker, travelling by bicycle. Unfortunately for her state of mind, his daughter has just undergone a panhysterectomy; if she is childless, he is to be grandchildless. His bicycle has a flat tire; and she invites him to dally, not even aware of her Freudian slips of language. She calls the bill-broker Mr. Rooney, and asks him to stop molesting her. But molestation she wants, from her husband or from a reasonable substitute even if the flatness of his bad tire clearly symbolizes the limpness of his sexual organ. As he rides off on his flat rim she calls to him to come and unlace her corset behind the hedges. Realizing that is out of place, she exclaims, "What's wrong with me, what's wrong with me, never tranquil, seething out of my dirty old pelt, out of my skull, oh to be in atoms, in atoms ... ATOMS."[7] The cooing of Venus birds exacerbates her condition, until she shouts in despair "Jesus, Jesus!"

Her third meeting is with a Mr. Slocum, the clerk of the racetrack, to whom she refers immediately as "her old admirer." He is driving a car, and offers her a lift to the station. But she has to be pushed up into the automobile. When he puts his hands on her back to lift her up, she tells him to put them lower, on the buttocks, one can only suppose, and push. On the way, they run down a hen, squashing it into atoms. Death and its surrogate, sex, are in the air.

At the station Mrs. Rooney is helped from the car by the station porter, Tommy. The porter and the clerk converse as if she isn't even there, and she is forced to call them back to the amenities. Mr. Barrel, the station master, then comes upon the scene, chastising Tommy for his absence. Following the usual pattern for passing encounters, Mrs. Rooney and the station master enter into a conversation about each other's condition; but, as befits their inauthentic personalities, their real states of mind are never expressed. Such is the virtue of inauthentic talk; it prevents the exposure of one's personal vulnerability. Mr. Barrel leaves her to her thoughts, inevitably about the sorrows of her own pitiable state.

In this condition she meets "the dark Miss Fitt," a maiden lady in her thirties come to meet her mother on the same train, the 12:30 "up mail." Except for Mrs. Rooney's insistence, however, the younger woman would have passed her by. Although they worship in the same church on Sundays, outside of that relationship they remain strangers to each other. As, indeed, they are. Miss Fitt exists equally in bad faith, but she rationalizes her absent-mindedness as a unique relationship with her Maker. Not even the sexton taking up the collection at the Sunday service will stop before her, because she is entirely wrapped up in the divine presence, too distraught to recognize that presence's vicar on earth. In this way, the good lady never has to pay her dues to the church. But it was "the protestant thing to do" to help Mrs. Rooney in her present predicament, even if she appeared to Miss Fitt now only as a "big pale blur."

The two begin to mount the cliff at the station, only to become bogged down in the attempt, the spectacle of all the curious eyes of the parish. By now Mrs. Rooney's state of mind has changed from her unfulfilled sexuality to the impending doom of the coming denouement.

The climb first of all reminds Mrs. Rooney of the Matterhorn, and all those newlyweds honeymooning there; but Miss Fitt is lost in a churchly hymn, a song about the "encircling gloom," which may be that of the coming night or the end of a purposeless life. Mrs. Rooney thinks it must have been the song intoned just before the sinking of the Lusitania, if it wasn't really the Titanic, but that ambiguous thought suffices to fix the prior ambiguity of the "gloom." Perhaps the song on the sinking ship, after all, was *Rock of Ages*.

The religious motif suffices to show a further ambiguity in the notion of transcendence. Religious transcendence, as personified in the personality of Miss Fitt, is as inauthentic as any human attempt to evade the necessity of facing up to one's ontological commitment, to determine oneself what one is to be, and the story will return to this theme.

As the two travelers continue to mount the cliff they continue to be the spectacle of all the characters encountered before, who are gathering for a day at the races. That too is an evasion of the problems of living. Everyone greets Miss Fitt and ignores Mrs. Rooney, who for a fleeting moment exercises the freedom of her consciousness, only to fall back once more into her ready-made state of self-pity:

Mrs. R: Do not flatter yourselves for one moment, because I hold aloof, that my sufferings have ceased. No. The entire scene, the hills, the plain, the racecourse with its its miles and miles of white rails and three red stands, the pretty little wayside station, even you yourselves, yes, I mean it, and over all the clouding blue, I see it all, I stand here and see it all with eyes . . . [8]

but her voice breaks, as she "falls" once again:

> through eyes . . . oh, if you had my eyes . . . you would understand . . . the things they have seen . . . and not looked away . . . this is nothing . . . nothing. . . .[9]

The conversation turns to the attention of the group on the lateness of the train, and to the rumors concerning what has happened to determine that incident.

At last the train arrives, and Mr. Rooney makes his appearance. An old man, blind, only half alive and obsessed with death, he cannot remember that today is his birthday, nor that he is wearing the tie his wife had given him to celebrate the occasion. The married couple trek their weary way back to their home, he occupied with his thoughts of retirement, and she, intent upon forcing her love upon him. He refuses, of course, to embrace her in public; that would be unseemly, but then so would any show of affection in the utmost kind of privacy, which Mrs. Rooney continues to imagine, excercising the only kind of freedom left to her.

As they pass the cottage from which the music of the dead maiden had been playing in the morning, the disc has not been changed. After all, maidens die a natural death, like the girl of the song and the Rooneys' own daughter, or they merely change their status into that of frustrated married women; and the older they get, the more frustrated they become.

Along the way, Mr. Rooney gives a broken account of what has happened on the train, losing the thread of the story on account of their deviating curiosities; and when he finishes, he mysteriously begs his wife to believe in the truth of his story. Reason: it is a total fabrication, false in every detail.

Their mutual curiosity is brought forward to the topic of the next day's sermon. He is disappointed that it is not the fitting "How to be happy though married," because the vicar given to such ironic topics has himself passed on to the great beyond. Instead a replacement has announced for "The Lord Upholdeth All That Fall," a title that provokes deep ironic laughter from them both. Whether Beckett himself expected such a divine approbation of his play of the derived title, of course, no one may know; but the irony verily ripples.

Each day of their meaningless lives the couple have both fallen into an unpardonable ontological state. Mr. Rooney, imagining the benefits of his approaching retirement, states the condition well:

> Mr. R: On the other hand, I said, there are the horrors of home life, the dusting, sweeping, airing, scrubbing, waxing, waning, washing, mangling, drying,

> mowing, clipping, raking, rolling, scuffling, shovelling, grinding, tearing, pounding, banging and slamming. And the brats, the happy little hearty little howling neighbours' brats.[10]

Such activities are horrors, of course, only if one finds them a meaningless deviation from one's personal ideal of self-determination, which would be the attainment of the authentic life. Mr. Rooney is characteristically blind to the fact that his description fits his wife's daily activities, and his disapproving statement goes far to explain the couple's estrangement. He has an alternative, equally inauthentic, which is to continue working. On his job he can shut himself up in a closed office from ten to five and imagine that he has already died (as indeed, in a sense, he has), from which condition no one need be lifted up.

When the fallen pair are obstructed by their own neighbors' brats, Mr. Rooney, completely oblivious to his wife's concern for their own dead daughter, asks her if she were ever tempted to kill a child. In her continual daily brooding over the death of her daughter, like some laic *mater dolorosa*, she is of course guilty of the continued murder of the girl. Loves, even maternal loves, must, like lives, come to an end once an essence has been achieved by the person loved.

At the last moment the boy Jerry arrives with an object Mr. Rooney has let drop: a ball or something like it, which at first he denies to be his, but later claims only as something he always keeps about his person. A ball, perhaps, to tempt a child. The train had been delayed because a child had fallen onto the tracks, where it had been passed over by the speeding train, at about the same time, one supposes, that Mr. Slocum's car was running down a defenseless chicken. Mr. Rooney's story of the delay was indeed a lie; he had wreaked his vengeance on a defenseless child, who had all these long years served as a symbol for the child, his own, who had stolen the affection of his wife, her mother.

All That Fall is therefore a story structured on at least two significant levels. It holds together as the naturalistic account of an unhappy couple, mismated and separated into two hostile units. Their child, the cause of their mutual unhappiness, has died; and they are both reminded, in spite of themselves, of this real cause of their misery by the music of their neighbor. But neither can accept the truth. A chicken is killed at the same time the young child is killed by the speeding train. And in the next day's sermon they will hear that the Lord Upholdeth All That Fall.

The final revelation is withheld until the last spoken words of the broadcast, when we learn that a child has been killed on that train. There would be irony in that Sunday's sermon; for if its message were true, Mr. Rooney's action would be annulled by the efficacy of the Lord. But that sort of transcendency rings hollow to the ears of the earthly couple. It was not only the chicken, not only the murdered child, but all the characters of the play who had fallen; and such a fall can not be lifted up by the miraculous, merciful Justice of divine efficacy. The real fall has been suffered by the tragic couple, who will spend the rest of their days in the pursuit of the emptiness of their own shallow lives.

For the second level signification of the drama, we need only hear the tone of their conversations with themselves and their fellows to understand that *Dasein*, in its fallenness, lives out everyday in its very everydayness: empty of hope unless consciousness becomes fully conscious of itself. The mode of inauthentic human awareness is, in its state of mind, uncontrollably curious; in its understanding, ambiguous; and, in its expression, riddled with idle talk.

Beckett, of course, breathes nary a word about all this ontology. He doesn't have to; his characters have manifested it all in living their concrete ontic situation.

In this play, as in all of Beckett's works heretofore, philosophy lends its structure to the created literary world. The old master sorcerer has merely moved his tricks into a new medium, where the contact of voice and ear is as direct as the airwaves permit. The hearer's response, however, is quite the same: the understanding is of a structure, philosophical and literary, of a completely, uniquely expressed state of mind; here, that of lostness in an alien world, felt as the uncanniness of an inauthentic existence. And the pessimism of the world view expressed is overcome by an act of the listener's consciousness. Human salvation is still mediated through the dramatist's art.

Embers (1959) is a radio play structured on the same idea. Once again there is a couple whose unity has been destroyed by the appearance of a child who alienates the mother's affections from the father. But so far have these two fallen that Henry, the husband-father, is almost totally alienated, from himself as from his natural surroundings.

In as clear a presentation of schizophrenia as Beckett has devised, Henry has only one last ray of hope: to tell a story clearly so that others may understand who and what he is. But the alienation and self-entanglement of his inauthenticity is such, and the turbulence it creates so overwhelming, that he shall never find either the end or the signifiance of his tale, to which only the sea is a fitting background. His confused understanding cannot even grasp the significance of the embers his imagination evokes. He is stuck there because what their symbolism reveals is too horrible to bring to consummation.

The play opens with Henry's mad fit of commanding the sea's movements. He knows that each sweep upon the stony beach is to be followed by a recession into the swirling deep, but that knowledge does not jibe with his present state of mind. To the background of the moving tide, the audience hears his commands: "On . . . On! Stop . . . Stop! Down . . . Down!" His mania is to talk, and talking, to evoke those images that have obsessed his subconscious mind from the time he was a young man.

His own father had disappeared one day, apparently a suicide who had expressed his dissatisfaction with his family life by slamming the door behind himself and undertaking the voyage that permits no return. Since the body was never found, the family had their troubles with the legal consummation of his deed.

The audience hears only the sounds of the sea and the voice of the demented son apostrophizing his dead father, a presence, for what it's worth, to respond to the son's need of communication. Nothing in this world is what it seems, and none of the words succeeds in expressing a clearly conceived thought. And the reason becomes slowly apparent, that the nature of the real world and of the authentic personality has become grossly entangled in the tissues of a totally imagined, not to say fictitious, world, constructed by the hero's tortured consciousness.

In his present state of confusion the sounds of the sea are interpreted as those of horse's hooves, as if a horse could be trained to mark time in place. Slowly that image too will receive its schizophrenic explanation, but as the schizophrene's mind is laid bare, the audience's attention must move from image to disconnected image.

Following his father's death the protagonist's family had moved from one side of the channel to the other. The all too human fascination with the sea is such that once its rhythms have entered the consciousness one is powerless to remove oneself from their source. The ultimate voyage of the human transcendence, on the evolutionary scale, is from the sea to the sea. Besides this natural fascination and the symbolic atavisim of the seaside dweller, Henry has in addition the memory of his father's last words to his wayward son, "You're a washout."

Indeed, the inference is inevitable, the son will end as his father ended, washed out of life in the endless deep. Below the waves, his wife tells him, all is as quiet as a tomb. The sea only does quicker what the world does to every inauthentic consciousness. Even should the son go to the mountains of Switzerland to get away from it, the sound effects of that resounding sea cannot be erased from the troubled consciousness.

As if the reveries of the inauthentic consciousness were not enough, Henry attempts his redemption by inventing stories:

> Henry: . . . I usen't need anyone, just to myself, stories, there was a great one about an old fellow called Bolton, I never finished it, I never finished anything, everything always went on forever. *Pause*. Bolton. *Pause. Louder*: Bolton![11]

Bolton ultimately makes his appearance as the old man who has called his doctor, a Mr. Holloway, to his residence for a nighttime visit. The two consult, if that is the word, since Bolton wants company more than treatment, before a dying fire on a snow-driven night. Whence the title of the play.

Henry grasps for the words that come in bits and pieces, as they did to the narrator of *How It Is*. But unlike that novel, this story is never finished as the narrator's consciousness is already too far gone. Consider:

> Henry: . . . Bolton at the window, his back to the hangings, holding them a little apart with his hand, looking out, white world, even the spire, white to the vane, most unusual, silence in the house, not a sound, only the fire, no flames now, embers.
>
> . . . .
>
> . . . great trouble, not a sound, only the embers, sound of dying, dying glow, Holloway, Bolton, Bolton, Holloway, old men, great trouble, white world, not a sound.[12]

The narrative thread is broken by the narrator's thoughts of his father, and then refections on his own lost condition:

> Stories, stories, years and years of stories, til the need came on me, for someone, to be with me, anyone, a stranger, to talk to, imagine he hears me, years of that, and then, now, for someone who . . . knew me, in the old days, anyone, to be with me, imagine he hears me, what I am, now. *Pause*. No good either. *Pause*. Not there either.[13]

The radio listener will hear, later, as Henry's wife, Ada, makes her appearance, that Holloway is the name of the family doctor who had treated Henry's father; so "Bolton" is the fictive reincarnation of the father, and the peculiar behavior of the fictional character is explained by the mystery surrounding the old man's death.

Ada is concerned with Henry's growing tendency to talk to himself aloud. She counsels him to see Holloway for that condition, suggesting another identification between father and son, before the dying fire of his consciousness has become completely extinct. She asks him to think of their daughter, Addie, and of what kind of explanation may be given her for that strange conversation between his father and himself. His suggestion, that she be told he is praying to God and all his saints, is only partially false, since a story is still the calmative by which the uneasy consciousness releases its tensions and fears. But Addie is now too old to accept such an explanation, for it is obviously false (she can hear); and she is not mature enough to understand the sense in which it is true.

Ada's physical presence is rendered in the same way as Mrs. Rooney's interlocutors. The couple speak to each other, but at cross purposes. And the existence of the girl plays the same function in this play as the dead daughter of the Rooneys in *All That Fall*. Instead of uniting the couple, she succeeds in separating them, and in further alienating the father from the real world.

Her mother insists that she learn to play the piano, and to ride on horseback, although the girl has no talent for either art. Reason: the mother never had the occasion to learn these things herself. Thus, while the mother may project her own life onto that of her child, the father is left both wifeless and childless. The hooves he hears at the seaside are now abundantly clear.

The girl has never enjoyed the walks, nor the conversations she was forced to have with her father, so she refused to him the same role she now reluctantly plays for her mother: the projected life of the parent in the child. As a child himself, Henry enjoyed his walks with his father, and could never understand the parental silences at home; but failing to recapture the past relationship with his own father in the present one with his daughter only serves to make his obsession with his father's death the more compelling. If he could tell that story in the Bolton tale, he might succeed in liberating himself. But he cannot, and does not. Only the scene of the dying embers comes clearly to his mind; and he, being deprived of the end of the story, cannot understand its significance.

Separated from his father by an unrationalizable suicide, and from his daughter by his wife, he is separated from his wife by their mutual inauthenticity. In his lost state of mind, Henry cannot relate Ada to his father, nor himself to her. As the fire of their own love had been reduced to an ember, Henry was driven inward to his own dreams and meaningless self-dialogues.

When he is brought away from the sound of his own voice and can focus his attention on the presence of his wife, he can see her only in terms of their past, when they loved one another.

Henry even succeeds in recalling an image of a scene, some twenty years ago, when he first seduced her, always fittingly demure and secretive. The "Don't, Don't's" she now uses only to dissuade him from indulging in his fancies, was then used to parry his advances, and when the evocation dies away, her attitude hasn't changed a bit. Her next words begin with the fatal negative prescription: "Don't sit there gaping. Sit down."[14] They can both recall the "hole" where they first did it, but that was a dip in the ground that hid them from the gaze of the curious. The hole is still there, but like all earthy holes it has become filled in by time.

The ambivalence of both their attitudes is clear by the *double entendre* of their words.

Henry: Yes, you were always very sensitive to being seen in gallant conversation. The least feather of smoke on the horizon and you adjusted your dress and became immersed in the Manchester Guardian. *Pause*. The hole is still there, after all these years. *Pause. Louder*: The hole is still there.
Ada: What hole? The earth is full of holes.
Henry: Where we did it for the first time.
Ada: Ah yes, I think I remember. *Pause*. The place has not changed.
Henry: On yes it has, I can see it. *Confidentially*: There is a levelling going on! *Pause*. What age is she now?[15]

The embers, then, are equally symbolic of a dying love.

On each of Henry's suggestions, to get up and go somewhere together, alone, for a hike or for a row on the sea, Ada can only disappoint his expectations: What would Addie think if she came here only to find her father gone?

Henry's father and Ada, had they ever met? The question follows Ada's warning: "You will be quite alone with your voice, there will be no other voice in the world but yours. *Pause*. Do you hear me?"[16] As if in answer he replies:

Henry: I can't remember if he met you.
Ada: You know he met me.
Henry: No, Ada, I don't know, I'm sorry, I have forgotten almost everything connected with you.[17]

That statement is true, and the ultimate revelation of love's dying embers.

Ada can recall her meeting with Henry's father, and recounts the incident. Henry cannot remember because he was not there, not having returned home the night before the day of the meeting. And prophetically she recalls a posture of the old man, sitting as if all breath had gone, stiff as a stone in contemplation of the sea. From the manner of the father's death the audience knows how that posture ended; and what's worse, Henry now looks exactly like his father.

Henry's final speech reveals the confused state of his mind. Imploring his wife to stay with him, recalling his father, hearing horse's hooves, recalling a scene of his daughter's earlier childhood, beginning his story of Bolton once again: nothing promises a satisfactory end, not even thinking about the last words in his father's little book.

Henry: ... Little book. *Pause*. This evening ... *Pause*. Nothing this evening. *Pause*. Tomorrow ... tomorrow ... plumber at nine, then nothing. *Pause Puzzled*: Plumber at nine? *Pause*. Ah yes, the waste. *Pause*. Words. *Pause*. Saturday ... nothing. Sunday ... Sunday ... nothing all day. *Pause*. Nothing, all day nothing. *Pause*. All day all night nothing. *Pause*. Not a sound.

Sea [18]

If Henry were capable of understanding the actions of his father, he might have understood his own self-entangled turbulence; but then the Bolton story was never finished.

The compelling image of this play is enshrined in its title, *Embers*. The dimming light and sound of a dying fire is first presented within the context of the Bolton story, in which the embers of an actual fireplace fill in the background of the relationship between Bolton and Holloway. But Bolton is Henry, is his father. Within Henry's own consciousness the light of reason and the heat of passion are likewise slowly dying. As a story teller he cannot finish his tale; it too will simply go out. Estranged from the father he loved, from his wife and daughter, both of whom he is prevented from loving, from any true connection with his real past, as from any meaningful future, Henry's end is predictable; he will fulfill either one of Ada's premonitions: to become simply a voice heard by no one, or go on forever stiff as a stone, lost in the fascination of the sea until, like his father, he ends it all.

What was the syndrome of the inauthentic, fallen consciousness? It lives day after day in its everydayness, in a tempting, tantalizing, alienating, self-entangling, and turbulent loss of connection with the commitment to be. That is what the voices of Henry, Ada, Addie, and her two coaches reveal to the disturbed audience in one more of Beckett's "philosophical" plays. And as always the reader's consciousness undergoes the "tragic lift."

Beckett's fondness for the multiply stratified symbol is nowhere as fully indulged as in *Embers*. He calls attention to the play's symbolism in the title of the piece in which a single surface vocable comes to mean (i) the physico-chemical occurrence in Henry's story, (ii) the socio-psychological event of Henry's and Ada's love, (iii) the psychological state of Henry's inner consciousness, (iv) the impending end of Henry's living self-transcendence; and all of these in their interrelatedness are themselves symbolic of the anagogical stratum of significance that provides the structure of the play.

By now the symbolism of the ontic conditions of characters has been amply commented upon; since it is the ontological structures of human disclosedness that make possible a certain kind of experience, the inference from the ontic situation to the ontological significance is almost psychoanalytical in its procedure. But in Beckett's plays the psychoanalytical procedures are, typically, existential in nature. The term employed for such reasoning by Heidegger is *Daseinsanalytik*, existential analysis.

In *Words and Music* (1962) the technique is applied to fictional characters who are themselves symbolic of an author's consciousness in the act of creation. If we continue to understand the fictional world created by the spoken word as a context of significance composed of surface and depth structures, the sensuous properties of the words as spoken and the meanings they intend, the "music" of any fictional world is the poetic tension of the sound patterns heard on the very surface of the expression. Whatever idea may be expressed in the words, as constitutive of the word's depth, the surface is never insignificant in itself; and should the words not purvey any meanings at all, they would approach the phenomenon of "white noise." It is precisely the task of the author to fit the vocal gesture to the intended meaning in such a way as to create a significant style.

Beckett uses three "characters" in *Words and Music* to dramatize this task of the creative dramatist. The actual music of the original production was produced by John Beckett. It is a series of musical motives whose expressiveness is gauged upon the themes of the ideas suggested by the words spoken by the character called "Words." Since Music and Words tend to enjoy their own virtuosity, they often work at counter purposes, and so need a third character, the conductor (author, dramatist), called in this play "Croak." Croak refers to Music as 'Bob,' and to Words as 'Joe,' abjuring them all the while to be "friends." To him they represent "comforts," and "balms." If they are to work together, they cannot continue to pursue their separate virtuosities.

The tone of the play is light and satirical, with the barbs of the satire aimed at the "philosophical" dramatist looking for an authentic self-expression; in Heideggerian terms, once again, what the creative writer seeks to achieve is the human disclosedness that reveals itself simultaneously as it lays out the structures of its uniquely defined world. But to do this within the play, Croak, the musical conductor, must work with the words and the music as every man must work with his own self-consciousness, partly master and partly slave, in order to achieve his own self-identity.

The project is to produce a musical poem on love for the aged. Irony upon irony, then, *Words and Music* will repeat the common theme of *All That Fall* and *Embers*: the tragic essence of recapturing a lost love, but this time as something to be expressed, not as a first order literary or dramatic expression. For this reason, the play is dramaturgical, i.e. dramatic to the second degree, like *Cascando*.

The play opens to the sounds of a small orchestra tuning and an actor reciting his words on the theme of passion. The topic is sloth, delivered in a prosaic, medieval philosophical style:

Words: ... Theme ... sloth. *Pause. Rattled off, low*: Sloth is of all the passions the most powerful passion and indeed no passion is more powerful than the passion of sloth, this is the mode in which the mind is most affected and indeed – ...[19]

It is as if the poet is rattling the memory banks of his brain, competing with the tuning orchestra whose charivari disturbs the process. The double redundance escapes the notice of the person looking for a clear idea; and, as if to clarify, the voice continues:

Words: ... by passion we are to understand a movement of the soul pursuing or fleeing real or imagined pleasure or pain pleasure or pain real or imagined pleasure or pain, of all these movements and they are legion sloth is the most urgent. ...[20]

The diatribe continues with two more redundancies on the urgency of the passion, until Croak makes his appearance.

Croak deserves his name, since his function is to let the other voices come to expression, without himself being heard in the final rendition of the projected work. Having nothing to say himself and breaking into the lines of the others, his voice is a constant croaking. He is, however, the man in charge, and will constantly goad his voices on to establish the work.

He arrives at the rehearsal late (illustrating the sloth of Words), having seen a vision of a woman's face in a tower, and announces the theme for today's production: love. On this cue, Words, now called "Joe," restates his diatribe on the passions, substituting 'love' for 'sloth.' To the uninitiate this process may seem absurd or a kind of playful nonsense, but the dramatist is deadly serious. Beckett's readers will recall the loves of the slothful Belacqua.

First of all, Joe's attitude must change from a slothful play with his words to a lively, loving concern for them; and then it may be perceived that the theme of any developing work may change in a moment's notice. In the end every theme must be reduced in importance to the starting urge for the developing composition, whose context will further reduce it to a single counter of the significant context.[21] The significance of that context will be controlled by the specific relations between the surface of the words (complemented by Music) and the images used to concretize the idea.

Joe continues his search for the theme's definition:

Words: . . . what is this soul that more than by any of its great movers is by love so moved? *Clears throat. Prosaic*: Love of woman, I mean, if that is what my Lord means.

Croak: Alas!

Words: What? *Pause*. Very *rhetorical*: Is love the word? *Pause. Do.*: Do we mean love, when we say love? *Pause. Do.*: Soul, when we say soul?[22]

Stymied here, Croak calls for the help of Music, as Words insists: "Do we? . . . Or don't we?"[23] Music enters with his same old refrain, completely out of tune with the effort of the words, driving Croak to distraction.

The conductor then proceeds to suggest the appropriate images, beginning with age; and Joe takes the hint:

Words: *faltering*: Age is . . . age is when . . . old age I mean . . . if that is what my Lord means . . . is when . . . if you're a man . . . were a man . . . huddled . . . nodding . . . the ingle . . . waiting –[24]

But Bob remains silent until Croak repeats his request for music expressive of the tensions of age. Joe and Bob continue their struggle with little unison, until the poem begins to take its imagistic shape:

Words: Age is when to a man
Huddled o'er the ingle
Shivering for the hag
To put the pan in the bed
And bring the toddy
She comes in the ashes
Who loved could not be won
Or won not loved
Or some other trouble
Comes in the ashes
Like in that old light
The face in the ashes
That old starlight
On the earth again.[25]

But the point is not just old age, where the fires of youth have already been reduced to cold cinders, the character of the loved one should be revealed in the face: more prosaic rumblings, followed by the poetic images, reinforced by the Music.

In the end, Joe (Words) defines his image in harmony with the music, beginning with the prose statement of the face's characteristics and ending with the lover's physical reaction to the beloved's entire body.

Words: – the brows uncloud, the lips part and the eyes ... *pause* ... the brows uncloud, the nostrils dilate, the lips part and the eyes ... *pause* ... the lips part, a little colour comes back into the cheeks and the eyes ... *reverently* ... open. *Pause.* Then down a little way ... *Pause. Change to poetic tone. Low*: ...[26]

Each pause, each new start brings the poet closer to his vision, as his words break the silence, until we hear the finished product:

Words: Then down a little way
through the trash
Towards where
All dark no begging
No giving no words
No sense no need
Through the scum
Down a little way
To whence one glimpse
Of that wellhead.[27]

But Croak can take no more, and leaves his voices to themselves.

Alone, they complete their task, and Words indicates his final approval with a deep sigh. In the process, an idea of the passion of love has been transformed into a list of suggestive images expressive of the urgency of this most urgent of all the passions motivating the human soul.

The very subject of the poem – the silent communication through an act of love – is demonstrated by the process of the poem's generation; and the completed poem unites poet and hearer in a similar process of communication according to which two minds fold into one as the separate consciousnesses intend the same unique universe created by the tension between linguistic surface and ideational depth.

If *All That Fall* and *Embers* are constructed on the philosophical idea of the inauthentic consciousness, *Words and Music* illustrates the process by which the authentic consciousness of the poet comes to its personal expression, revealing its own self-identity as it creates its imaginative world. The apparent strife between Words and Music is the price one must pay to discover the expressive idea, and one will never know whether one has succeeded or not until the context of significance has been laid out for hermeneutical verification. That is the point where creator and his audience achieve unity in communication, hearer responding to speaker in the same way as the speaker responds to self.

The calmative effect of the well-told story has no other philosophical explanation: the labor of creation is the pain one goes through to relieve the pain of an inauthentic existence. And in projecting themselves as creators of a novel's significance, literary artists lay the foundations for an enriched society; for the fruit of their travail is eventually transformed by others into a new cultural entity. In order to achieve this kind of renewal of the general culture, they must continue to turn inwards upon themselves, to put up with their own, and the limitations of their medium, to suffer the loss of easy, but fruitless companionship, to accept the uncertainties of self-determination against the certainties of the hollow determination by others, and thereby obliterating all that is non-essential to their creations, to engage their senses along with their minds totally in their tasks, no matter how schizophrenic they may be made to appear in the process. That is the poet's plight.

The self that speaks and the self that hears are indeed at the moment of creation the same self responding to different inner voices. Call the one "Words" and the other "Music" if you will, the third voice of that "pretty trio" is the one that decides, he who in a moment of reflection regards what he has done and sees that it is good – or, *horribile dictu*, irretrievably bad. Should that occur, art could not succeed in transmuting life's pain into aesthetic pleasure.

To live authentically is to accept the risks of self-determination, to create a better world if one's own self-projection forces one to esteem that the universe we all live in floats under us on a ball of muck. The price the poet pays is the oddity of his appearance; but the value we all gain is a newer, more humanized world.

This means only that one must give up the easy lure of an inauthentic life if one is to achieve fulfillment in the authentic. And nothing guarantees success in the attempt; for the master poet, whatever the fruit of his efforts, is always submitted to the condescending gaze of his servant readers. That too is part of the poet's unhappy plight. Beckett, the artist, does not speak discursively about this plight; he shows it in the plights of his "characters."

CHAPTER X

ENGLISH VOICES FOR THE STAGE

When the trilingual version of Beckett's dramatical works was published in Germany (the French plays, 1963; the English, 1964), the principal English works were *Krapp's Last Tape* (1958), *Happy Days* (1960–1), and *Play* (1963). Each is structured on the idea of a self-consciousness entangled in existential inauthenticity and desperately incapable of working out a solution to what Sartre has called "our concrete relations with others."[1] As is well known to historians of recent continental philosophy, Sartre's phenomenological ontology was derived from Germanic sources; principally the works of Husserl, Hegel, and Heidegger.

Where Sartre's version of fundamental ontology places emphasis upon the strife between two absolute consciousnesses seeking to dominate the factual limitations of their freedoms, he is treating of Heidegger's *existential, Mitdasein* (our being with others), in a peculiarly Hegelian manner. It is as if the dialectical progress of the consciousness as treated by Hegel in the *Phenomenology of Mind* had been arrested at the stage of the conflict between master and slave. And in the plays of Beckett, *Godot* and *Endgame* may give the same impression.

The impression would be false, however, because Beckett never forgets what seems to have slipped Sartre's notice: that the master and slave consciousnesses are only external symbols of the dual aspects of the single self-conscioueness as it interrelates with all the phenomena of the world, including the existence of others, ordering them willy-nilly into some kind of personal significance. Thus, the meaningful strife for consciousness is internal, and the source, ultimately, of a profound personal unhappiness.

In Hegel's account, of course, the dialectic of self-consciousness moves beyond the conflict of master and slave, as in its search for identity consciousness goes on to liberate itself in stoicism. But stoicism gives way to skepticism before self-consciousness achieves its ultimate apotheosis, in another return to itself, as essentially unhappy. What before was felt as a profound malaise, and what is experienced in all true cases of schizophrenia, now becomes the definition itself of the human self-consciousness. In this condition it must return to the world in search of whatever significance it may experience in life. And as Heidegger has shown, the ordering of the phenomena of the world may take place either authentically or inauthentically; either one determines oneself what one is to become (Dasein's existentiality), or one's existence is determined by the restrictions of the world of everyman.

Moreover, whether one's existence is authentic or inauthentic, Dasein exists in a world as itself a phenomenon of disclosedness. In its states of mind, its understanding, and its discourse, it reveals itself simultaneously as it reveals the nature of its world. For this reason, once Beckett had discovered the philosophy (either by reading the masters or by consulting his own experience) his heroes embody the notion of existential disclosedness (*Erschlossenheit*).

In the later novels and in the dramaturgical plays the heroes are in the throes of disclosing their own imaginative worlds, indefinable except through the words used to create them. In the earlier novels and the later straight dramatic plays, the plight of the hero is to be cut off from the authentic expression of the true human disclosedness, and to find himself lost in the pull of Dasein's *Verfallenheit*, i.e. lost in a meaningless present, cut off from the past by the flow of time and incapable of projecting a future that would be the rewarding sense of the living present. Such, it should be recalled, is the interpretation of human existence from the existential analytic point of view. The term 'fallenness' is, however, an ontological and not a theological concept.

How each person, each character, falls depends upon the whims of his curiosity, which clouds all issues in ambiguity and comes to expression in idle talk. Whether that talk is with oneself or with others, the difference matters little, since all speech is a dialogue, and at the heart of self-consciousness in the most intimate "monologue" speaker and hearer are the same. Indeed, when inauthentic personalities attempt to communicate each speaks only to himself of things the other already knows. Whence, the disjointedness of their "conversation"; the one may talk, but the other refuses to listen. The same is true for the inauthentic monologue.

In the novels and plays embodying inauthenticity, the hero functions as an active human consciousness that finds itself ontically lost because it has ontologically fallen. Where the mode of existence attributable to the authentic human personality is such that in its being its being is in question (what one is to be is not yet determined), the inauthentic personality exhibits a curious forgetfulness of Being, absorbing itself in the immediate world of its surroundings where curiosity, ambiguity and idle talk have replaced the resolute decision to be oneself. Result: self-alienation. In this way Beckett's characters give testimonial weight to Heidegger's claim that, under the influence of philosophical positivism and modern technologies, contemporary life has developed a characteristic separation from its sources. But their lives nonetheless exhibit the ontic conditions of the ontological structures that make them possible.

For this reason the ontic conditions easily symbolize the ontological structures. Indeed, as novelist and dramatist Beckett uses these ontological *Existentialia* as structures for his own creative works, and thereby supplies the understanding found lacking in the suffering heroes, who, upon superficial analysis, seem to illustrate nothing more than the absurdity of the human condition. But that too would be an inauthentic understanding.

The more careful reader, supplying in hermeneutical expansion these same ontological structures in his interpretation of the works, comes to an understanding of the all-pervading symbolism, arriving at authentic communication with the author. In this way two self-consciousnesses bridge the gap of essential separation. Writing is itself the initiation of an act of love; and should by accident that writing be about love itself, it would do no more than to illustrate the essential moves of consciousness: to proceed outward from itself and then to return to itself, enriched in authentic cases but destroyed in the inauthentic, for its journey.

Returning to itself, consciousness exhibits its essential move inward; but that move may be meaningful only if it has not lost its bearings with what it has found on the outside. In *Krapp's Last Tape*, the hero attempts to move inward, but his consciousness has lost its worldly bearings, both in space and in time. The play was written for a man and a tape-recorder: a single consciousness, present and past.

It is curious that neither the French nor the German translation of this play contains the name of Krapp in the title. *La dernière bande* and *Das letzte Band* both fail to capture the shitty suggestion of Krapp's name. In the stage directions he is described as a "wearish old man": very nearsighted, but not wearing spectacles, hard of hearing, having a cracked voice with a distinctive intonation, and walking laboriously; no mention is made of his chronic constipation – a condition he constantly exacerbates by an excessive consumption of bananas. That condition is revealed in the reminiscences recorded on the tapes, and in his current reflections on them.

Krapp is as pleased to hear "slight improvement in the bowel condition" on his tapes as he is to relive the moments of his lost loves. Everything comes to Krapp, slowly; even the word 'spool' is rolled from his tongue as 'spooool,'[2] the vowel held so long that one may hear 'stooool' instead. The word is to be delivered "with relish," as befits the constipated man. That's how far inward Krapp's consciousness has turned. The French 'bobiiine' captures the rhythm of the moving bowel's effort.

The space of Krapp's world is his den, brilliantly lighted about the desk, but in total blackness around the periphery. Nothing to be seen, therefore, save the basic black and white contrast that symbolizes the least determinate form a consciousness may be aware of. This night, some night in the distant future, there are no sounds except those made by Krapp past and present. He is alone, with his bananas, his whisky, and his tape-recorder.

For years he has had the habit of recording the highlights of the passing year: along with the condition of his bowels, his resolutions, the state of health of his father and mother, his amorous conquests. Indeed, the deaths of his parents are duly recorded along with his vain resolutions to consume fewer bananas and alcohol, and to slow down his all too active sexual life. That is found in the record of his past; now, some thirty years later, he continues to indulge his passion for bananas and alcohol, and has largely lost his ability to have any but the most vicarious kind of sexuality that is the recall of his past exploits. With no future other than the curious designation of the play's first stage direction, "A late evening in the future,"[3] Krapp is lost in the present attempting to establish a link with his past self.

Had his past resolutions not all been empty, of course, he would have established the continuity between the moments of his self-consciousness; for it is by a present resolve that one determines what one is to be in the future. The present moment of Krapp's existence which is a future to his past is lost in an empty effort to relive that past. And not even the tape-recorder, one of those technological achievements of the modern age that makes man forget his commitment to authentic expression, can turn the trick. When Krapp has gone on the tape, only Krapp can come off.

Krapp rummages through his boxes of tapes, consulting his catalogue for a specific spool that cannot be erased from his memory. Box three, spool five; always "with relish" on the 'spool.' That is the one recording the death of his mother, the black ball whose significance he can no longer recall, and a "memorable equinox," no longer remembered. It is the notation, "Farewell to love," that clinches his decision to replay the spool.

From the tape the audience hears the voice of a younger Krapp, recording the events of his thirty-ninth birthday. As usual he has spent his day at the local winehouse, alone before a fire. He has eaten three bananas, and realizing that they are bad for a man of his condition, has refused a fourth. Already then he has made his resolution to give them up – to no effect, since before listening to the recording of the resolution the old Krapp has just eaten another. The way he eats them sucking the ends, leads one to suspect a vicarious act of fellatio. His love affairs are referred to at age thirty-nine as "post-mortems," for the act of love is for him primarily a release of physiological tensions in which the orgasm, pursued for itself, becomes the natural symbol for the end of a life.

His essential disclosedness is expressed in the words that reveal his character as they reveal his world:

These old P.M.s are gruesome, but I often find them – *Krapp switches off, broods, switches on* – a help before embarking on a new ... *hesitates* ... retrospect. Hard to believe I was ever that young whelp. The voice! Jesus! And the aspirations![4]

The voice speaking, Krapp at thirty-nine, comments on hearing Krapp's voice as a still younger man – exactly the same relation as that between the old Krapp listening to his voice at thirty-nine. Thus, three moments of time in the same life: the present, a future for the voice of thirty-nine, the past future for the voice of the younger man released from the tape. The character of the voice has changed, but now the old man of sixty-nine can only look back to his former aspirations as totally senseless; otherwise he might have believed that he was that young whelp. As it is, his present moments can only be projected toward a future that will be lost in the search of his past.

According to Krapp's voice at thirty-nine, Krapp was at the crest of his transcendence: no place to go then but down, and what we hear proves him to have been right. His future remains in the past. The tape is incoherent, and its playing is interrupted to allow Krapp to indulge his unbridled craving for drink. That too was to be avoided according to a past resolution.

Slowly the passage of interest approaches:

When I look –

. . . .

back on the year that is gone, with what I hope is perhaps a glint of the old eye to come, there is of course the house on the canal where mother lay a-dying.[5]

but it is broken by the tape's reference to his mother's "viduity." Krapp has forgotten the word's meaning – the state or condition of being or remaining a widow, and other associated meanings, so Krapp searches down the word in his dictionary to satisfy his curiosity. Switching back on to the appropriate spot, he listens to the voice of his past consciousness:

... after her long viduity, and the –

. . . .

bench by the weir from where I could see her window. There I sat, in the biting wind, wishing she were gone.[6]

Around him at that moment were only a few habitual denizens of the place, "nursemaids, infants, old men, dogs."[7]

Krapp came to know one of the nursemaids well: "One dark young beauty I recollect particularly, all white and starch, incomparable bosom, with a big black hooded perambulator, most funereal thing."[8] It was her eyes, like chrysolite, that attracted him, always looking at him, but immediately diverted when he took notice of her. Such an interplay between two consciousnesses, the one female, the other male, will not long be without its consummation; and it is duly recorded:

– upper lake, with the punt, bathed off the bank, then pushed out into the stream and drifted. She lay stretched out on the floor-boards with her hands under her head and her eyes closed. Sun blazing down, bit of a breeze, water nice and lively. I noticed a scratch on her thigh and asked her how she came by it. Picking gooseberries, she said. I said again it was hopeless and no good going on, and she agreed, without opening her eyes. *Pause.* I asked her to look at me and after a few moments – *Pause* – after a few moments she did, but the eyes just slits, because of the glare. I bent over her to get them in the shadow and they opened. *Pause. Low*: Let me in. *Pause.* We drifted in among the flags and stuck. The way they went down sighing, before the stem! *Pause.* I lay down across her with my face in her breasts and my hand on her. We lay there without moving. But under us all moved, and moved us, gently, up and down, and from side to side.[9]

The "Let me in," given the sexually symbolic description of the nature surrounding the trysting couple, may have been too strong; certainly too brutish, in view of the lady's decorum, to persuade her to part her legs. In his French translation, Beckett specifies that the entrance is of the image of his body into her eyes, just released from the glare. Krapp wanted his love to look at him, to be as fascinated by his presence as he was by hers; indeed to find him as lovable as he had found her. And their union, one can only suppose, was as tender as the natural movements about them. The ambiguity is, once more, characteristic of the inauthentic personality's understanding.

In rehearing his old account, Krapp loses patience with his thirty-nine year old experience and places on a new tape, his last, in a gesture to disavow his former clumsiness. But that gesture is hollow: nothing comes but a series of incoherent statements, beginning with the familiar repudiation of his past. The thirty-nine year old is now, with the passage of time, nothing more than a new young whelp; the tryst recorded happened thirty years ago, and since, a lot of water has flowed beneath the bridge. There were other loves, no good to be sure, but better than masturbating, as the following "post mortem" indicates:

Fanny came in a couple of times. Bony old ghost of a whore. Couldn't do much, but I suppose better than between finger and thumb. The last time wasn't so bad. How do you manage it, she said, at your age? I told her I'd been saving up for her all my life.[10]

But that lie must have been apparent even to Fanny; at any rate, their encounters had no significant sequel, for Krapp's next immediate remembrance is his having attended Vespers, in old age as in his youth. Characteristically, he had gone to sleep and fallen off the pew.

Precisely, we note, as he had thirty years ago when he loved and might have won his heart's desire, whom he rather fatuously rejected as his social inferior. From that time on he could easily conquer, but not bring himself to love, others – mere pretenders to that tender nursemaid's hold on his life. Oh, to be again! We hear the voice of the sixty-nine year old pining over his lost love:

Be again on Croghan on a Sunday morning, in the haze, with the bitch, stop and listen to the bells. *Pause*. And so on. *Pause*. Be again, be again. *Pause*. All that old misery. *Pause*. Once wasn't enough for you. *Pause*. Lie down across her.[11]

The words of the thirty-nine year old voice refuse to be obliterated from the old man's consciousness. He tears off the new and last tape, to replay the words recording the memory of his lost love, whose "funereal" perambulator at their meeting might well have been laden with too much significance: a child that would have been theirs might have gone on to kill their love, and he failed to take the risk.

No matter how long he lives, Krapp will bear the regret of not having committed his freedom to the love of his life. Having convinced himself that it was no good to go on, given the disparities between their conditions – and his own fear of the consequences, he gave up his last chance for earthly happiness, a happiness that would have been his had he resolutely chosen it. But lacking the resoluteness, he lost his grip on life; his understanding was diverted, and confused by the platitudes of idle talk. On that day thirty years before, Krapp had fallen into the inauthentic world of everyman, there to remain for good. His transcendence ceased, and the rest of his days were spent indulging his appetites and reflecting on his past reflections on a still more remote past, with the help of his tape-recorder.

We can be sure that his last tape will bring him back to box three, spool five, to relive once more that day he wished his mother dead without redirecting his oedipal drive to the girl who might have allowed him to become a man for all that. The flesh indeed was willing, but the spirit was weak.

The salvation from existential, ontological fallenness is the response of conscience, that call from the future to the present self, to commit oneself to a personal decision. But alas, the call of conscience can only be heard in the silence; it cannot be recorded on a tape. Krapp's tragic fault lay in his never having had the courage to be, and of not knowing that salvation was at hand, in the silence.

In *Happy Days* a second unhappy consciousness, made that way by a love gone astray, almost constantly attempts to escape its condition in two dramatic acts of barely interrupted idle talk. There are two characters, Winnie, the fifty year old magpie who is well preserved, but to no purpose – since she has renounced all save the external trappings of her sexuality. Having successfully killed her husband's love, she can now do nothing but continue to indulge her bent for male-castrating conversation. Her ontological condition is symbolized by her relation to the earth and her possessions.

In the first act she is buried above the waist in a mound of sand; and her possessions, mostly the feminine regalia for making herself more attractive – toothbrush, mirror, comb and brush, lipstick, are stored in a bag she had been given for shopping. Thus, once intended for the use of them both, the bag now lies beside her mound to preserve her personal articles, which she uses to increase her ambivalent hold on the masculine world, now that her *mons* is covered over with sand. Her parasol protects her from the sun, and the Browning revolver she has confiscated from her husband to prevent his suicide lies prominently on the ground for everyone to see. "Brownie" it is called. Will Winnie win? Or will Willie exert an act of his will?

The entire play grinds by in the immediate present. Winnie is awakened by a bell to begin her morning prayer. For her it shall be "another heavenly day,"[12] but only because she is oblivious to everything except her self. She too has become entangled in the world of everyman. Her interest in the events around herself is expressed in a fruitless curiosity by which her consciousness jumps from one thing to another without ever being brought to fruition in the least kind of knowledge, and which always comes back to her own well-being, the very question of which denies the stability, the equanimity her opening words falsely proclaim. Completely alienated, she cannot perceive the hollowness of their ring. Feigning happiness, and believing in it herself, she is the living incarnation of Sartrean bad faith.[13] But bad faith is bad because it is never wholly convincing, and Winnie will demonstrate to Beckett's audience that she has failed to convince herself of the goodness of her life.

She fails, of course, because her speech and her understanding reflect the empty curiosity of her state of mind. She needs another to hear her words; and her husband is there, as he had been in the past, present at least to give her confidence to continue her monologue, and at times responding to supply the meanings of her words that demand some kind of relationship with the events going on in the world. He too is fallen, but not quite as profoundly as his wife. He crawls on all fours behind the mound in which Winnie is half-buried, and sleeps in a hole in the ground. For his own connection to the world he continually reads the newspaper.

Following her opening prayers, Winnie begins by directly speaking to herself: "Begin, Winnie. . . . Begin your day Winnie."[14] And there follows the daily round of everyday's events: toothpaste, toothbrush, inspection of teeth, inspection of toothpaste, toothbrush, inspection of teeth, inspection of toothbrush: curious, that; made of "guaranteed genuine pure" hog's setae, but she will need a magnifying glass to read that advertisement. The inspection of her world continues: spectacles, too weak for the fine print, the condition of her eyes, parasol, revolver, tonic, handkerchief in and out and back into her bodice. She goes on to contemplate her own condition: no complaints, too much to be thankful for, slight migraine, but relieved by the many merciful gifts of providence; never any better, never any worse; no pain. In handling her parasol, she lets it fall behind her. It too was a "wonderful gift" – from Willie, her husband, no doubt, who hands it to her again without letting himself be seen. For that she is thankful once more, and admonishes him not to curl up in sleep again. The overtones of that suggestion are freighted with naturalistic significance: no doubt he has failed her more than once, when she needed his sexual attention. He preferred to slink from his duty into Belacqua's unmoving position.

The tonic is good for "loss of spirits . . . lack of keenness . . . want of appetite,"[15] and Willie needs such a remedy as much as she does; so she drains off the rest of the bottle in one swig, and throws the empty bottle behind her. It strikes Willie on the head, as she rummages for her lipstick. If the gestures were not so deeply tragic, they would appear as low burlesque comedy. She looks for an old saw, which she refers to as "that wonderful line" from the classics to cover the situation: "Oh fleeting joys . . . oh something lasting woe,"[16] never removing her attention from the rouging of her lips. For Willie, their power to attract never needed any enhancement.

His head appears, bloodied by the flying tonic bottle; and he covers it with a handkerchief, and the handkerchief with a boater decorated with his club ribbon. She apparently misses this, for she tells him to put on his drawers before he becomes singed by the sun. Instead, her words inform us that he is rubbing in some ointment, and she tells him to work it in well, on both sides of some member or other of his body. Since he continues doing what he wants to do anyway, she concludes, in apparent bad faith, "Oh this is going to be another happy day!"[17] as she continues to paint her lips.

It is a happy day for her when Willie pays her some attention, but he is apparently more interested in the events of the world. His newspaper appears, and he reads: about the death of the Most Reverend Dr. Carolus Hunter, dead in his tub. That notice puts Winnie off chasing the glories of her past. To her the reverend was Charlie Hunter, a former beau; from him, her awareness recedes to other masculine acquaintances, to her first ball, her second ball, and to a man whose name she has forgotten: a Mr. Johnson, Johnston, or Johnstone; she can remember the scene of their embrace, but the name is beyond recall.

Willie continues to read, and what he reads sounds like a regret on his own past: "Opening for smart youth." "Wanted bright boy."[18] Had he himself been brighter, his present predicament might never have occurred. But the heart has it reasons.

Unconscionably, Winnie's curiosity is still struck by the hog's setae. The connection, of course, was Mr. Johnstone's mustache; but she is not aware of this, being too concerned with the platitude that covers all situations:

> That is what I find so wonderful, that not a day goes by – *smile* – to speak in the old style – *smile off* – hardly a day, without some addition to one's knowledge however trifling, the addition I mean, provided one takes the pains. . . . And if for some strange reason no further pains are possible, why then just close the eyes – *she does so* – and wait for the day to come when flesh melts at so many degrees and the night of the moon has so many hundred hours. *Pause.* That is what I find so comforting when I lose heart and envy the brute beast.[19]

Accustomed to her habits, Willie is not listening, but looking at a pornographic postcard instead. As she asks to see it, he shows it to her: genuine pure filth; enough to "make any nice minded person want to vomit!"[20] But Winnie is trapped, her curiosity won't let her drop the scene: "What does that creature in the background think he's doing? *Looks closer*: Oh no really!"[21]

Finally she lets the card drop, but it brings her attention back to the hog's setae. What is a hog? A sow she knows, having been one; but a hog? Willie blows his nose. Uncouth of him; it would have been better for her to let him sleep.

Once more centered upon herself, she realizes, not in a mock horror, that she has not yet done her hair. Where is the comb, the brush? Nowhere to be found. How does one refer to hair – on one's head, that is? It or them? To a person losing his hair, of course, one will think of it as "them," but Willie replies, "It." Another sign of a happy day; he has deigned to reply, and she has put on her hat. Now Willie may crawl back into his hole.

But she will need him again. She sees an ant crawling, and through her magnifying glass she notices that it is carrying something that looks like a little white ball. What in the world could that be? An egg, says Willie. Eggs, eggs, eggs; "formication" in process. Willie laughs at his joke, and she joins in the laughter. Still another sign of a happy day; but were they laughing at the same thing? Yes and no, as Willie has finally scored his first point for the day; for he was actively contemplating what his wife will not admit she is capable of thinking of: her own blood fulfilling fornication. There are too many other phenomena, natural and other, that tug at her curiosity: her seeming weightlessness, the parasol again, the passage of time, her listlessness, the parasol's bursting into flames. Why not? Of course, it had never happened before; she might herself be consumed, all her visible flesh "charred to a cinder."

Her conversation changes into an insightful description of the world of the everyday, there where she has been lost so long, without the slightest realization:

> It is no hotter today than yesterday, it will be no hotter tomorrow than today, how could it, and so on back into the far past, forward into the far future. *Pause*. And should one day the earth cover my breasts, then I shall never have seen my breasts. . . .
>
> . . . .
>
> Yes, something seems to have occurred, something has seemed to occur, and nothing has occurred, nothing at all, you are quite right, Willie. The sunshade will be there again tomorrow, beside me on this mound, to help me through the day.[22]

On this day, like all the others, she will have to while away the time.

She takes out a music-box that plays the waltz duet from *The Merry Widow*, 'I Love You So.' She sways to the rhythm, and he hums along. Oh, happy day. But much to her sorrow, Willie refuses to sing the words; for one can sing, she muses, only what stems from one's heart. While filing her nails, she imagines that a couple has visited them in this condition – the Showers or Cookers, what's the importance of the name? Willie is eating the mucus scraped from his nostrils, like the symbolic hog he is. Mr. Shower-Cooker finds himself in the position of the play's audience, as he asks:

> What's she doing? he says – What's the idea? he says – stuck up to her diddies in the bleeding ground – coarse fellow – What does it mean? he says – What's it meant to mean?[23]

And his wife replies, as if for the dramatist,

> And you, she says, what's the idea of you, she says, what are you meant to mean? Is it because you're still on your two flat feet, with your old ditty full of tinned muck and changes of underwear, dragging me up and down this fornicating wilderness – coarse creature, fit mate. . . . [24]

Anyone who is a healthy, adjusted human animal would be a coarse creature for Winnie, who fails to see herself and Willie in the fictions of her imagination. No one ever means anything, for whatever significance a human being achieves in his life is the result of a resolute decision to act.

In her condition, Winnie is incapable of acting with any effect on others or on the world, and so asks for the definition of a hog. "A castrated male swine . . . reared for slaughter,"[25] replies Willie, realizing that he has long since been castrated by his magpie, self-alienated wife. She will not even grant him the distinction of crawling as well as he used to. If all males were castrated like the swine they are, it would be only merciful to slaughter them as the hogs they would be. The definition, now well fixed in her mind, occasions her last exclamation of the day: "Oh this is a happy day! This will have been another happy day! *Pause*. After all. *Pause*. So far."[26]

Willie turns to his paper, where he reads, as before, "Opening for smart youth." "Wanted bright boy." And he announces to Winnie that it is time for her nightly prayer.

So ends the first act of *Happy Days*. The irony of the title has by now begun to cut, and no unhappy couple will fail to see themselves in Winnie and Willie's condition. The pessimism may even have begun to hurt; still the play's effect is yet to come. We have seen once more what lostness in the world of everydayness means for the inauthentic personality; the mood is entirely set in the rambling disclosedness of Winnie's world: her chaotic discourse, the refusal to recognize what she really is, her attempts at fleeing what at bottom she knows herself to be; and the very smokescreen of the words reveal the "tempting, tranquilizing, alienating, entangling, turbulence" of inauthenticity, as described in Heideggerian ontology.

After his study of Heidegger, Sartre renamed the syndrome "bad faith," the state of a consciousness, divided against itself, that has successfully lied to itself about its essential condition. When Winnie refers to the passage of time "in the old style," she refers to a time when she and Willie both were capable of projecting an end to their activities, and of measuring the phases of its accomplishment. Those days might not always have been happy, but they would have been normal and the striving couple could have been proud even of their failures, as being their own. The "old style" language reveals the life of a speaker in touch with the conditions of his life, and a continual reference in the present to the old style of speech underscores the change that has taken place.

All this seems clear. What does not seem clear, from the purely naturalistic point of view, is why Winnie and Willie have fallen to the depths they have. Act II will supply the answer.

For, in the second act the curtain opens on a changed universe. Winnie is now buried up to her neck in the sand, and Willie is no where to be seen. Her parasol, the bag and the revolver still lie beside the mound, where she can only see them. Thus, the preceding day's burning of the parasol happened only in Winnie's feverish imagination. After all, it had been a gift from Willie one day when they went punting on a lake, amongst the reeds (like Krapp and his beloved). Its apparent burning therefore was her vision of Willie's lost affection. But that, too, was an error of her imagination, since Willie is still there, in spite, and because, of her narcissistic masochism, bent over double in pain from desiring her; his newspaper is only a means for escaping her inevitable rejection of his advances.

Now that she can no longer manipulate the things of her environment, they too have lost their utility (their readiness to hand, *Zuhandenheit*) and have sunk into the least kind of being possible, being present at hand *Vorhandheit*), as the objects of Winnie's perception. Her world therefore becomes restricted to the activities of her mind unfettered by natural law; instead of stopping, time is measured and given value only by the content of her immediate consciousness. It is as if Winnie has been lying on the psychoanalyst's couch, exposing the innermost secrets of her inauthentic life.

The session continues with her awakening, and her salutation of the light. She imagines that someone is still looking at her; she is or was, after all, an object to be desired, and her enjoyment of that rôle is the fullest measure of her masochism. Her subjective freedom functions erratically, as her chaotic discourse reveals the condition of her present restricted world: another forgotten unforgettable line of poetry; her need of Willie, remembering the old style of a time when their love was functional, time in general, her being alone, the surrounding wilderness; her lost hope, she no longer can pray; the independent life of things, there in the bag, her arms and breasts, her Willie, the dead Charlie Hunter, the universe without an atmosphere.

A universe without an atmosphere is like a man without opinions:

It's understandable. *Pause.* Most. *Pause.* The earthball. *Pause.* I sometimes wonder.

Pause. Perhaps not quite all. *Pause.* There always remains something. *Pause.* Of everything. *Pause.* Some remains. *Pause.* If the mind were to go. *Pause.* It won't of course. *Pause.* Not quite. *Pause.* Not mine. *Smile.* Not now. *Smile broader.* No no.[27]

The very words deny their content; for her mind is incapable of establishing the connection between the pauses, where her consciousness is reduced to nothingness, a nothingness of awareness except her own anxiety, which is expressed in a smile that should have been a prolonged frown or earthshaking sigh. Still another evasion: she begins to manipulate the muscles of her face, looking out upon the parasol of fateful memory, and the revolver that would have allowed Willie to put an end to his own misery, since he was not allowed, by more concernful means, to do the same for her.

She hears voices, bare sounds that help her through the days as they were measured in the old style. Has she lost her reason? No, she says; the things are there still in her bag: again her words deny their intent. If all else fails, she may give the story of her life. Ironically, it is about another, a little girl, Mildred, who will remember life in her womb.

The child was given a doll, with a full panoply of clothes, even a tiny picture book to delight the little china-blue eyes that opened and closed. One morning the girl rose before day break, and descending the stairs from sleeping quarters on her hands and knees, took her doll under a table of the nursery and began to undress it. But the thread of the story is lost as Winnie's mind, avoiding the pain of its recall, shifts to other topics: Willie, was he stuck in his hole? Was it time to sing her daily song? Why does one become sad after singing a song, as one does after sexual communion? Is there a post-chantal triste? The classics might help, but fail to be revived. Why does one lose one's classics? But only a part? The Shower-Cooker's again make their appearance to her, as if to answer the question. The classics are an escape from reality; the love of man for woman is hardly ever exclusive and always badly requited; and one chooses to forget whatever brings up an unpalatable truth. But this only the analyst knows, as he hears Winnie's report:

Standing there gaping at me. *Pause.* Can't have been a bad bosom, he says, in its day. *Pause.* Seen worse shoulders, he says, in my time. *Pause.* Does she feel her legs? he says. *Pause.* Is there any life in her legs? he says. *Pause.* Has she anything on underneath? he says. *Pause.* Ask her, he says, I'm shy.[28]

That scene could only have happened yesterday if Winnie was correct in reporting that the Shower-Cookers were the last living people to pass by. She had missed the significance of its recall yesterday, and does so again today. The wife replies,

Ask her yourself, she says. *Pause. With sudden violence*: Let go of me for Christ sake and drop! *Pause. Do.*: Drop dead! [29]

The appearance of another, seen as desirable, can do nothing but turn one's mate away in jealous anger. Winnie's subconscious mind approaches an awful truth.

The other that tore apart Willie and Winnie was the little Mildred. Winnie's attention returns to her unfinished story, and the epiphany of the play:

Suddenly a mouse ... *pause*. Suddenly a mouse ran up her little thigh and Mildred, dropping Dolly in her fright, began to scream – *Winnie gives a sudden piercing scream* – and screamed and screamed – *Winnie screams twice* – screamed and screamed and screamed and screamed till all came running, in their night attire ... to see what was the matter ... *pause* ... what on earth could possibly be the matter. *Pause* Too late. *Pause.* Too late. *Long pause. Just audible*: Willie.[30]

Winnie's own screams betray the cause of her own alienation: her daughter has died or lost her reason, and now there is only Willie, already alienated from her for having been cut out of her affection – or herself. With the trauma to her daughter, Winnie's transcendence ceased; and her predicament came to be aptly symbolized by the mound of sand, that will henceforth obscure from view her *mons veneris*. She still wants to be desired, but only on her own terms.

Sleep is her escape from the desires, the terms of others; but the bell has not yet rung. Her curiosity comes back to time. And her old style notion of that concept, which was naturalistically true but psychologically false, comes to be psychologically true:

I say I used to think there was no difference between one fraction of a second and the next. *Pause*. I used to say ... *pause* ... I say I used to say, Winnie, you are changeless, there is never any difference between one fraction of a second and the next. *Pause*. Why bring that up again? [31]

Answer: because now, at last, it is true.

She has a pain in her neck, which she eases by moving it. What differentiates one second from another in psychological time is the quality of one's state of mind, but even the pain fails to bring Winnie back to reality. There are other avenues open to the imagination for the avoidance of pain: the classics again; and the memories of a past happy moment when she and Willie had taken a last drink together, a pink fizz, when all the guests had left their party. It is time for Winnie's song.

Before she can decide what it is to be, Willie makes his appearance, dressed fit to kill: top hat, tails, striped trousers, white gloves, but still crawling on all fours. Winnie cannot move, of course; but she cheers her lover on. At times his strength gives out, and he falls into the sand, but only to lift up his face to hers. She cannot stand the look of his eyes:

Don't look at me like that! *Pause. Vehement*: Don't look at me like that! *Pause. Low*: Have you gone off your head, Willie? *Pause. Do.*: Out of your poor old wits, Willie? [32]

In spite of the refinement of his trappings, what Winnie sees is that old desire on her husband's face. He had dressed up to please her, in a last effort to meet and win her on her own terms. He has lived up to his name, but she cannot, to hers. His just audible "Win" touches her deeply, but his gesture is futile; for his use of her endearing name only makes her exclaim as she had the day before: "Oh this is a happy day, this will have been another happy day! *Pause*. After all. *Pause*. So far." [33] And to forestall the change in her condition she intones the words of her music-box's love song:

Each touch of fingers
Tells me what I know,
says for you,
It's true, it's true,
You love me so! [34]

But after listening to the song she is struck by that curious post chantal triste; her happy expression disappears; and the unhappy couple look at each other as the curtain comes down on their separation, freezing them into their plight.

The happiness Winnie feels is a figment of a dement's imagination, and he who suffers most is her husband, whose love she has refused. Winnie has, ironically, lost.

Play is an experimental piece of theater, a dramatic round presenting a round-robin love affair between a man, his wife, and his mistress. The three characters, referred to simply as W 1 (First Woman), W 2 (Second Woman), and M (Man), are enclosed in urns about a yard high. Stuck fast by the lips of the urns, their necks support their unmovable heads. They speak their lines only when flooded by a spotlight, but without expression, from a stage that is totally dark, except for the play of the light.

The symbolism of this stage setting is partially repetitive, as far as the *oeuvre* of Beckett is concerned, and partially original. In an ascending order of symbolism, the urns are wombs, are closed worlds of aborted transcendences, where each character is separated from the events he or she recounts by the expressionless tone and rapid tempo of the discourse that illustrates nothing so clearly as the lack of real communication in their lives. We already know from the tradition that *tota mulier in utero est*, but are left unenlightened as to what transpires when a man attempts to dispossess a woman of her rightful totality. In his urn, the man is stuck up in some womb or other, but that womb (since the three urns are identical in size and color, an equally expressionless grey) contains equally well another woman, the wife or the mistress, as the occasion demands. Each of their wombs is blocked to him, filled, as it were, by self, and haunted by the image of the other.

The broken circuit of love's requital is reminiscent of the way in which Neary's love for Miss Dwyer came back to him in a complicated chain of intermediaries. Here the chain is simplified into three, just enough points for the short-circuiting of love's natural path. And that is not the only theme of *Murphy* to be restated in *Play*; the import of the third party, superfluous to an authentic act of love, is to break the tension of a significant gestalt, the relationship between figure and ground. It will be recalled that Murphy learnt at Neary's feet that "all life is figure and ground."[35]

What the audience sees on stage is a revolving sequence of shifting figures and grounds. In the opening chorus, in which Beckett reduces that ancient stage convention, a shuffling group of omniscient commentators, to a shiftless, motionless presentation of words and the barest visual image, all three characters give tongue to the words expressing the states of mind peculiar to the person trapped in its unique existential situation. The events recounted are of their pasts, which, once understood, explain their stage positions: a man, motionless in an urn, flanked on either side by two women in similar urns; he, the victim of wombs in general, and they engorged by their own. Prevented from acting as a result of their past actions, their present disclosedness is restricted to an inauthentic speech.

As, in the opening chorus, the spot falls on all three, each begins to express his own state of mind, but simultaneously with the other two. Thus, there are three concurrent temporalities, which symbolically never achieve significant, sympathetic, synchronic resonance, as they would if one or the other fell into phase with the third in an act of reciprocal love.

Woman 1, who later becomes revealed as the wife, speaks of the strange darkness (her reaction to having been abandoned by her husband); Woman 2, of the possibility that she may have lost her wits (hers, to the same phenomenon); and the Man, of his astonishment at his failure to achieve peace in leaving them both.

On the first presentation, the audience will receive the message in bits and pieces; but unless one's attention is geared for receiving three messages at once, as the musical mind may grasp a series of related notes as a chord, the impression will be that of an intended confusion. Once the initial chorus has been followed by a second and third sequence of single representations of the various points of view, however, the states of mind will be understood; and the audience will be ripe for a second presentation of the round.

In this way, the dramatist directly presents his image of the eternal return of consciousness to itself. But having lost its bearings with the realities of the world, each of these consciousnesses will complete the return in an alienated, unhappy state. And the repetition of the round will underscore the universality of the message: for the solipsistic consciousness, love is nothing but a labor lost.

Thus, where *Krapp's Last Tape* dramatizes the consequences of the lack of courage in love, and *Happy Days*, those of a love gone sour, *Play* sets its sights on the loneliness involved in an act of playing at love. Yes, the simple title is not quite that simple; the hero is the man, who, in the first sequence insincerely plays at being in love with two women at once, and who, in the second, equally plays at being indifferent to them both, tragically finding himself no longer even the object of perception for either of his false loves.

His wife has become enamored of the darkness, and his mistress, concerned with the loss of her reason. If one were looking for a moral to the tale, there would be none better than de Musset's *On ne badine pas avec l'amour*; but that would tend to take away the reader's attention from the philosophical structure of Beckett's *Play*: a round robin of love, three conscious temporalities running summatively in counterpoint, but each returning back into itself, there to suffer the consequences of its own inauthenticity.

Following the opening chorus, there is a five second blackout that encloses each consciousness back into itself. The new light opens once more on all three characters, who begin to intone their separate tales as if to continue the simultaneity of the disclosedness:

W 1: I said to him, Give her up –
W 2: One morning as I was sitting –
M : We were not long together – [36]

but another five second blackout cuts off the simultaneity of their discourse, in which the difference of the point of view threatens all comprehension. When the light hits its target again, it will fall on one or the other of the faces; and its owner will continue with the story told from his or her personal point of view.

For example, in the bridging chorus, the wife begins with her account of ordering the husband to give up the mistress; the mistress with the intrusion of the wife in her quarters to demand that she too desist from the illicit dalliance; and the husband with the short time necessary for the wife to discover his deception. As the spot lights one, then another of the faces, each continues with the story from the point of his or her personal engagement.

In brief, the whole story is as follows: without lessening the physical ardors addressed to his wife, the husband has taken a mistress, who accepts his attentions under the assumption that his relations with his wife have ceased. The ploy is so apparent, it is a wonder that it ever works on the second woman: a tale of dissatisfied, misunderstood husband at home, looking for fulfillment elsewhere. The mistress is duped because she needs no such explanation of the man's attentions, gratified to find herself the object of the errant lover's "true" love.

When the two women meet, the mistress will console herself with the judgment that the wife's portraits were too kind to her; and the wife will reciprocate with a list of deprecating judgments on the mistress's physical characteristics. Both, then, have already fallen into inauthenticity, full of bad faith, appealing to the value of their own subjective transcendences, but reducing their respective rivals to the phenomenal appearance of their objective, physical characteristics. The self-deception allows both of them the luxury of underestimating their foe: indeed, of misidentifying the foe, since that one is properly the unfaithful husband and lover.

Although she cannot judge that her husband is unfaithful because his attentions to her have slipped, the wife smells the odor of the mistress in contact with her husband. She orders him to give up the girl, puts a private detective on his tail, and goes to pay a visit to her rival to let her know that she is on to their little game.

Both the husband and the mistress deny there is anything between them (and this would be true, if sincerity were the only gauge of the significance in their contacts), and the triangular relationship continues for a time as it was. The husband is capable of buying off the gumshoe, but not his own conscience. When his wife threatens to kill herself, he comes sneaking back to her and confesses. He is forgiven, provided that he give up the tart. In asking for his wife's forgiveness, he claims that he cannot live without her; but he says the same thing to the mistress, who confronts him with his betrayal of their secret after the wife has made her second appearance in her apartments to gloat over her "victory."

To settle this situation, the husband decides to leave them both, without telling either of them. The wife can only decide that he has gone off with the mistress; and the mistress, that he has returned definitively to the wife. The husband, having played the game of a double love, now decides to play that of the disinterested hero, giving up both, each one to please the other. But the ambiguous words of his declaration of love to them, that he cannot live without her, turns out to be true; for he too can only imagine them as now communicating directly with each other: drinking tea together, and discussing their past "happiness."

About this he is mistaken, and even his imagination cannot sustain the fiction. So he moves on to consider the situation in which he would possess both of them at the same time and in the same place. Whoever would wake up first would awaken the two others; punting on the waters, he could contemplate both their expectant faces in the bow of the boat.

In reality, however, both women have turned to their own form of purely imaginative life. The mistress burns his clothes, and begins to contemplate the loss of her reason; the wife imagines her worthless husband in the company of his mistress on a distant pleasure trip. But that image too becomes dark, and she is reduced to the words by which she currently expresses her own confused state of mind.

As far as he is concerned, his game has been revealed. He now realizes that he was playing at being in love, and then haplessly playing at sacrificing himself in the best interests of the wronged women, ultimately to find himself wondering if he still exists – if only as an object of one or the other's bare perception. Not even that kind of *Vorhandenheit* existence is his to be enjoyed – the masochistic reveling in his objective state, since both his loves are now too concerned with the hurt he has occasioned them.

No gloss of the story, however, will do justice to the manner of its stage presentation. The whole story may be pieced together only from the incidents reflected in the individual points of view. Once the tale is understood, the opening chorus is repeated, with the simultaneous presentation of the three states of mind, ordered here in such a way as to emphasize their isolation:

W1	*W2*	*Man*
Yes, strange	Yes, perhaps	Yes, peace
darkness best	a shade gone	one assumed
and the darker	I suppose	all out
the worse	some might say	all the pain
till all dark	poor thing	all as if . . .
then all well	a shade gone	never been

for the time	just a shade	it will come
but it will come	in the head	*hiccup* – pardon
the time will come	faint wild laugh	no sense in this
the thing is there	just a shade	oh I know . . .
you'll see it	but I doubt it	none the less
get off me	I doubt it	one assumed
keep off me	not really	peace . . . I mean . . .
all dark	I'm all right	not merely
all still	still all right	all over
all over	do my best	but as if . . .
wiped out –	all I can –	never been – [37]

Each of the vertical columns presents the continuity of a single consciousness, and each horizontal line, delivered by the three characters simultaneously, expresses a single moment of love's alienating isolation when engaged in with anything less than the full commitment of one's resolution to be with and for the other.

Whether the playgoer's mind is capable of taking in such a complicated expression of a star-crossed love only the individual can attest. But if not, there would always be the blinking of the lights. They do go on and off, and symbolize a human awareness in the minimum of its conditions. And if one has missed the significance of the words the first time around, there is always the second, in which the hapless playgoer is given his second chance.

Like the man of the story, however, his second chance is conditional upon his giving up his love for a merely traditional form of theater. Should he fail, to give up the experimental or the traditional, he might give up both; but as the story insists, there can be no peace in that direction either.

Krapp's Last Tape, Happy Days, and *Play*, all exploiting a consciousness inauthentically turned towards a past love, depict once more the unhappy state of living without an authentic future. But, within context, each of the characters living that situation has chosen the inauthentic state. The depth of the tragedy merely strengthens the percipient's hold on the truly authentic condition: to find oneself one must lose oneself in the love of another. Such a love is not that which unites a pseudo-couple, but a real couple in which the two "I's" become a 'We.' Failures all, these couples are fair game for a new, more contemporary treatment of tragedy.

CHAPTER XI

THE LIMITS OF THEATER

In the traditional criticism Beckett's plays have, from the beginning, been dubbed "experimental"; and Godot's failure in Miami proved that a profound tragic sense can become confused when it is expressed in surface ironic humor. That experience should have proved only that Beckett was never the victim of the inauthentic plights of his heroes. Although the Miami fiasco was in a sense tragic, it was a natural consequence of an inept reading and of the ridiculous commercialism of the American theater. In spite of his multifarious uses of the "convenience" of philosophy to structure his fictional worlds, his own world view remains a matter of conjecture, unimportant for the interpretation of his works; and *Godot's* reception in Miami[1] touched his actors more than it did Beckett himself.

In the theater, as in the novel, he shows himself an accomplished magician, continually reworking old themes into new resonances. A man of his times, he had learned through practice that the medium is the message, and unceasingly changed the surface qualities of the particular medium in question, in an effort to enhance the tension of each new expressive vehicle. The "experimentation" of his writing, then, is to work for the maximum effect with the minimum effort.

Once he had learned the trick of effacing himself from the context of his heroes' concerns, he discovered the true power of words: to evoke images, to tell a story, to capture the conscience of the narrator telling his story. The minimum characteristics of the literary medium, those underlying all those meanings that go together to form the literary world, are sounds and silences; the meanings are found in between, in that tension where the barest consciousness is awakened. Even the surface characteristics of the verbal medium, then, may serve to symbolize the developing theme of a story. Beyond that last "ping" of his own creative consciousness, it would be difficult to introduce into fiction a further abstraction: all perception, if not life itself, is figure and ground.

Looked at in this way, the stage became a new medium, a new area for the achievement of the maximum in the minimum. What before could only be referred to by the words of his medium could, in this newer mode of expression, be presented by the settings of the stage; that too could be symbolic in ascending orders of signification; and *Cascando* and *Words and Music* show that Beckett had not lost interest in bringing the creative consciousness to explicit dramatic representation. Once that lesson, too, had been thoroughly learned, the magician could return to the dramatization of consciousness' unique unhappiness, ironically in that one project in which man seeks his happiness most, in the love of his fellow creatures. But there is a reason for the human failure: to find oneself in an act of love one must lose consciousness of oneself.

If Beckett chose the inauthentic over the authentic patterns of existential commitment to life, his readers and his audiences could come away from his words with a deeper appreciation of those fleeting moments in their own lives when they find themselves in a positive relation to others. In this way, the positive existential mode is enlightened, in Beckett's plays, by contrast with the profoundest display of human failure to achieve the happiness each of his characters seeks.

On the surface, then, the maximum in the minimum; and in the depth, the positive through the negative. And neither the abstraction to a surface nor the inversion of the depth qualities of experience is anything new to art.

The miracle of Beckett's achievement is that a single philosophy is responsible for so much of the structure. There have been other examples of the "Theater of Ideas," e.g. in G. B. Shaw, another Irishman who took pleasure in plaguing the intellects of the English with their own language; but he talked too much. On the continent, Pirandello worked more successfully with philosophy in the theater, but he was obsessed with the illusion of reality, which he invariably treated as fixed in Platonic essences. In Beckett, the philosophy in drama became thoroughly existential. If a play represents an illusion, its writing does not; and the experience of every play is a conscious response to the tensions of its organized medial counters.

Symbolism was not new either. That too was found in abundance in Pirandello. Beckett's use of it, to increase the tension of the dramatic medium, may make the traditionalist think of Wagnerian opera, itself "experimental" in its day, since Beckett's characters and situations express related leitmotifs of the developing theme. And when stage settings themselves have become symbolic of the overall philosophical structures of the plays, one might think of Brecht, who used his props as props the way Wagner used motifs as motifs and Priandello, the play as a play.

But Brecht's use of the prop is naturalistic in spite of his theory. In calling attention to the prop as prop and to the actor as actor, he could achieve a double alienation – of actor from character and character from his natural environment. Such goings on may serve the purpose of preventing a too eager audience from being drawn into the action of the play, as Brecht intended, but his props, after all, indicate a natural setting for the play's action; and if actors step out of their roles to comment, like the ancient chorus, upon their predicaments, they merely intensify the audience's awareness that they are being told a story – with the dramaturgical counsel to cool it. But both the props and the actors were meant to mean something, and the action of the play made clear what that meaning was. As often as not, Beckett's actors give the impression of meaning nothing – but that is the meaning they should portray, the nothingness of human aspiration when a consciousness has become alienated from its world.

Nothing in Beckett's theater, then, was absolutely new, or even experimentally original to himself. He merely continued to abstract, to reduce the number of surface counters in the expression of the same philosophical ideas in order to increase the tension and thereby the intensity of the individual works. Hamm's parents are placed in garbage cans – for a representational purpose, the direct visual presentation of an ontological condition; Krapp is locked into his den, for the same purpose: he had already locked himself up in his past. Winnie is buried in the ground, and the man and two women are stuffed into urns, in *Play*, for similar purposes.

If the move from the medium of words to those of the stage and film opened up the possibility of exploiting the visual aspects of experience without the mediation of words, it likewise permitted a new opportunity for artistic abstraction: one could disavow the word and yet express the same or similar ideas with none but a changing series of visual representations. That is what has happened in Beckett's *Act Without Words, I* and *II*.

These plays were originally "written" in French, for production in the French theater. And the reason he used this language to write his plays without words is not difficult to imagine; just as he had found French publishers more open to the experimentation of his prose, there having been in France a longer tradition of literary experiment, so too he found in France the conditions favorable to an open theater, and a long tradition of respect for the mime. But rather than have his actors perform by imitating natural phenomena in their bodily gestures, Beckett introduced an ontological dimension of signification to his characters' interactions with their environments. And none of the themes was in anyway new to his established philosophical repertory.

Act Without Words I (1956) is a representation of a man in his changing environment. The action opens with the man "flung backwards" from the right wing onto the stage. The scene is a desert, bathed in a dazzling light. For each change in the environmental setting a whistle blows, calling the man's attention to a passing phenomenon. The whistle blows first from the right and then from the left wing of the stage, but in each case the consequence is the same: the man goes out, and is thrown back into the desert. As he did on his first entrance, he picks himself up, brushes himself off, and begins to reflect. At the second whistle from the left wing, he is tempted, begins to go on that way, but hesitates to reflect, changes his mind, and gives himself away to a longer meditation.

While he meditates, a tree descends from the flies. The man can't see it, and takes no notice. Whistle: man turns, sees, goes to sit under the tree, which offers its shadow to protect him from the heat of the sun, and contemplates his hands. This process repeats itself for a pair of tailor's scissors (to cut his nails); a tiny carafe (to slake his thirst); three cubes, large, medium and small, (for bringing far objects closer, by mounting them); a rope, with knots (to climb, in order to reach the carafe, which is continually pulled out of reach). Since the man has fallen from his scaffolding of the two cubes (large and medium), he refuses the gambit of the third; and it is pulled back up into the flies of the theater, unused. It is then the rope makes its appearance. Whistle breaking reflection, reaction of the man.

Whereas the carafe is pulled up out of reach when he has climbed his cubes, it remains stationary as he shinnies up the rope; but the rope itself gets longer, until finally he finds himself at ground level once more. Another reflection: the scissors to shorten the rope. As he starts cutting, the rope is pulled up, but he clings to it long enough to cut off the rising end. When he succeeds in cutting it, he falls without reaching the water and drops the scissors. He brushes himself off once more, and reflects, as the longer end of the rope disappears into the flies.

With the shorter end he makes a lasso. Such is the fruit of his reflection. But when the lasso is finished, the carafe is pulled up into the flies with the rest of nature's paraphernalia. Nothing now, except the tree, two cubes and a lasso of his own fabrication. He approaches the tree with the lasso, makes the connection between it and the overhanging branch; looks for the cubes, brings them next to the tree, places one on top of the other, tries out their stability (small on top of large, after recognizing that the large could not be fit on small and still retain his weight in balance). When the scaffold is prepared, he turns to look for his lasso; but the boughs of the tree sink to the vertcal. He drops his lasso, and resumes his reflection. No suicide today.

Result of reflection: he straightens out the cubes, places the lasso on the smaller one, reflects, decides to go out, does so, and is thrown back into the desert. Same response: picks self up, brushes self off, reflects. Whistle, stage left. The man ignores the signal.

He inspects his hands, finds nails too long; looks around for scissors, finds them and begins to trim his nails. Reflecting the while, he runs his hand along the blade to test its sharpness; finds it adequate, and turns to lay down the scissors, while he prepares his neck for the cutting. But he has placed his scissors on the smaller cube, alongside the lasso, and it is now drawn out of sight into the flies. Surprised, and with his second attempt foiled, he sits on the larger cube, which is pulled out from under him. He falls, lies on his side, staring before himself into the audience.

The carafe descends to a few feet from his face. Whistle; no response. It comes closer, dangling before his very eyes; still no response. But now his thirst has been forgotten. Rejected, the carafe disappears once more into the flies. But nature is benign. The boughs of the tree resume the horizontal, offering their shade. Whistle, no response. As the tree disappears above, he looks at his hands. Curtain.

Like most of Beckett's plays with words, the audience may find in the above sequence of images whatever it wishes to discover: either the surface quality of ironic humor in the inadaptability of the ape-like man, trained by circumstances to adjust means to ends, or profound tragedy, in which even the easy solution of suicide is foiled for the lack of an adequate means.

We are all familiar with the psychologist's trick: hang the banana too far out to reach, give the chimp a stick too short to reach the banana and enough time for him to see the stick as a means of harvesting his food. Then another stick, shorter than the other, with enough time for the chimp to make a single tool of the two sticks. The trick works for the psychologist because the chimp is not a chump; he puts his intelligence at the service of his instinct, and never is deceived by his own symbol system. In this way he never puts himself in the way of the experiment's success. Psychologists have patently less success with human subjects on the other hand because a man is often a victim of his symbolic systems, and almost always gets in the way of his own, and his experimenter's designs.

The difference in the two modes of conduct are striking. For the non-human animal: hunger, for motive, food for incentive, and objects for making a tool; instinct and intelligence work together for changing the situation to the chimp's advantage. The man, on the other hand, loses himself in his reflection, and begins to brood over his plight. Instinct and intelligence are separated in the pursuit of their common aim; and, as a result, motive goes unsatisfied, and a new chain of purely subjective reactions takes place: pain, continued wish to avoid the pain; new imagined incentive, self-destruction; foiled, more pain. New situation, same sequence of events; loss of contact with the means for any activity; return to self in an empty, hostile environment; no change in the man's condition. He looks out upon attentive audience seeking sympathetic understanding.

We may laugh at this display of nature's taunting the hapless hero; or we may tend to see ourselves in that suffering humanity, begin to feel our own anxieties, even weep, depending upon whether we interpret the scene as a man reduced to the condition of our simian forbears, with a *deus ex machina* playing the rôle of the psychologist who pulls the wires in the flies, or as the bare essentials of a human life subsisting in an environment that continually changes without regard to human purposes and which supports human life or not, depending upon man's ability to bring instinct and intelligence into an effective harmony for the fulfillment of his needs and desires. Both interpretations lend a meaning to the mime; but only one suggests a deeper, anagogically structured meaning, and it presupposes the other. The clue to this ascending symbolic significance may be found in the hermeneutical expansion of "the world."[2]

Man and his world, indeed man in his world, is not an unambiguous concept. It all depends upon whether we take the "scientific" or the "ontological" world view. Ontologically, whenever anyone begins to reflect, making a problem of his worldly situation, he always finds himself already situated in some kind of world – existential thrownness, or *Geworfenheit* in the Heideggerian philosophy; and significantly, Beckett's player is thrown onto the stage, backwards to be sure, but thrown nonetheless, to pick himself up and dust himself off. There now in the desert world, he begins to reflect. Where now? Who now? When now? The Unnamable's three ontological questions are as apposite to this situation as to his own.

All human reflection becomes complicated when it is addressed to man's situation in the world; for what a man is, is as ambiguous as the essence of his world. Indeed, neither concept can be understood in isolation from the other. A man finds himself in the world as a form of disclosedness by which he reveals himself simultaneously with the revelation of his world. Whence the natural ambiguity in the concept of truth, caught up in the contending intentions of the objective world and the subjective mind.

Heidegger chose to show that what a man is conditions the kind of world each human being lives in, rather than pursuing the usual tack of explaining what man is by virtue of the kind of world he happens to live in. Both routes are possible, depending upon what the interpreter takes as problematic; but the one gives an ontic, and the other, an ontological explanation.

Even ontically, however, there are two descriptions for the human condition. One either decides to build a universe of significance to oneself – one's own world, or one decides to live a life in accordance with the unchanging laws of the nature that surrounds one – the so-called "natural" environment. And either way, one will have lived through an experience. The determination to exist either way has its source from within or from without. This ontic dimension of human existence is therefore a choice: to be authentic and self-determined, or inauthentic and naturalistically determined.

Explanations of these ontic conditions, whether of authentic or inauthentic determination of the self, are ontological. And they too may be authentic or inauthentic, in a derived sense of these terms. If a man is distinguished from an animal because in his reflection he may pose the question of what it means to be, that distinction is still ontic – it can happen to anyone, anywhere, as a matter of course. Recall Clov's repeated observations, in *Endgame*, that something is taking its course: time runs through all nature.

In the course of a man's existence, however, the question arises as to what, exactly, takes place when something takes its course; so, ontically, he is the ontological being, posing the question and sometimes giving an acceptable answer; in his being there is question of his being, and the property of Dasein's existence is to relate understandingly to his world. Whenever the question of its being arises, the man finds himself already thrown into a world as a projection to some indefinite future.

Moreover, since he has yet to be what he is to be by virtue of his projection, his existentiality is future-determined, whereas in his facticity, he is related to a fixed past, where his actions have already been determined, either by himself or by his world. The present is therefore either authentic or inauthentic: either the resolute anticipation of his end or the state of falling into the world of everyman, where his disclosedness assumes the alienating forms of endless curiosity, ambiguous understanding, and prattling discourse. In this mime, of course, we have no discourse; but the state of mind and the understanding are apparent in the player's actions.

Traditionally, ontological explanations have been the same for men and things. Ontically, every entity is present to someone's act of perception; its being is its presence, its being perceived. And the concept of "the world" fitting such beings is merely the summation of all things that may be perceived. Ontologically, corresponding to this being present at hand, is the concept of nature, that according to which entities present-at-hand come to be and pass away, the laws of which it is the business of the natural sciences to describe. And therein, a man may be very like a monkey, as indeed in some respects monkeys are like men. Perhaps that is what makes them so amusing, as the viewers of this play, ensnared by this surface, inauthentic appearance of comedy, may contend.

Does Nature play with the foibles of man? Or does a man show his weaknesses in the failures of his strength, in the abortion of his efforts to work out the condition of his own destiny? The difference is ontologically founded.

The attitude by which investigators establish the laws of nature is called, by Heidegger, "inspection," itself a deficient mode of human concern for the entities man always finds himself alongside when he is thrown into the world, whereas "circumspection," by which these entities are ordered as means for the accomplishment of his ends, establishes another kind of world, more closely related to man's being the animal he is. Ontically, this world consists of things ready-at-hand, adjusted to man's own purposes by a personal assignment of values, depending upon the immediate task to be performed.

In this peculiarly human world,[3] the task is primary, and things make their appearances as variables for the blanks in the expression " . . . in order to. . . ." The things are merely an "*x*" or a "*y*," the "*x*" being close to hand, and the "*y*" further away. The purpose of the action is to bring the "*y*" close. In this way, man establishes his spatiality, with respect to his body as here, by casting his circumspective glance around himself; the there, or thither, is away, where he must go to establish his region. For this reason, Beckett's heroes are on a journey, somewhere, anywhere, since the completion of that journey is a return to oneself. That towards which the entities ready to the traveler's hand point is the end – either the self, which will complete its circuit to itself in the successful journey, or to various way stations thereto, where the traveler's ontological commitment may be forgotten. That, it should be recalled, was Molloy's predicament, as he went through the process of creating his region ontically, and then gave the ontological explanation in his pages.

But the nonexistence of a fulfilling end may short-circuit the conscious return to self, allowing the traveler to turn his attention inward prematurely, as it were, where he finds only a blind, suffering striving – a volition without corresponding object, motive without incentive, and pain. Man then finds himself in separation from nature, which refuses its entities to his personal uses.

Such a situation, as despairing as it is, is to be contrasted with man's authentic disclosedness. If an object, x, points towards some fulfilling y, the relation between them is the assignment of a use value. And the y itself may become an x for further relation to some other y. Obviously there can be no mere summation of such entities into a single totalizing collection. Since their being is to be ready-to-hand, they can be related only by the total context of assignments of their use values. In this world, where the entities fund into a context of assignments, only the circumspective concern of the acting, existent being can grasp the significance of the whole. And that whole will disintegrate and become a mere collection of entities present-at-hand whenever nature refuses its support to the human project of elaborating a personal world. Man will then have fallen into the state of being present-at-hand – a bare datum of his own alienated self-consciousness as he is to the consciousness of the assembled audience.

To reconstruct a world view on such data, the consciousness re-adopts the attitude of science. But scientific inspection can never bring into consciousness the worldhood of the natural world as a single phenomenon (since its composite phenomena are infinite). Only existential circumspection permits the appearance of a singular world in the equipmental totality of the entities ready-to-hand at any given moment; for it is the active human being itself which is the ultimate value assigned as that for which each other assignment is made. The end of man's conduct is man. Thus, that on behalf of which an action is contemplated is the welfare of the acting being, Dasein itself. And being finite, Dasein's world is fraught with tragic consequences.

The first and most important of these consequences is that Dasein, the active human being pursuing its own ends in nature, exists, for the most part, inauthentically. Losing sight of his own potentiality for existence and ignoring the call of conscience, the average human being piddles away his substance in the world of everyman, seeking what everyman seeks in nature: love, honor, wealth, or the bare wherewithal for continued existence, which may be symbolized by the shade of the branches of a tree offering protection from the heat of the sun, or a drink of water that would slake one's thirst. In this reduced condition, man's overriding concern becomes self-preservation; his morality, convention; his knowledge, a commonplace. If suicide under these conditions appears the easiest way out, the reason is that the self has already been lost in the wrong kind of world. Such is the ontological fallenness with which Beckett has already made us completely familiar.

In the throes of such a life, one is alienated from self as that self would find itself, if it acted authentically, on the other side of its present projection in some fulfilling future, and pressed back in upon itself as it is now, lost to its possibilities in the never changing present, where it becomes adjusted to what is right and proper for every man.

Thus, having fallen from an authentic concern with his own being in a world of his own making, Beckett's player seeks refuge in the world already there, that in which objects make their appearance in accordance with natural laws, and where everything that happens has been determined by something else that happened in the past. Whence, the comfort, and the entangling temptation of living in the inauthentic world – until the agent opens himself to the call of conscience and responds to it in the silence of his reflective nature.

That call, it should be remembered, is from Dasein's being-in-the-world at some future time to Dasein's being lost in the world of the inauthentic present. Failing to respond to the call or lacking the imagination to envisage the future that would fulfill his projected being, the hero can only be driven further into alienation from self, until even suicide becomes an impossible possibility.

Act Without Words I plays on a man's ambiguous ontological states: ideally to establish his self-identity in the revelation of his personal world, but factually reduced to a reaction towards the stimuli of nature. The first, and most important, step in the player's alienation is, indeed, his lack of speech. That avenue of self-disclosure is not open to him. His reflections are therefore internal, out of all connection with his surroundings until nature imposes her will upon him. The whistle symbolizes that.

But the nature of Nature is to change, and to force men to change in order to adapt themselves – to the obstacles or to the opportunities offered by their environments. To do this, they must fashion tools, projecting their own changes to coincide with those of nature. Failure, under normal circumstances, is only a stimulus to readjust one's efforts; and this can be done, provided that one continues to look for the self where it is – on the other side of an existential projection; on this side, man can only be lost in the haplessness of his efforts, and haunted by the survivals of his past.

In such a condition it is natural that he appear as an atavism, a lower form of life which on its own terms, ironically, is well enough adapted for survival provided it responds to nature's whistled call with a direct action. The existence of inauthenticity in the human subject makes such a response impossible: the self has gotten in the way of the self. The player reflects instead of acting, and therein lies his tragedy; for his circumspective concern has broken down into an inauthentic, scientific inspection. That's the wrong attitude with which to approach the entities of one's environment if one is looking for a self.

In this way, the action of *Act Without Words I* encompasses a simple dramatic evolution: a man is thrown into his environment to work out his own survival. The place is a desert, and conditions are hard. The hardness of the conditions and the frailty of his imagination drive his attention back inward to a realization of his own weaknesses. But self separated from its world is a bare thing, determined by the whistle calls of nature, in which the agent becomes still less a self-conscious self. The conditions become worse until only suicide seems a possible way out. But there transcendence would cease, and with it, the significance even for the sufferings of this life.

As if in accordance to the absurdity of this act, Nature withdraws the means for its accomplishment, and man stays where he was, is, and always will be, in the passage of the days in their everydayness, each like the other until the authentic call of conscience may be heard in the silence. The inauthentic whistles of the wind need replacement by the resolute decision to be – oneself. Otherwise, one is reduced to a symbol of fallen humanity, definable only as the negation of what man most authentically is.

Thus instead of the rising action of a success (self thrown into the present, struggle with environment, newly developed adapted self), this mime presents a continually falling action: self thrown into present, struggle with the environment, self further reduced to dependence upon the environment. If the work is a tragedy, as I think it is, the reason is that the falling action intensifies our understanding of the missing transcendent rise.

As a whole, the play is constructed with an abstracted surface medium presenting a philosophical depth in which human value significances have been inverted. After all, we can call a pattern of life inauthentic only if we can recognize the authentic; otherwise we must resort to the inauthentic discourse concerning the normal and the abnormal, the rational and the absurd, where statistical averages are used to construct the concept of the normal, rational, human being. But that attitude will not take us beyond an ontic description of the events presented in this play.

With the ontological hypothesis, according to which inauthenticity enlightens the conditions of an authentic existence, the ontic dimensions of the sequenced images stand as signifier to an ontological signified, and the play becomes a triply layered construct: first of all, the images the audience perceive; next, their ontic, naturalistic interpretation; and finally, the ontological interpretation, the anagogical dimension, that sets the whole in order.

What the audience sees occasions what they think, and what they think is brought back into the interpretation of what they have seen. Such is the natural order of hermeneutical expansion, which in its development presents in mirror image the way in which the work has been elaborated. The context before the artist is his experience; that before the audience is the artist's completed work. And in the audience's response to the entire dramatic structure one is brought to an understanding of what lostness in an alien world means. That meaning is nothing more than a feeling expressed in the tense relationship between abstracted surface and inverted depth significance.

Once again, the magician comes nearer his objective, to represent the maximum in the minimum, the better to quicken the attentive consciousness of his audience.

If *Act Without Words I* is an abstraction performed in the dramatic medium, it still represents the complicated relationship between a man and his environment in a rising order of significance: the sequence of the perceived images, their naturalistic (ontic) interpretation, and their anagogical summation in an existential ontology. And in conterpoint to the rising order of significances, there is a simultaneous descent of the "hero" into a deeper state of haplessness before the onslaught of his environment that translates ontologically into total self-alienation.

Act Without Words II (1959), likewise "written" in French, performs a further degree of abstraction; but an explanation of this contention would be meaningful only after a description of the events depicted in the mime.

It was written for two players and a sting, representing an external goad on the consciousness of the players. All the action takes place on a low-lying platform at the back of the stage, where there is a violent play of white light; the rest of the stage is in total darkenss, representing the unknown beyond.

In the contrast of the light and dark a world opens up, as the consciousnesses of the audience are presented with a double image of an act of awareness: first of all, a bare contrast is still the smallest possible form susceptible to human perception; and accentuating that contrast in the violence of the lighting calls attention to the consciousness. Then the light itself, in which all the action takes place, vibrates at the center of the initial gestalt and reveals a series of conscious activities wherein a human presence projects a field of existential significance.

To express this idea in Heideggerian terms, we should have to say the light symbolizes *die Lichtung des Da* – the light of a consciousness's being there,[4] "clearing" itself a world, either by inspecting its environment or by expressing its circumspective concern in the creation of its own meaningful universe; or, failing this, with escape from one's personal commitment to be oneself. Whichever, something is being disclosed.

When the curtain rises the audience sees two sacks in which the players, A and B, are sleeping. A's sack is on the right, and B's on the left; next to them, to the left of B, is a neatly arranged pile of clothes. The three recognizable figures are lying on the platform about two yards from the right wing of the proscenium. The first action is that of the goad, which makes its entrance, plunges into A's sack, then withdraws. Nothing happens. Repeat the action: A stirs in his sack, and the goad exits right.

A's day is accomplished in a symmetrical performance of his daily round. His motions are to be "slow, awkward . . . , and absent."[5] Between actions he broods. As he climbs out of his sack wearing only a shirt, he prays before getting to his feet, takes out a bottle of pills, swallows one, puts back the bottle; puts on clothes, takes out a half-eaten carrot, bites off a chunk, finds it distasteful, spits out the cud, puts back carrot; picks up both sacks, carries them to the middle of the platform, takes off clothes, except for shirt, lays them down in a disarranged heap, takes another pill, kneels to pray; finished, crawls back into sack and sleeps. The sacks are now arranged with B's to the right, the clothes heaped to the left of A's, in the center (right to left) of stage.

Pause.

Re-enter goad, on a single wheel (something is taking its course); prods B's sack; B stirs immediately, exit goad. B's movements are all "brisk, rapid, precise,"[6] the exact antithesis to A's. Between his actions, however, he consults his watch. He, too, gets out of the sack wearing only a shirt, but gets to his feet immediately; takes out watch, consults, puts it back; does exercises; takes out toothbrush, brushes teeth; rubs scalp, takes out comb, combs hair, puts back comb; takes out clothes brush, brushes clothes and hair, rearranges brush; takes out mirror, inspects appearance; takes out partially eaten carrot, bits off chunk, eats with relish, puts back carrot; takes out map and compass, consults, puts back; picks up two sacks, carries them to left wing of proscenium, about two yards distant, sets them down; takes off clothes, arranges them in a neat pile to the left of the two sacks, A's on the right; does exercises, rubs scalp, combs hair, brushes teeth, consults watch for last time, winds it and crawls back into sack for sleep.

Pause.

Re-enter goad, now on two wheels (a second round). Prodding of A's sack: no response; second prod: A stirs. Goad exits. A gets out of sack, stops, broods, and prays as he had the day before. Curtain.

Where B's actions are about twice as many as A's, they are accomplished in the same amount of time; and where B consults his watch, A broods. To what effect?

Merely to make the time pass. Both players represent an inauthentic personality, lost in everyday, habitual activities, merely filling up the time of the real world instead of creating a temporal significance of their own determination. Let us not forget that temporality is the meaning of the being of consciousness. But instead of a humanly significant temporality, the play presents a uniformly divided day, each moment of which is determined by the performance of the habitual acts, each of which in turn is separated from the other by a personal quirk, the brooding silence of A, turned inward upon himself, and the compulsive consultation of his watch by the equally self-regarding B.

Since both sets of actions are inauthentic, their substance matters little; both represent an equally empty form. And it is this form that is a clue to the mime's ontological interpretation, pointing towards the temporal development of the action.

That action is presented in two equally enduring moments, like movements of a musical piece, followed by a short coda, the projected prolongation of A's second day. Whence, a three part division, in mock sonata form: A-B-A′. The whole orchestration, however, is not a bare repetition; the initial contrast between the brilliant light of the platform and the total darkness of the rest of the stage is repeated in the balance of the attitudes of A and B, but in their differences they exhibit a high degree of similarity: each acts to make the time pass, and each makes pass the same amount of time.

Repetition of balances is an old formalist's trick, and would tire without some perceptible change. This change occurs in the particular movements: A's actions establish an identity, call it a "self"; these actions give way to their opposite, in the actions of B, where the difference is noted as both "matter" and "form," what is done and how. In the bridging pause between A's and B's performance, the self has passed over into its own Other, and, as might be expected from the dialectical pull, returns to itself in the tail piece. And as the self returns to assert its presence in the coda, the format of the sonata comes to closure. Thus, the "matter" of the particular movements, the themes, illustrate the same "form" as the sonata's abstract orchestration.

We pass, then, from the beginning repetition of contrasts (light-dark of the proscenium and self-other of the players) to the perception of the like in the unlike (the duration of the disparate selves), to the change in the selves, culminating in a return that repeats the abstract form of the play's development, both coming to closure at the same instant: a circular matter enclosed in a circular form. In the world of the everyday, everything comes to repetition, and the habits of the characters guarantee our perception of the phenomenon – provided we pay close enough attention.

Each movement, as it is considered in itself, bears moreover a double relation to the other; each is both like and unlike the other: like in so far as the duration of the activity is concerned, and unlike, with respect to the substance. But they are similar in another formal respect in that both unfold in almost flawless symmetry of a rising and falling action, climaxing in each case with the food sequence. And symmetry is the simultaneous occurrence of a repetition and a contrast in a single gestalt configuration. So each separate movement repeats the overall configuration of the players' activities.

What the audience sees, then, from the strictly abstract point of view, is a contrast of the two movments related as symmetries internal to the symmetry of the sonata form: a second inclusion of a form within a form. And were we formalist critics, our analysis would stop here.

But we are not formalist critics, and since the play is symbolic, we must pursue the symbolism. From that point of view, the only difference between this play and *Act Without Words I* is that the initial object serving as symbol in that play is a concretely significant act, whereas in this one abstraction is made from the concrete significance of the act (being alike in their differences, the differences are reduced to irrelevance), and the formal relations between them become the symbols of, the pointers toward a higher significance.

The argument runs as follows. If it is accurate to claim that the lighted world of the stage symbolizes an act of consciousness opening up a region, that region is presented in these circumstances as a one-dimensional horizontal project, later accomplished by concrete activities. These activities present the conscious temporality of the characters, whose spatiality and temporality structure a world. Check, but how? By the funding images of A's and B's transcendent action – moving from the right to the left of the stage, in a third symmetry, by the way, which will reveal still a fourth: the self-other-self return in the closing of the spatial region opened up by the initial act of consciousness.

It is as if Beckett were taunting formalist critics. Let us review the argument. First symmetry: the whole play is developed in an occult symmetrical pattern, the A-B-A′ structure of the sonata format. Second: each movement develops a "vertical" symmetry of rising and falling action. Third: the "horizontal" symmetry of the positioning of sacks and piles of clothing. Fourth: the "substantive" symmetry of the self-other-self relation, creating its world by a constant projection of a single activity within the universe described by the overlap of the vertical and horizontal symmetries.

In its abstractness, however, the activity of the players represents or symbolizes the spatializing and temporalizing habits of a human consciousness (thereby disclosing a world), that was presented at the outset by the contrast of a blinding light in an otherwise totally dark stage. And if we were merely "ontological" critics, our analysis would stop here; for we would have explained the ontological significance of the symbolism in a second degree: the blindness of the light reinforced by the inauthenticity of the personalities.

Since, however, we are not ontological critics, having been convinced by Beckett's own criticism that the use of any given philosophy is only a literary device for structuring a story, if our method of structural analysis is to be found fruitful in the present instance, we must continue until the last level of significance has been made clear.

The clue for our interpretation, once again, is to be found in the equal emptiness of A's and B's worlds. We have already shown how this phenomenon permits an abstractionist's analysis of the mime; but, ontologically, it signifies, at the same time, the inauthenticity of the human personality spending its days in meaningless, habitual activities.

Yet, behold. The magician has waved his wand: that meaninglessness has become significant at a still higher level of experience. For even inauthentic worlds may be significant to others. All one would have to do is to show, by a formal analysis, that the emptiness of those worlds is repeated in the abstractly significant dramatic structure. And, indeed, this is what has been achieved so far.

But now the original image is open to further interpretation. The region that has been lighted up in that initial *Lichtung des Da*[7] was initially the world of the creative dramatist. He enjoyed himself playing with the sensitivities of formalist critics and aestheticians, but that prestidigitation hid from view a more sinister intent: to structure one more play on an existential-ontological base. And that he has done once more with eminent success.

Nevertheless that initial "lighting up of the there" also represents the consciousness of the audience, which finds itself once more contemplating itself on a voyage toward meaning: moving from a presentiment of the truth in the initial visual image to its confirmed realization in the explicit awareness of the play's overall concrete structure. That, too, is a return to self, authentic this, mediated by the contemplation of a workable hermeneutical hypothesis. And the relation between the conscious activities of dramatist and critic creates a last symmetry, the fifth, as the two separate sets of activities are developed in mirror image around the interface of the work produced in the one and appreciated in the other.

In creating this dramatical world, Beckett has shown that the abstractions of a dramatist may be as thorough, and as profound, as those of any other artist-magician. To perform such a trick, the dramatist merely has to learn how to manipulate the counters of his medium to create semantics out of syntax. Such is the lesson, a dramaturgical one this, of *Act Without Words II*. Thus we learn in this second mime, first, that human activities inauthentically projected may form the substance of a tragic representation, but second, that if they are they must be expressed through the formal constraints in the dramatist's art.

Come and Go (1965) is still another artistic attempt to reduce semantics to syntax. The first impression a prospective director of the play receives is that this English play represents the coming and going of the three female characters, Flo, Ru and Vi, as they shuffle positions on the stage, talking about the past from which they have come and to which they cannot return.

In their youth, they had been students together in a religious school run by nuns, and are now reunited in this indefinite place to relive their old spirit of comradeship. They are seated, in the opening of the play, at center stage on a bench that is almost hidden from view. The central configuration is illuminated by a weak light, and surrounded, once more, by total darkness. The feeble light indicates a feeble consciousness, at this stage of the play's development shared by the characters and the audience.

Each of the players is clothed in a sombre dress: violet, for Ru, red for Vi, and yellow for Flo. They wear large-brimmed hats that almost conceal their faces, and their hands are heavily made up to be as visible as possible in the weak light flooding them. As far as can be noticed, neither of them wears a ring. They speak in a tone barely audible, except on a single occasion each, when an exclamation is in order. Although each of them leaves the stage, they make their exits by disappearing into the shadows surrounding the bench and reappear in the proper position. Moreover, their stage comings and goings are inaudible, since each wears a pair of rubber-bottomed shoes.

At first glance, the action appears to be a round-robin break-up of the threesome into three more intimate duos; and after the exchange of their dialogue *à trois* and *à deux*, their positions on the bench have been reversed.

Vi opens the conversation by addressing the other two by name in turn.

Each of the others replies with a laconic "yes." And when she asks when it was, the last time the three were together, Ru requests, in order to avoid the subject, that they remain silent. The silence ensues, and Vi exits right.

The trio being thus reduced to a duo, Flo takes up the word, asking Ru how she is struck by Vi's appearance (except for their distinctive colors, they all appear the same). The answer is "I see little change."[8] Since Vi had been seated in the middle of the other two, that space is now vacant and Flo moves over to whisper in Ru's ear something the audience does not hear. Ru exclaims, "Mercy"![9] and asks whether Vi is aware of the whispered secret. Flo responds with her own exclamation: "Heaven forbid!"[10]

Vi reappears and takes the old place of Flo, as Flo informs the audience that the three are now on a bench as in the good old days in the court of the convent school; and when she enters into the spirit of the reminiscence, she is cut off by Vi's "hush!" The subject of the relations between the three is thus refused for the second time. Since her initiative was blunted by Vi's refusal to dig up their past, Flo exits, left.

There follows a sequence between Ru and Vi similar to that between Flo and Ru. Ru moves to the center of the bench next to Vi to whisper her secret confidence, following the same initial question: how does the absent one strike you at present? The responses to the question and the revelation of the confidence are very similar: about the same, more or less; and two sequential exclamations, one of alarm and the other an invocation of the deity to prevent Flo's learning the secret.

Flo makes her re-entrance, and takes Ru's former place on the bench, with Ru now in the middle. Ru takes up the broken conversation about their past, recalling a peculiar manner of taking each other's hand in a closed circle – and, adds Flo, dreaming of love. An embarrassing silence ensues during which each character turns her attention inward, and Ru exists right.

Thus, the third attempt to bring the group's collective consciousness into focus on their common past is inadvertently stopped by Flo. The pattern of the openings and closings of the conversations *à trois* is as follows:

	Openings	*Closings*
1.	Vi	Ru
2.	Flo	Vi
3.	Ru	Flo

Each broaches the subject of their past, and each prevents successful completion of the initiated project, Ru and Vi intentionally, and Flo inadvertently.

Where Ru and Vi feel the embarrassment, indicating that their pasts have changed with the passage of time, Flo does not.

There is, however, a third sequence *à deux*, now between Vi and Flo: question concerning present appearance of the absent one, repositioning on bench, whispered confidence, exclamation of dismayal, and prayer for the absent person's continued ignorance of the condition.

The significant difference between this confidential scene and the previous two is that Flo, still different from the others, pleads darkness as an excuse for not noticing anything at all concerning Ru's appearance. Thus, to Flo, Ru is what she is, an old friend. Note that while it was true to say it was dark, that truth implies the unawareness on Flo's part that Ru has actually changed: it was Ru's embarrasment that blocked the first revocation of their past.

Ru makes her re-entrance, taking Vi's former place on the bench. Vi, now once more in the middle, initiates conversation *à trois* once again. If we can't speak of the good old days, and of what has happened since, let us at least join hands in the manner of our youth. They do. Vi, in the middle, takes in her own right hand the right hand of Ru, on her left, the joined hands resting on Ru's knees. Vi's left hand is joined to the left hand of Flo on her right, hands reposing on Flo's knees. Thus Flo's right hand becomes joined to Ru's left, resting on Vi's knees; and Vi's arms, in the middle, rest on the left arm of Ru and the right of Flo.

Following this final positioning, Flo exclaims: "I can feel the rings." Silence. Curtain. The silence, of course announces another split-up of the triad into three distinct units, with each person facing her own embarrassment.

But what has happened in the play?

On the first level of significance the answer is simple. Three old school chums meet some time distant from their school days, and take seats on a bench as they did in the past. The person sitting in the middle begins a conversation about the past, only to have that initiative blunted. As if affronted by the implied exclusion from the circle, that person leaves and the other two engage in the intimacy of a confidential revelation, the nature of which is unknown to the audience. The process is repeated three times, until they reassume their closed, criss-crossed arm positions, when Flo, in the final position, has each of her hands joined to the left hands of the other two. Although no ring can be seen on any of the players, she exclaims that she can feel the rings that would indicate that the other two had fulfilled their former wishful thinking about a love to come, either for having been betrothed or married. But to all appearances, there being no rings, she is either hallucinating or displaying an astounding feat of extra-sensory perception. The question is, which?

If the dramatist is truly engaged in abstraction – assimilating the semantic to the syntactic functions of his construct, further analysis should reveal the answer.

Since each is capable of gossiping about the other's appearance, it is obvious on second thought that their coming together does not represent an authentic mode of being-with-the-other, the *Mitdasein* of a human transcendence in solicitous concern for the other. One would have to know the other to assume such a relationship. Nor is their behavior representative of a negative, inauthentic hatred.

True solicitude, which would be authentic, is merely feigned on all three of their parts; and the feigned solicitude covers over an actual indifference to each other's present condition, if not each other's insincere curiosity and embarrassment. Whatever their feelings for each other were in the past – motivated by mutual opposition to the restrictive morality of the nuns, presumably, since their habit was to join together in an act of defiance, enjoying a collective furtive dream of their future loves – in the ensuing time each has become lost in her own world of inauthentic solicitude for the others. Such an interpretation is borne out by their inability to converse freely about their pasts and the apparent bad faith of their present relations *inter se*.

The world of their today is nothing more than a repetition of everyone's world of the everyday: when old friends meet to relive the past they had shared, they engage in ritual. But in the past, they had gone their own respective ways, only to come together by chance once more. In this meeting each goes out only to come back and find it impossible to speak an authentic word; and Flo's final expression of an imagined or indirectly perceived present merely serves to emphasize their disconnectedness from reality. They had always been thus, beginning with their daydreams on the bench of the convent school.

In this way it becomes apparent that the ontic events of their lives bear an ontological significance, and Beckett has told that story many times over. If this recounting of the story is to add to his *oeuvre*, there must be another clue for the hermeneutical expansion of this unique situation.

We might, for example, begin with the title: *Come and Go*. A reference to the coming together and parting of the three characters? Obviously. But the French translation offers a further clue: *Va et Vient* would express the same surface idea, but suggests as well the distinctive movement of partners engaged in an act of sexual intercourse; and that indeed is what these three were day dreaming about in the convent court. The naughty girls. At least now we know what brought them together: the absence of a fulfilling masculine intention. But why should that send them spiraling around each other now, on a bench in the dimly lighted dark?

As for the lighting – because all three are acting in bad faith, aware of what ails each one of themselves in particular and of their group as a whole, yet feigning to believe otherwise: that's a consciousness, multiplied three times over, dimmed by self-deception. As for the activity – latent homosexuality? In the past, indeed, prevented by the convent ambiance to find a heterosexual outlet for their childish eroticism, they were forced to express that imperious drive covertly – in a daisy-chain on the court bench. Thus, if their spiraling around each other now is to have any significance, it is for the purpose of illuminating the difference time and a change of circumstances have wrought on their relationships. But the relative motions around and through each other's positions have a surface significance as well, and that surface significance is of a form, appreciable for itself, as well as for whatever it may add to the depth already revealed.

Let us consider first the significance of the form in the abstract. Such a motion would have the nicety of being closed, like those of a mobile sculpture, emphasizing the internal changes that alone are possible in any closed world of shifting circumstance, such as that described in the players' present situation. Together they form a human mobile.

Thus, staying close to the ontological significance of the playlet (Beckett refers to the play, in French, as a "dramaticule"), this clue offers a further set of analytical procedures. For, if the motion of the players is as significant as the discourse of the characters they represent, the worldhood of this world would be traceable in the patterns of their changing relative positions, even if the ambiguity of the characters' language tends to preserve the obscurity of their closed world. The audience, in this respect, is like the nuns who might have surprised the girls in the past.

Beckett emphasizes the changes in the players' positions by marking out, in his "notes" to the play, the seven patterns of relative positions. They are as follows:[11]

1.	Flo	Vi	Ru	:	opening sequence
2.	{Flo		Ru}	:	first confidence
	{	Flo	Ru}		
3.	Vi	Flo	Ru	:	first return

4.	{Vi		Ru}	:	second confidence
	{Vi	Ru	}		
5.	Vi	Ru	Flo	:	second return
6.	{Vi		Flo}	:	third confidence
	{	Vi	Flo}		
7.	Ru	Vi	Flo	:	third return, final tableau

It is apparent now that after all their comings and goings. the three players have changed positions on the bench in a distinctive pattern: $\{a, b, c\}$ at the start to $\{c, b, a\}$ in the end. Are Ru and Flo in contention for the affection of Vi? No more so than the other possible relations.

Nor no less. The distinctive pattern of their changes in position is a chiasm. By chance or intent the same chiasmatic figure is found between the meanings of the words in the English and the French titles of the play as

Come and Go

Va et Vient

Drawing a cross between the associated words, we realize two symmetrically arranged triangles, base right and base left, with apices sharing a common point and the bases left undetermined, open, as it were, to variation.

Oddly enough (again by chance or by intent) the conversation follows a similar pattern, beginning always with the central player. Vi opens the conversation, since she is seated in the middle of the other two in the first sequence. Her gambit is refused; and Flo, first to Ru's left, then in the middle, delivers her secret confidence. Ru responds with dismay, and Flo prays that Vi be kept in ignorance of the confidence. Direction of conversation: triangular, apex left, with return to middle, now occupied by Flo.

Flo's attempt at conversation *à trois* is stymied; and as if on cue, she exits, leaving Vi and Ru together. Ru delivers her confidence to Vi, receives her exclamation of dismayal and prays that Flo be kept in ignorance. Direction of conversation: triangle apex right, with return to middle, where Ru attempts to bring all three to the topic of their former comradeship. That too is blocked, and she exits, leaving Vi and Flo together.

Vi whispers her secret to Flo, who responds with dismayal and Vi prays that Ru be kept in ignorance. Direction of the third dialogue: triangular, apex left again.

One would expect here that Vi, now in the middle once more, restart the conversation; but that would only repeat the process and force her to leave again. Instead, they join hands in their peculiar manner; and Flo, who has all along remained in the dark (not perceiving any changes in Ru and not understanding the embarrassment of the two others), ends the conversation with her exclamation of sensing rings on the others' fingers – and on her own as well.

Thus, in the pattern of their directed conversation the two overarching chiasms (traced between relative positions of the players at the beginning and end of the action, and between the associated meanings of the English and French words of the title) are broken up into constituent parts, as the players come and go, into three changed, individually manifest, segments: swing left, swing right, swing left again.

Vi, at the center, had started the conversation; and Flo, on the right, ends it. The present action, then, given over to the inauthentic speech of the characters, is suspended at the center where it begins and returns for each assembled group of three. The turning point of the "mobile" changes from left to right, back to the left, and then to the right again. Thus the motive force of this human mobile is the consciousness initiating an inauthentic discourse.

As the fulcrums of the turning points change, a new inauthentic consciousness continues the impulse of the motion, but only to pull the entire arrangement hither and thither, without for this reason destroying the overall configuration of the inauthentic world so described: the chiasm generated by the comings and goings – nowhere. All are lost. Vi and Ru wear no rings, and Flo discloses the deception of the other two in her words as she discloses her own in her actions.

In this way, the spatiality of this inauthentic world is joined to the temporalities of the three consciousnesses to form a universe in which nothing happens of consequence, outside the disclosure of the inauthentic springs of the action. For this reason it is not terribly important, at this point in time, to inquire about the substance of the confidences revealed in private within the three separate pairings. Any group of three is an unnatural arrangement. Someone's hair might have begun to turn grey, as would be normal for an aging group of women; they might all have been seduced by the same unfaithful lover; or, as is more likely, have become married or betrothed without telling their soul sisters.

All that an analyst would be able to deduce from the patterns of the conversations is that, although with respect to the persons revealing their secrets the activity is transitive – from Flo to Ru to Flo – and that the directions of the conversations from speaker to listener have switched from the right to the left and back again.

But tarry a moment on the colors we see switching position before our eyes. Ru's violet dress harmonizes with Vi's red, since violet is a composite of blue and red; but the blue in that violet contrasts with the yellow in Flo's dress. From the point of view of the colors, too, Flo is the odd-woman out.

In the bridge to the first confidence, Vi in her red dress appears to Ru as being about the same as she was in her youth; and indeed in a sense they are now alike (as they were in the past) with respect to their harmonious reds. But there is that hint of blue in Ru's violet dress, which would indicate a difference between them now. That's true; the appearance of sameness was only approximative.

When, in the bridge to the second confidence, Ru asks Vi if she has noticed Flo's appearance, the answer is that Flo appears the same, more or less. But since she is now wearing a yellow dress, Flo is as different from Vi as one primary color from another, and the difference could hardly be passed over even if Vi were color blind (since that phenomenon occurs in the simultaneous perception of red-green or yellow-blue). Now appearing the same to Vi, Flo still looks different – from Vi. Their contrasting primary colors, set off against Ru's violet would only produce an orange. So be it. The result may be significant.

In the sequence of changes called above "the first return," Vi, Flo and Ru are seated in that order from left to right. Vi's and Flo's dresses would combine into an orange next to Ru's violet. The two composite colors break up into the following sequence: Red, Yellow, Blue, Red, a pattern that will re-occur, abstraction made from the things related, in the position of the six hands in the final sequence or third return. But of that, more later.

To continue the description of color pattern changes, we may finally note that in the bridge to the third confidence, Vi asks Flo if she has noticed Ru's appearance; but Flo answers that it is too dark to see. Either her vision has slipped, or she wishes to hide the fact that the blue in Ru's violet dress contrasts with the yellow of her own. Judging by appearances, anyone can see they are now different from each other, not alike. Flo is thus aware of her oddity, or of Ru's change in external aspect, but feigns not to notice it; and in this way she confirms her status as the odd-woman out in the intimate threesome.

As they are seated in the final sequence (third return), the pattern traced by their arms is itself a double chiasm:

as Vi's two arms cross over the left arm of Ru and the right of Flo, while the right arm of Ru and the left of Flo trace an arc that, completed, would constitute a closed circle. An apt enough symbol for the closedness of this directionless, love-barren world, where long ago dreams had taken the place of reality; wish, that of self-conscious perception. The appearance of the closedness of this world is both true and false: false, since other circles have intervened in time to short circuit this one; but true in its symbolic separation from reality. In their present inauthenticity, each is like the other, even in their responses to the revelation of a personal secret. Does she know? Heaven forbid! For if she did, she wouldn't be there, and there would be no play.

It is not difficult, however, to establish what those three secrets were. One must come up with a statement that would justify the exclamations and prayers of each respondent. Such a statement might concern the existence of a co-respondent, such as "I've been married!" To a person with whom one has enjoyed a homosexual attachment in the past, the response "Horrors!" would be perfectly justified. And since there was a third partner to the lesbian play of the past, natural, too, the question whether the other knows. Finally, since each of the three is now engaged in the reminiscence of their past, trying to restablish the old ties, the prayer likewise is perfectly understandable.

But why, then, reveal one's own change to one of the others as each of them do? Again: a natural response to that person's having recognized a change in the absent third. The reappearance of the third in each instance prevents the receiver of the confidence to reveal her own secret. It is the third that blocks the free flow of information, since in their embarrassment before the judgment of the third all the individuals are thrown back inward upon themselves, where their threeness is irretrievably lost.

Flo is the least embarrassed by her own change of condition. She is the first to reveal her secret to Ru, after perceiving a change in Vi. She discloses their past habit of daydreaming of love. She does not notice any difference in Ru, and is equally in the dark concerning her own difference from Ru (the color complementarity). Her last differentiating characteristic is her final position on the bench. Let us continue to follow the pattern.

In the closing tableau, in which their arms are doubly criss-crossed, there appears still another chiasm, as the double cross now becomes triple. Attention should be focussed on the hands. Ru's right and Vi's right on Ru's knees at the left; Ru's left and Flo's right on Vi's knees in the middle; and Vi's left and Flo's left on the knees of Flo at the far right. If the scene gives the appearance of a seance, the pattern is interesting: two right hands on the left; two left hands on the right; and in the middle, a right hand on the left and a left hand on the right of Vi as the central axis.

The configuration is complicated. It may be clearer in schema, as follows

This new pattern, it now appears, is the same as that of the earlier mentioned color scheme of the first return: likes at the extremes, surrounding unlikes in the middle.

Now focus attention on the knees of Vi. By the order of previous speeches, she, being in the middle, should initiate the conversation. She doesn't; instead, Flo gives her revelation. Looking at Vi, but surprisingly hearing Flo, the audience's consciousness is presented with its final visual revelation. The left hand of Ru grasps the right of Flo; but with respect to Vi on whose knees they are placed, Ru's left hand is on the right and Flo's right hand on the left: a final chiasm, then, as follows

If we now amuse ourselves by counting the criss-crossings formed in the comings and goings of the three players, we find a total of six. They are: English and French titles, end positions, color scheme in first return, two sets of crossed arms, and the final right-left transference of Ru's and Flo's hands with respect to Vi's centrality. If the chiasmus of the two titles of the play is thought to be too far-fetched to be applicable to a single performance of the play, there is still a sixth chiasm in the real positioning of Ru's and Flo's hands as presented on scene, with Vi the center of perception, and the apparent positioning of their hands to the audience, with the bodies of the audience as the center of the configuration. What is to the left of a person facing him is on the right of the perceiver, and through a mirror one's own right appears on the left-hand side of the image.

Six crosses, then; or three double crosses, as indeed each has double-crossed the other, having changed sexual orientation in the interim between their past and their present, as well as in their motions in the present, which symbolize still another betrayal in the exclusion of one of the partners (an action repeated three times).

But a transitive double-cross is equivalent to the statement of a truth. Although Flo does not feel the rings on the left hands of her comrades (there are none to be felt), their attempts to recapture their past by reassuming their characteristic daisy-chain are foiled. She senses the truth. The past refuses to be recaptured; not even the magic circle can perform the Proustian recall: their present affective memory is a bare shadow of the past reality, which has been irrevocably changed since they gave up their collective day dreaming of love for the fulfillment of a true heterosexual relation.

However indirectly the word is received, it is the power of the word to give birth to the image. Flo's final report was therefore not a hallucination after all. She did not feel any real rings, although she is in contact with the left hands of her two partners; but she now at least has felt the effects of their presence in their very absence. Such is the ambiguity of the word "to feel," to indicate an object of concrete tactual perception (here false) or an affective correlate to any sensuous contact. In this latter sense she does feel the presence of those rings – at long last.

That is why the present, in Flo's consciousness, is filled with a true symbol of all their conditions. They are playing at the game of recapturing an irretrievably lost past. And the images of those *three* rings – Flo's own, Vi's and Ru's – are created for the consciousness of the audience through the power of her words alone.

In spite of their presently linked hands and arms, the friends had indeed grown no friendlier and could not relive the friendship of their past in the same way now as then. If the play is Proustian at all, it is anti-Proustian. In an inauthentic world such as theirs no one has a friend; for there no one is capable of giving oneself to another. The appearance of the contrary (in the final return sequence) is therefore deceiving, and assuming that position is only the means the characters use to mask the differences that separate them. Each exists, each acts, in bad faith, deceiving herself by deceiving the others. The world disclosed in such a triple deception merely turns around a shifting fixed point of central reference.

Does the significance of the play's abstraction stop here? Hardly. All this abstract concern for the chiasmatic configuration reveals a search for a concrete personal identity. The clue for this interpretation is found in the philosophy of Maurice Merleau-Ponty, the French companion of Sartre who was attracted to Heidegger's philosophy late in his life.

In his *The Visible and the Invisible* (a title itself significant of an abstract meaning, the invisible, portrayed by a visible sign), Merleau-Ponty describes self-perception as the chiasmatic crossing of the projections of selves into the other.[12] In the temporal development of any two persons, as indicated in the abstract, dialectical schema of Hegel, the concrete content must be generated by an interaction of the two transcendences. No consciousness, first of all, may be taken in abstraction from the body, so the communication between self and other is most clearly manifest in carnal knowledge – the object of the three characters' dreams. Past circumstances prevented such a knowledge except via their dreams, and the substitute form of an unfulfilling homosexuality. Their most recent pasts changed the order of dream and reality, making it impossible for them to reconstruct the distant past of their youth, for figure and ground had switched positions in still another chiasmatic crossing.

When we attempt to schematize the paths of two projections of unity in a single temporal fulfillment, the following results:

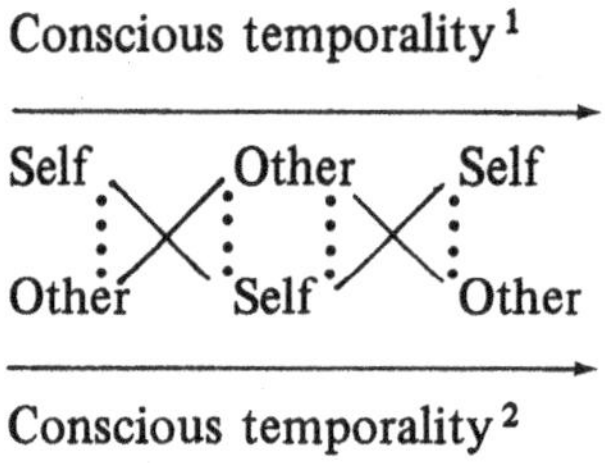

In this scheme the synchronic cross-section of the two on-going projections represents a moment of real time. So each self, at first separate from the other, moves into contact with the other, only to separate, in time to be joined again, and then separate once more. But, *mirabile dictu*, that is precisely the trajectory of the individual transcendences represented by the three players, developed sequentially in the three separate repetitive conversation scenes; and the final tableau does nothing but collapse them into a single presentation that contains all the triadic relationships of the changing triangular forms: in each sequence one talks to the other about the absent third. And even that third is representable on the schema above, as the missing bases of the four triangles. Being missing, it is a variable, and the variable may be exchanged for a value that is either masculine or feminine.

The meaning of that obscure title then becomes clear: the configurations of the triads are the same, whether we look at the final tableau or understand the peregrinations of the convergent and divergent conscious temporalities described in the changes of their ontic conditions.

The representation of self-consciousness in the figure three is nothing new to Beckett. Recall the three voices of Molloy, and the "pretty three-in-one" of the narrator of the *Texts for Nothing*. Indeed, with respect to the world, consciousness is nothing. It is the denial that the world exists in itself, but only for it, as an object of perception. Consciousness, on the other hand, exists in itself, and can come to exist for itself only as it perceives itself in relation to some Other.

The dialectical swings of the play's conversations, then, do nothing more than represent in the abstract the internal structure of self-consciousness. Vi, Ru, and Flo are personifications of a functioning self-consciousness, partly self-determined and partly determined by the Other. And there is still further confirmation of this now almost horrifyingly complex hypothesis.

Beckett has dedicated this play to his own friend and editor, John Calder. That fact, too, may be significant. For there was another Calder (Alexander) who delighted the world with his invention of mobile sculpture, which presented in physical fact a similar construction of suspended forces in motion. If a sculptor can turn the trick, why not a dramatist? And if a dramatist, why then not a philosophical dramatist?

Indeed, *Come and Go* represents a story of three friends whose ontic comings and goings are structured on two principles at once: a dialectical, existential ontology, and the principle of the balance of forces in a mobile construction. Or perhaps it would be better to say that the ontic events of the three characters are revealed in a pattern of ontological disclosure that is figured on the visual surface of the play's abstracted motion. If so, there would be once again a simple ascending order of meanings to be apprehended: the ontic event, the abstract pattern, and the ontological structure.

The motions of a mobile have a double determination, the balance of their weighted positions and the wind of the environment; they do not move of themselves. Nor do the characters in this play. Although they illustrate the pattern of dialectical relatedness with an Other, they are moved by the forces of convention and habit that have little to do with a resolute decision to commit oneself to the welfare of another.

In disclosing his fictional world in the way he has – matter, form, and substance – Beckett has given a lesson to formalist critics; he has pointed out still another direction for the development of the dramatic medium, retaining the device of structuring the concrete whole (matter and form; ontic appearance and abstracted spatial relations) on the basis of a single philosophical idea (the substance). In Merleau-Ponty's terms, the audience attains to the experience of understanding the "invisible" idea when they have correctly perceived the visible events in their proper perspective.

One can achieve more with less. The test is the situation in which sense is overlaid with non-sense – not to say 'nonsense' or 'absurdity.'

If the earlier plays of Beckett were considered "experimental," because of their multi-layered symbolism, the three plays analyzed in this chapter show a true spirit of dramatic experimentation. The limits within which the dramatc representation operate are the same as in any other artistic medium, even two as traditionally disparate as music (nonobjective) and sculpture (representational). The history of art has shown, however, that the most nonobjective of the arts may symbolize some object or idea, and that the most representational may be turned into a nonobjective abstraction. What are the determinate limits of artistic creation? On the objective side, the nature of the surface materials that must be arranged into a meaningful context; and on the subjective, the force of the artist's own knowledge and imagination expended in the arrangement.

If the critics of Beckett's drama wanted experimentation Beckett would gratify them, and give them a lesson in theoretical aesthetics as well. He abstracted from dialogue entirely in the first two plays of this chapter, and reunited it, in the last, with the significance of the spatial relationships assigned to the players on stage.

In *Act Without Words II* he used a musical format to lend a significance that surpassed the concrete meanings of the mimed actions, achieving the feat with a play of stage positions. Being the same in their differences as signifying the unchangedness of an inauthentic world, the mimed actions of the two players were reduced to irrelevance; and the concrete reference of the individual mimed acts was attached to the forms the audience had perceived: the indefinite repetition of symmetry, that closed perceptual form of simultaneous likeness and difference.

Something similar happens in *Come and Go*, but here the format is sculptural. Dialogue, which once was thought to be definitive of the dramatic medium, prior to the development of mime (itself an abstraction from dialogue, or a replacement of the centrality of intrigue by that of the ancient shuffling chorus' dance in a simple reversal of figure and ground on the stage), is here used to make abstraction of the content of speech, which must be deduced from the abstract relations presented. As in the earlier transformation of the Greek chorus in mime, only the form of the discourse remains significant – and that signification turns out to be the same philosophy Beckett had used in his earlier plays, i.e. existential ontology, which once again lends its structure to the events of the created world.

It still remains to be determined, however, whether the same effect may be registered in other media, and whether the philosophical structure will remain the same or change. Beckett's plays for cinema and television will provide the examples.

Philosophy reflected off the silver screen? Projected through the tube of blinking cathode-rays? We need only look to see. Such will be the case if "the invisible" turns out to be the in-visible. Pushing the theatrical medium to its limits in these three plays, Beckett has called attention to the artistry necessary to turn pain and suffering into successful tragedy. And, as always, he does not lecture the point; rather he shows it.

CHAPTER XII

SOUL MADE LIGHT, AND SOUND

Writing for the filmic media has its hazards, some of which are shared with writing for the stage. Producers and directors inevitably come between the artist and the object of his imagination, and sometimes shortcircuit their connection. While the producers are charged with the problems of financial backing, they usually insure close cooperation with their directors by exercising their personal choices in selecting the personnel to fulfill this function; and the directors themselves may be given a free hand in the choice of performers.

The Miami production of *Waiting for Godot* was from this point of view an enormous comedy of errors from the start. Upon the recommendation of Thornton Wilder, the producer, Michael Myerberg, chose Alan Schneider to direct the play, and cast two well known comic actors, Bert Lahr and Tom Ewell, to portray Estragon and Vladimir. In the New York production Schneider was replaced by Herbert Berghof, and Ewell by E. G. Marshall; and the play caught on.

According to John Lahr's account of "The Fall and Rise of Beckett's Bum: Bert Lahr in *Godot*,"[1] Schneider's direction of the Miami fiasco was hampered by Myerberg's choice of a stylized set, and his own deference to the play's philosophical structure led him to underestimate its comic potentialities, which the producer had hoped to capitalize on by hiring established comics for the major roles. On the other side of the picture, there was Lahr's insistence on his (Estragon's) rôle as "top banana," and his almost insensate pursuit of laughs.

But the play was a success in New York because its promotion was brought into line with what might be expected from the genius of Beckett, with the surface comedy brought into ironic relationship to its tragic depth, under the direction of Berghof. In general, the surface antics of the clown merely serve as a mask for hiding the underlying humanity that suffers as much as anyone's. If Estragon is only an inadvertent clown, Bert Lahr was, in the original American production of *Godot*, too self-consciously such. The actor, then, is as much a hindrance to the realization of the dramatist's fictional world as the producer and director; and if all three of these rôles cannot be brought into a cooperative comprehension of a play's potential, there is little hope for maintaining the artist's vision.

What is it about the life of a man looking for himself and a place in the world that appears comic? The way he does it, perhaps; knowing the truth but fleeing from it, looking for himself as he flies away from what he is looking for: an absurd project, absurdly pursued; an abstract logical impossibility – were it not for the inner scission of a man's psychological structure into two, or three, as the case may be.

Such is the subject of *Film* (Evergreen Theatre, 1964), Beckett's first excursion into the filmic medium. It too was directed by Alan Schneider, with Buster Keaton impersonating the two roles of a single protagonist, the man seeing and the man seen. Once again, therefore, the surface comedy of the situation is offered over to the ironic antics of a well-established comic that run about twenty-two minutes off the projector.

And in this new medium there is the additional problem of the camera. How does one photograph a sequence in which a man sees himself seeing the objects of the world, or himself aware of being seen by others as an object in the world? Beckett admits the impossibility and suggests that the simultaneity of the two conscious processes not be directly represented. A difference in background color, perhaps, for the man seeing and being seen. It is, of course, possible to film an entire drama from the point of view of a single person engaged in an action. When this technique is used, what the audience sees is what the character sees; it focusses consciousness on a single datum, concentrates attention, and establishes a point of view.

The technical problem of *Film* is to establish a point of view on something ordinarily considered to have no external point of view, since it (the consciousness) is the point from which other things are viewed. Where the seeing consciousness may be presented as the camera's reproduction of the objects it brings into focus (a plus, for the realization of this film), there can be no camera to reproduce the intersection of the two points of view. Beckett merely suggests to his technological experts what he thinks might be done, and gives them a free hand for the realization of his plan. Drama and the film may only be performed in cooperation, and the writer must avail himself of technological assistance.

No sound is to be heard in the film, except a woman's "sssh!" as she prevents her male partner from commenting upon what he sees. That is to be left to the camera, which works from only two points of view: O's, the man fleeing from the sight of the other, and E's, the man pursuing the other, attempting to keep the other in sight. There is the surface comedy of the situation's unreality, and the stylized actions of the fleeing man, frozen into immobility by the gaze of his tormentor. Although these two points of view represent two functions of the same consciousness, the audience does not discover the identity of the protagonist until the final sequence of the film, where the surface comedy attains to the deepest kind of tragedy.

What the director and cameramen had to work with was a "score" composed in three parts: general remarks; an outline of the action; and the author's notes.

Unlike any other creation of Beckett considered thus far, the general remarks explicitly state the philosophy used to structure at least one aspect of the film: Berkeley's principle of subjective idealism, *Esse est percipi*, which states that to be is to be perceived (or to perceive). The expressed part of the principle holds for the objects of the world; and, indeed, the camera brings such objects into existence for the perceiving audience. The unexpressed part of the principle is the source of the film's paradox: the perceiver himself cannot be perceived qua perceiver. Not even an eye seeing itself in a mirror sees itself seeing, but only another eye in which itself is reflected; if it could see itself it would only do so as reflecting the other reflecting itself in an infinitely closed circle of the comings and goings of the reflection. And in this activity of the come and go, there is no love of any kind: only the relentless activity of mutual self-reflection.

Should the camera be able to represent the impossible, that is what it would show. As Beckett states it in his general notes: "All extraneous perception suppressed, animal, human, divine, self-perception maintains in being."[2] In order to compensate for the impossibility of the situation, the protagonist engages in a symbolic suppression of those other forms of perception. He has fled all human contacts into the empty room (Beckett supposes, without elaboration) of his mother; puts out the cat and dog; covers the mirror; covers over parrot cage and gold fish bowl; destroys a picture of the all-seeing eye of God the father, in order to sneak a look at pictures from his past, when he is cornered by his alter ego, his perceiving pursuer. In this symbolic way self-perception is maintained in being.

Thus, the original project, which Beckett states in the general remarks as a "search of non-being in flight from extraneous perception"[3] breaks down "in (the) inescapability of self-perception."[4] Unfortunately, however, the inescapability of that final self-perception is not understandable in terms of the Berkeleyan hypothesis, unless the second and unexpressed portion of the principle is made explicit: the being of a man is to be a perceiver.

I shall therefore attempt an alternative reading of the film's structure to show that Beckett's intuition – the inescapability of the final self-perception

– is accurate, and that it is interpretable in terms of the same existential ontology used in the prior chapters of this work.

Beckett is aware, of course, that no truth value attaches to his, or any other author's use of a philosophical world view to structure the world of a given fictional context; and he says so explicitly: "No truth value attaches to above, regarded as of merely structural and dramatic convenience."[5] That, after all, was the lesson he had learned from Joyce, and practiced so well in his own prose and drama.

The literary artist need not argue his points; if he did so, there would be a confusion of his task with that of the philosopher's. The value of the dramatist's work is its expressiveness, not its truth; to argue otherwise is to confuse categories, to insert 'truth' for usefulness, and 'beauty' for truth. Only a romantic, befuddled with Platonic idealism, could argue, like Keats, that truth is beauty, beauty truth. The literary artist's job is to show, and not argue at all. All this Beckett understands fully well.

But what he has lost sight of, for once, is the complete structure of his symbolic context. For the order of his images may very well be achieved by the structure of one philosophical idea, here Berkeleyan idealism, and yet point towards still another level of philosophical signification. Although in the film the usual words of the writer's script have been replaced by photographic images, and the sequence of these images signify the first order philosophical depth, nothing prevents the first order philosophy from signifying still another, more profound, depth significance; one, indeed, in which the inevitability of self-perception follows from the very attempt to search out non-being in the flight from extraneous perception.

Instead of the usual ascending order of significance, brought to closure by the tension between expressed depth and expressing surface, and controlled in a strictly literary medium by word, image, idea (world), and worldhood of that world, the film dispenses with the word. But the silence of the filmic medium allows a more direct presentation of the being of consciousness, gathering itself together in the silences that surround its multifarious voices, and shows directly what is going on in the consciousness that sees: the organization of a spatial world.

In *Film* that consciousness flees being seen even as it is relentlessly pursued by itself in reflection. Instead of the word, then, images; and these images are indeed of the world; in their being, constitutive of the being of that world. And, finally, that world points to the worldhood of that world as the self-fleeing consciousness is trapped in its own reflection. Berkeley's philosophy is, then, only an ontic description of the events taking place in the "worlding" world of a human consciousness fleeing its own facticity.

The clue for the interpretation is given in the temporal references of the images. The first two sequences, a street and a stair scene, are photographed from E's point of view: that self left behind in the inauthentic flight of the present. The past was the moment of decision to escape the vision of others. And the future is the final tableau, where the fleeing self is caught in the vision of the self pursuing – an indication at once of the futility of the initial project of flight from self, and of the tragedy of the human condition in which all projects are defined as a project from past to future where, if any kind of identity is to be achieved, the self must be able to discover itself as the realization of its own project. Transcendence, it's called; the being of a human being, thrown, as it were, into its situation and forced by its nature to decide the question of its existence by an act of its choice.

To flee from oneself, under these conditions, is to fall – into the inauthentc world of everyman, where the anxiety of authentic self-knowledge may be experienced as the fear of being seen, of being the object of another's judgment. Here, as before, then, Beckett's insight is into the tragic circumstances surrounding a man's attempts to be himself in a world determined by the perceptions of others, whether it be those people he disturbs in the street, the animals in the room, or the all-seeing eye of God.

To indicate the self-diremption of the man who has decided to flee himself, Beckett adopts the convention of having E (pursuer, perceiving) follow O (pursued, perceived). The camera instantiates E's point of view. Always behind, to indicate the pursuit, E remains within an "angle of immunity," where he is unperceived by O. O may then continue his flight unhindered by his own perceptions' being perceived. When that angle (45°) is exceeded, E enters the perceptual field of O, who then perceives himself being perceived, and cringes from the visual contact. He covers his face, hides against a wall, etc. When E recedes back into the angle of immunity, O's flight may continue.

E is perceived by O twice, as E inadvertently exceeds the angle of immunity; and the final revelation shows E and O face to face (like eyes in a mirror) in the secrecy of a closed room.

The outline of the story given in the score is varied somewhat in the production of the film. In the score Beckett establishes the unreality of O's world, by calling for the photographed setting of the surrounding scene from E's point of view. That was to be the first clue of the identity of their two worlds. Adopting the director's point of view, let us imagine that scene.

In a period setting, circa 1929, of a small factory district, workers come and go, in pairs. Bicycles go by, with men carrying women on the crossbars.

A carriage, with horse and whip swinging driver. All look conspicuously at each other or at the objects around them: poster, shop window, and the like. The *esse est percipi* theme is established.

O makes his appearance, hugging a wall to avoid the glances of the others, but is caught up by E's vision, fixed by the camera. He goes counter to the current, wearing winter clothes in summer, collar of his greatcoat high to hide his face. E exceeds the angle of immunity; and O freezes, cringing against a wall. Released from perception, he continues his furtive flight.

Hastening along, O bumps into a couple reading a newspaper: a woman and a man, she with a lorgnon and pet monkey, he with a pince-nez. Recovering from the shock of the contact with O, she brings lorgnon to eye (she is farsighted), while he takes off pince-nez (he is nearsighted) to look at O. Then changing the focal distances of their eyepieces, she lowers her lorgnon and he puts back his pince-nez. As he is about to vituperate, she quiets him with her "sssh!" Feeling E's gaze, they change focal lengths once again to look into the camera. Their faces assume an anguished expression Beckett calls "the agony of perceivedness," which is repeated in a later scene on the face of a flower woman descending the stairs O and E are then climbing.

The woman's pet monkey passively regards the passing events, and sets the stage for O's reactions to the animals in his mother's room.

The filmed version of the story contains only the sequence of the man and woman, indispensable, says Beckett, for establishing the *esse est percipi* theme, and the setting is a run-down, decaying district of a large city. To be seen seeing is to have one's transcendence transcended, one's possibilities made probabilities, one's subjectivity made object. That indeed is the Sartrean interpretation of the look of the other.[6] And if this scene is indispensable to the film's entire structure, it is hard to understand why Beckett has not himself seen that Berkeley's principle is transcended in the structure of his film. It is only the perceiver's objectivity that is determined by the look of the other; his subjectivity is his activity of seeing or of pursuing his projected goal.

The flight up the stairs is marked by two incidents, both registering the transcendence of a transcendence. E, for a second time, inadvertently exceeds the angle of immunity, and is seen by O, who cringes against the wall; but when the camera switches to the flower woman descending the stairs, O is released from the anguish of his perceivedness, since E now looks at the woman. She in turn feels the anguish, and falls to the floor with her face in the scattered flowers: confirmation of the objectifying look of the other.

In the meantime O has continued his upward flight. By the time he has reached the second landing E has caught up with him again, and stays well within the angle of immunity, as they both enter the room. Here once again there are technical difficulties for the camera. Part of the sequence is shot from O's point of view, part from E's; and when E deliberately exceeds the angle of immunity and comes around to face O, the two points of view must be shown in montage. When O is viewed by E, the headrest of the chair is found in the background of O's head; and when E's presence is felt by O, O perceives a background of tattered wall paper (the shattered world of O).

The room sequence takes place in three sub-routines: the preparation of the room, all external perception and all symbols thereof destroyed; O in his rocking chair, looking at a series of photographs of himself –

first, as a six month old infant in his mother's arms
second, as a four year old child, praying
third, as a fifteen year old schoolboy
fourth, as a twenty year old graduate
fifth, as a twenty-one year old fiancé
sixth, as a twenty-five year old service man, holding a little girl
seventh, as a thirty-year old veteran, black patch over left eye –

the record of a life, then, to which O reacts emotionally in sequence; and, finally, the slow fading of the two points of view into one.

Within the second sub-routine, O takes out each picture, looks at it, and places it on his lap. He studies pictures 1–4 for about six seconds each; 5 and 6, with trembling hands, running his finger over the little girl's face, about twice as long. After looking at picture number 7 about six seconds, he tears it into four and throws the pieces on the floor. He proceeds through the pile in order, throwing all the pieces to the floor. Number 1, made of sterner stuff (his relation to his mother), resists being torn, but he finally succeeds in destroying it too.

Following the destruction of these symbols of his past life, he sits at rest in his rocking chair, where E (camera eye) moves in. O awakens, stiffens under the look until E removes self into the angle of immunity. O's rocking, started at his release from the anguish of perceivedness, comes to a halt as O dozes once more. E circles his subject, making an object of him. Startled in his sleep, O with same black patch over the left eye seen in his photograph as returning war veteran, looks up at E, in an identical black patch. With an expression that is neither "severe" nor "benign," E continues to look at O with acute intensity. In the background, a nail in the wall. Cuts back and forth; O with the agony of perceivedness, rocking nervously, E still acutely intent. O covers face in hands; then shoulders bowed, head sunk into hands, gently rocks. Fade out on the rocking coming to a standstill.

So much for what the audience has seen. In capsule, a world brought to life by a sequence of objects brought into focus by the camera: a man fleeing from himself into the safety of his mother's room (womb); destruction of all symbols of external perception; deliberate destruction of the man's past, most recent to most remote; and final confrontation with that past he has fled.

The traumatic circumstance: the loss of wife and daughter while he is away in the war. From that moment on, the man has been fleeing himself, a flight from the loneliness of a life without his loved ones, during which flight, that he is, he cannot escape the anguish he is; his project to flee is itself an act of bad faith.[7]

The fact that he cannot escape from the awareness of his anguish is symbolized by the final investiture of the dual aspects of a single consciousness. No one may successfully flee from himself – unless it is in the pursuit of a self in a world where to be is not to be perceived, but which is the result of one's own resolute decision to be that self. That too would be a flight, but from oneself towards oneself through the organization of a personal world. When the protagonist's old world had been crushed by the effects of the great war, he had made the mistake of defining himself by a fixation on the ruined past. His transcendence became aborted, and he lived the impossible project of fleeing the anxiety of his existence.

Thus, the clown has two faces: the exposed one that revels in the absurd, but only to cover up the tragedy of the absurdity of his personal existence. If there is a philosophical idea grounding the total phenomenological structure of this film, it is not the device of Berkeleyan idealism. That device is fitting for the necessity of the camera's intervention between the author and his imagined world; for in a film whatever exists is brought into existence by the camera's eye. But when those images are related into the structure of a global significance, the critic must find still another philosophical device – the existential ontology represented by the meaning-relations established in the film's Berkeleyan, ontic world.

It remains a matter of useless conjecture, then, whether Beckett would hesitate to defend the philosophy that structures the entire fictional universe of this film; that too is an artistic convenience. But the created universe, it should be recalled, is the context of the expression, i.e. a stratified system of signs and signifieds. In the film, the first level signs are the photographic images. They signify the world of their relations, and these relations themselves signify a still further philosophical system.

Author Beckett's and director Schneider's success at communicating the ontological worldhood of the Berkeleyan ontic world displaying itself as a phenomenon may be judged by the film's international reception. Unconcerned for the film's intellectual meanings, Buster Keaton was perhaps a more versatile clown than Bert Lahr; and all three performances – the author's, the director's, and the actor's – were crowned with two film prizes: the *Prix Filmcritica* at the Venice Film Festival in October, 1965, and the Special Jury Prize at the International Film Festival of Short Subjects at Tours, in January, 1966.

Let us suppose, then, that the reason for this success was the juries' appreciation of the philosophical structure they had perceived in the images reflected from their screens. That, after all, is the only place a philosophical film may be truly perceived. The author's "score" contains only a set of directions to his director.

To run into circumstances that may be fortunate or unfortunate (fortunate, as with *Film*; and unfortunate, as with *Godot* in Miami) is the risk all authors of drama, directors, actors, and audiences must accept if live or filmed drama is to exist. The creation of a play is still a cooperative enterprise. The only guarantee of communication, wherever it is one finds himself in the chain of communication, on whichever side of the camera's eye, is the authenticity of the participation.

The deeply felt tragedy of the situation is that no one may flee from the flight that one is. Destroy your past and all its symbols, and you still must face your personal anguish.

Eh Joe (1966) places still another consciousness on the other side of the camera's eye. Filmed for television, and first shown by the BBC, this drama takes place entirely within the subject's experience of conscience. The voice he hears is and is not his own; its timbre and tone bespeaks the voice of a former love, which has become internalized to the protagonist's consciousness as the haunting, taunting voice of his own bad conscience. Try to flee from oneself, and one will be brought up before the bar of one's own self-judgment. Self-consciousness, which is still achieved by externalizing itself in the Other, and then returning to Self, cannot escape either from self or from the Other. The theme is repeated: one cannot flee from the flight that one is.[8] But how does it feel to try? One need only look – and listen – to come to an authentic understanding.

Eh Joe is a multi-media portrait of a man, of middle age, haunted by his bad conscience. The television camera registers the outward signs of his inward disgrace, revealed by the poetry of the words spoken, as if by the lost, dead, loved one (either wife or mistress) who in her life had been driven into infidelity by her lover's own faithlessness; and it is her voice the audience hears, lending the words that reveal her point of view that has been internalized within the haunted man's inner self.

The camera, which never loses its focus on the man, catches him unawares; first, in a series of actions that indicate a man under considerable emotional stress; and then in two close-ups, in which his internal agony is reflected by the appearance of growing tension in response to the words. When the words stop, his face decontracts, leaving an expressionless physiognomy, until the taunting voice begins once again to apply its punishment.

Visually, the action is presented in three parts. An opening sequence shows a grey-haired man, Joe, in his bedroom wearing slippers and a dressing gown. In spite of the relaxed circumstances, he shows all the signs of his inner tension.

The camera catches his back, shoulders haunched, as he sits on the edge of his bed. He gets up, goes to the window, opens, looks out; closes, draws curtains, and becomes immobile once more. Still from the back, the audience sees him leave the window for the door, which he opens, looks out, closes and locks; pockets the key; draws portal curtain, and freezes into his immobile tense position. From there he goes to the cupboard; opens it, looks inside; closes and locks; pockets key, and freezes into same tense stature. From the cupboard he returns to the bed, gets down on all fours, looks under the bed, gets up, and sits in the same position there as in opening sequence.

Having found nothing in his immediate surroundings and nothing inside the room, he knows he is alone and begins to relax. The camera catches the relaxation in the first full front portrait, when Joe hears the voice of his former beloved calling him by name. His response is to open his eyes, and listen. Alone, he is not alone after all; his self carries with it its own other. That is the nature of self-consciousness.

And, as the philosophy explains in the abstract, the play shows in the concrete how self and other have come together in the past, separated in time, and now inevitably re-converge into a single struggling moment in which the former slave exerts its freedom to enslave the self-deceiving master. The opening series of camera fixations itself symbolizes that third consciousness, the phenomenologist's own, to whom the whole process is revealed. In his fear, Joe has already lost all appearance of the masterful male; and that feminine voice will explain his enthrallment, as Joe's tension grows in response to each revelation.

This inner drama takes place sequentially in a series of nine movements of the camera. In the preparatory scenes, the voice has begun by taunting its victim: nothing forgotten? The light is still on; the lice might see. Can't get in that bed? Did you change it? Your heart falling to pieces, eaten with worms? Eh Joe. . . . The camera moves closer, to show each response to the taunting, haunting voice, and will continue to do so until the whole story of Joe's faithless love has been made clear.

The steps of the revelation are as follows, beginning with the revocation of the past love between Joe and his mistress. On their parting, Joe had promised her happiness for tomorrow.[9] There was no happiness in that bed, then, only Joe's momentary pleasure. But his parting words were prophetic; his loving mistress has died, achieving the only happiness possible in this life of love for someone who is capable of loving only himself.

The second revelation exposes his pathetic present. What's going on inside that head, behind those expressionless eyes? Memories, no doubt, of all those dead – Joe's father, his mother, whose love for him was full of pity, the girl whose voice now speaks, refusing him the pity he deserves. Only one passion is left to be registered in that skull – the drive to kill those who are already dead, and that's a manifest impossibility, not only because they are already dead, but, being the others with respect to which his own self has been determined, they cannot be silenced, and their images cannot be erased from his memory, until he himself is dead. He might do it by suicide, but he had no stomach for that; he prefers his life of fear – fear of being seen, fear of seeing, fear of living, and fear of dying. There might be an escape in loving, but he has never known how to love. The camera moves in once more, in a relentless exposure of its subject: three seconds of silence, in which Joe may relax his tension; camera; four seconds of silence in which his face is studied by the audience. Then the voice, which tightens the face muscles once again.

This time the subject is the living present. The one who comes in now, every Saturday, she makes him pay doesn't she? No love there; better not be caught short of change – or of memories of the dead to be erased from his mind. If that happened he would have to listen to himself. At least he has always been in love with himself; no problem there, except that he is growing old, and will disappear like the others into the silence of the tomb. Silence; camera; silence. Joe returns to Joe, audience not far behind.

The voice, on this fourth time around, brings the audience from Joe's future demise to their relations in the past. Why the tension now? He always said she had a voice clear as flint glass; could never hear enough of it, then. Indeed, in the disturbed breathing of their copulation, he could only hear the breath; the words were lost to the music of their dance. Breath, words, strain; breath, no words, stress; loss of breath. It all ends with coming to silence. But if one didn't come, the silence would be onesided. The breath would go on, exhaled in words, the flint glass now broken; but there in Joe's head where the sense of the words escape him, he can still hear the breath and will continue to do so until he too is dead. Eh Joe. . . . Silence; camera; silence.

On the fifth round, the voice taunts once more. How goes his Lordship? As well as can be expected, no doubt. The voice he hears now is his own (conscience), speaking in order to kill off the dead, in the silence of that still room – a tomb without the worms. There's another dead one to be killed, the voice of the one speaking in the name of the other. Ever thought of that, Joe? When you finish with that one, you will be with the rest of us, if you ever do finish, that is. . . . Silence; camera; silence.

Sixth: let that voice tell you what really happened. There was another love, you know, a real one this: tenderer, stronger, finer, more beautiful, cleaner; it was loyal, faithful, healthy, all that Joe's love wasn't. The speaker had saved herself with a lover, not like that other one. . . . Silence; camera; silence.

As the voice takes up on its seventh round, Joe's guilt is rubbed in. Joe seduced the second love with the same technique as he had his tormentor. "Happiness tomorrow!" Always tomorrow. It was winter, and the poor, duped girl could not get the horned buttons of her coat through the button holes; she didn't realize the horns were her own, or that Joe already had the tickets for his departure in his pocket. Those pale green eyes, so beautiful to behold! In sexual embrace, brighter still: soul made light.[10] But that was in the winter, and winter's lights go out in the summer. Silence; camera; silence.

On its eighth round, the voice returns to its taunts. She never said anything, did she, Joe? But the announcement appeared: an innocent taken away before her time. Pray for the repose of her soul, that light, now extinct. The least I can do is tell you what happened, Joe. When you yourself have had it, I shall stop. Why flinch? The truth will out. She was the last to have loved you, unless that Saturday's bitch by some freak of circumstance happens to love you, too. We have all the time we need to befoul the air. Then you can have your silence, all to yourself. Listen to that story of that love of loves which began on a winter's night. Silence; camera; silence.

The voice begins its ninth round, and Joe tenses once again. The audience has had its interest in Joe's story broken, as its consciousness was focused above on his present. The story of that damning past continues, as the voice of his conscience continues its taunts. Never knew what really happened, eh

Joe? Let me tell you. That summer, clothed only in that lavender slip you loved so much, not being able to sleep at the seaside, she slipped into the lapping tide, face down; no luck. She went back to her room for her razor – the same you recommended for removing the superfluous hair; came back to the tide, and scratched her wrist. No stomach for that route; you know how much she feared pain. All this new, eh Joe? The third time's the charm. Up, and back into the room, her superficial wound bound with the silk of her own slip, she took out the pills, swallowing them all, gulping them down on the way and finishing off the bottle in the water. That was a love that was, Joe. Could you ask for anything more? Dig her a grave, Joe, in the rocks; a grave for her narrow whatzit, waist or what have you; for those pale green eyes, that soul made light. That's what you always said, not Joe? Her soul made light, by your love. . . . Silence; camera; silence.

Joe's figure is now fixed in magnified position on the screen, and the voice continues the punishment, delivered in a rhythm imitative of the moving sea and the evanescence of a fading consciousness.

In a low voice, hardly audible, his tormentor describes the scene of the ruined girl's death. Those are the objective details. The desired response is given in recurrent stronger accents. The tide goes out with: "the face in the rocks . . . the lips on a rock . . . ," and comes back in with IMAGINE'[11] Each time it recedes it comes back with another image: a rock; Joe, Joe (the desperate cry for help, unheard); the rocks; the lips. IMAGINE: a rock, the eyes, the breasts, the hands; IMAGINE: in the rocks.

Joe's portrait is now effaced from the screen, but the voice continues: that's love, eh Joe? And the verbal tide goes out, approaching silence, as Joe's tormentor urges him to compare such a love with their own, and with Joe's love for himself, the Lord and Master who before could know no other because he did not know himself, and could not see himself for looking in the wrong direction.

The voice comes as if from another, but what we have been hearing all along is the inner voice of Joe, berating himself with his guilt. So guilty is he now, he fears seeing and being seen.

But Beckett has exposed Joe's internal torments to the external gaze of the camera's all noticing eye. Joe's, too, is soul made light: from the initial fidgeting with his exits on the world, to his first exposure in enlarged projection; then, in a series of nine inward moving shots, to the quick of the matter, back to that enlarged projection of the face that sees nothing – that self which is as nothing in relation to the reality of the world.[12] The world at least feels nothing, but Joe must go on feeling his guilt.

The structure of this internal drama bears comparison with that of *Krapp's Last Tape*; it too, a disclosure of a world of inauthentic love. Joe, like Krapp, is locked up in a closed world symbolized in this instance by the shut off bedroom instead of a den, but both den and bedroom are themselves only symbols of the activity enclosed therein: Krapp, looking to the past in the regret of a true love; and Joe, haunted by his past of failures to give himself to any love other than that of self. Both characters personify the arrest of a human transcendence, presently incapable of functioning in the world – Krapp listening to the recording of his own voice of the past, and Joe being tormented by the voice of a former mistress, his present conscience. Where Krapp suffers from regret, Joe suffers from guilt; but each of these responses represents an ontic configuration of the same ontological *Existenzial*, personal responsibility. Neither has learned that, in an authentic project of love, one's concern is solicitude for the existence of the other.[13] Krapp could not bridge the gap of social class; Joe, that of sex; and each learns of his failure when it is too late to change the sense of the past into anything but what it has always been, the constant reflection of their respective failures.

The first theme of *Eh Joe* is the familiar dialectic of Self and Other. All selves must feel the pull: no one is self-conscious without consciousness of the other. The authentic self-consciousness knows of this dependence on the other, and accepts it as the condition of its own existence. There is no Master who is not, in this dependence, a slave to his slaves;[14] no dominating lover who is not dominated by the one he loves. Failure to recognize this dependence makes the inauthentic master a slave to himself, who in his sexual relations replaces a true, reciprocal solicitude by a senseless pursuit of physical pleasure.

On this point, *Eh Joe* is more explicit than *Krapp's Last Tape*, since Joe's conscience brings to light the conscious consequences of Krapp's numerous *post-mortems*. All of Joe's loves are actually dead, and one of them in particular returns to haunt him in the very scene of his crimes: not the closed bedroom so much as his own interior "monologue" that defines the scope of his current self-projection. Failing to transcend himself in the normal passage to the Other, his own Self has become frozen in inauthentic indecision.

His pursuit of sexual pleasure is the negation of the other that would affirm his being. Looking to the past, he feels guilt; around himself in the present, an irrational fear. The reason is obvious to the existential analyst: he has substituted an empty concern with self for a solicitous concern with the other, and now fears even being seen. The camera merely catches him at his fear, and the inner voice of his conscience exposes the causes of his guilt.

Had he ever led an authentic life, he should have felt responsibility for his own being (the authentic response to having to be what one is to be, of which the feeling of guilt is but a circumstantial, ontic exemplification), and his inner state of mind would have been anxiety for the consequences of his personal decision. But failing in this human responsibility, he feels guilt; and instead of anxiety, which might have saved him, fear.

In his present self-conscious state one of his former love slaves performs the function of the Other in the definition of his self-consciousness. She has sought her happiness in another love, and claims to have succeeded. Perhaps she did. Her task now is never to let him forget his own responsibility, his own guilt over the death of the girl with the luminescent green eyes, who was forced to kill herself to escape the domination of this Lord of the passions, this slave of self.

Happiness tomorrow, he promised each of his loves: the same technique, with the same results. Each of them has died, making it impossible for him to despoil them further – the first kind of happiness, that will be prolonged by the guilt their death produces in the silence of his own inner world.

This must be the only denouement possible for the solipsist's love, projected so long ago in *More Pricks than Kicks*. There is still more pain than pleasure to be found in love, but the difference between that novel and this televised play is the dialectical, and existential, ontology Beckett had perfected in the trilogy and the *Texts for Nothing*.

What the camera does to bring the audience closer to the inner torment is to symbolize the understanding of the audience, whose consciousness is focussed in ever enlarging detail on the understanding of this inauthentic world – of the unhappy consciousness seeking its reprieve in love. But those inauthentic loves are a false happiness, a bare physical pleasure in which one self destroys another. Result: initial unhappiness unchanged. Joe has merely exchanged the unhappiness of his past loneliness for the loneliness of a one-sided love, and moves on to the loneliness of a guilt-ridden future.

The self-other-self projection that was his past is recorded in the voice of his conscience. The camera catches his present reaction to that maelstrom of existential inauthenticity, and presents it, in essential form, to the phenomenological analysis of the audience, which projects it against the norm of an authentic love for its understanding of the tragedy of loving another too little for having loved oneself too much. Could that have been the reason Joe closes himself up in his bedroom? His fear of the consciousness of an authentic Other? One has every reason to suppose so. Eh Joe?

Being alive, under such conditions, is punishment enough for any crime –

provided one never lose one's self-consciousness; and should Joe try to flee being aware of himself, he will catch himself. Compare the analysis of *Film.*

Breath (1967) is a short televised interlude of a very high degree of abstraction: no words are heard, only the muffled breath of an adult and the inarticulate cry of a new born baby. It was originally a play first presented in 1969 at the Close Theatre Club in Glasgow. The visual line, establishing the consciousness of the viewer concentrated by the camera on a world of light, moves horizontally through the temporality of that consciousness from black, to weak light, to moderate light, and then descends symmetrically back through weak light to complete darkness. As the light increases, consciousness is focussed upon a desolate world, a nondescript mass of junk scattered around, with nothing left standing upright.

Within the opening and the closing of this visual world, synchronized to its diastolic and systolic movement, there appear two other consciousnesses. Following five seconds of silence, in the feeble light, a baby's cry is heard, sharp and piercing; and as it ends, the breath is heard, now as a deep, mournful sigh, first in its mounting diastole, which reaches its apogee after ten seconds, and then, after five seconds of total silence, when the light is at its maximum (only moderate in intensity), begins its systolic descent into total expiration, followed by the same piercing baby's cry.

Silence succeeds the cry, as the attention is focussed once more on the desolation of the world in a five second interval of feeble light. Then, the darkness extinguishes all.

The entire interlude lasts less than a full minute's time: fifteen seconds of silence, a few seconds for each cry, ten seconds of inspiration and ten of expiration, framed by the opening and closing images of total darkness.

Within this formal, abstract composition, there are two concurrent symmetrical processes, a larger englobing the smaller. The larger is the visual line: darkness, through feeble, to moderate light, and reversal; and the smaller is the audile: six seconds of silence, the baby's cry, mounting sigh, silence, descending sigh, baby's cry, and another six seconds of silence.

Both processes are rhythmic, systole following disastole in the pulse of life. Combined, the two lines define the essence of the enduring moment: black silence, grey silence; sound piercing lessening gray; sound of inhalation concurrently with the descent of gray to white silence; apogee held for five seconds before the descent and the ultimate return to black silence.

No comments are made, because none is necessary. The images show it all as the phenomenon shows itself, from itself, as it is in itself.[15] It was Neary who taught Murphy that all life is a figure on a ground, and in this interlude the world of sound is played against a world of light. So what one hears must be interpreted against what one sees on the tube. The symbolism of the outer and inner worlds is likewise clear, but less discursive than in *Murphy*. The worlds of the inner selves are presented in tension against the outer world of the Other as the gestalt forms an uneasy synthesis of their opposition.

In the beginning and in the end there is nothing to be seen, but as the light begins to intensify we come aware of the desolation of the outer world. Thrown into that world, the baby gives sign of its life-giving breath that to others is a sign of distress. Whence, the sigh of the adult: either parent or both taking in the breath that permits their life to expire, whether from weariness, grief or resignation. At the height of that breath, in the moment of syncopation: the highest degree of illumination on the desolation of the world. With the completion of the sigh, the world begins to darken, and the baby cries once again, whether from the fear of the darkness or the presentiment of its abandonment. Now truly in distress, at being thrown into such a world, alone and abandoned, the child's consciousness is directed to an uncertain future. But the rest of that fading global image reveals what lies ahead; it too dies, as it fades into the surrounding darkness of the outer world, without so much as the saving grace of a "Calmative" story.

The symbolism is managed, as in Beckett's abstract dramas, by transformation of surface syntax into depth semantics. The relations of the visual and audile counters of the context represent, in their abstractness, the essence of the human predicament. The artist has said, "Let there be light," and we perceive an illuminated world, where the sounds and the silences of the inner world alternate to create the image of a life. Thus, instead of forming discursive words, the voices are reduced to the essence of a feeling, the physiological pulsation of an organism responding to its environment. Soul and body are united in that pulsating tension. Indeed, in the movement (the sounds) within a movement (the visual images) of *Breath* Beckett has achieved the maximum in minimal art. Less than a minute of real time to depict, in non-objective fashion, the rhythms of conscious temporality in a single universal image.

The philosophy is by now familiar. In human transcendence, space and time are lived as spatiality and temporality, the ontological foundation of a man's ontic experience of body and mind. The spatiality of the viewing consciousness is defined by the scope of the enlarging vision; its temporality, a projection, modulated by the sounds of the breath of life, towards a future, where all is to be dark again. The same ontological structures articulated in the drama are brought to bear upon it by the viewing consciousness engaged in the act of perceiving the global image. In this way, the play is a symbol of self-consciousness.

For within the world of that image, as in the real world, one must breathe to show the signs of life, whether it be the undisciplined wailing of the infant, or, at the other extreme, the muffled passage of air into and out of the lungs of a sighing, expiring adult. Those are two minimal expressions of the human "spirit," loaded with significance: the birth and the death of a living transcendence, the past and the future of every viewer.

Between these two extremes, within the living present in which consciousness comes to know itself, spanning its past and future in the moment of its self-transcendence, the human being discloses itself – as consciousness, disclosing the world. The poet engages in the disciplined use of the human vocal apparatus, breathing the air he finds in the desolation of the world and exhaling it to utter the creative words for expressing what that consciousness, what that significance is; but on first glance, it appears that this middle ground of a human transcendence, in full self-consciousness, lacks representation within the depth of the piece.

And, indeed, on second glance as well; for the poet is present in his work only in the artistry. We may, if we wish, refer to the artist's presence, as the consciousness, the mirror image of our own, that has organized the contrapuntal surface of this work in such a way as to represent the extremes of a human consciousness in the very pattern perceived. But the inference would be otiose.

For every consciousness is an opening on the world, both a negation of that world, as per Hegel, and, at the same time, as per Heidegger, that disclosedness by which we know the world for what it is. The consciousness of the poet is there, on third consideration, like ours, in the world he has created, as a feeling brought to an understanding by the expression. Such a consciousness is not represented in the interlude; it is presented in the total expressiveness, the signification of the entire context of the work: the tensional relationship between the surface and the depth counters of this artistic construct.

The process of expression here is indirect, to be sure; for the poet is not in the interlude using his own breath to utter the creative word. Instead, he has played the role of the "Opener," controlling the electronic gadgets of his assisting technicians: one who manipulates the lighting, another the recorded sounds, and still another the camera that sees for us all. The play is not in the opener's head, but there on the tube and the sound tracks of the tape. And the result of their cooperative behavior is the conscious image described above.

That image, too, is a figure on a ground, standing out against the booming, buzzing confusion of the natural world (Neary, it will be recalled, borrowed William James' expression, too!), wherein every living being may lose the sense of the direction of his life, unless in a very rare moment of reprieve he is led back to self-consciousness by the work of the creative poets in our midst.

In sum, then, Beckett has in this interlude made abstraction of the more traditional first order signification of an artistic visual and audile surface, and thereby presented the patterned whole of a life in a fifty second sequence of synchronized images. Result: the abstract representation, in a more direct fashion, of the usual second order significance. Although the context is less specific, with respect to the individuality of a protagonist's character, it has gained in intensity and clarity as the universal idea becomes readable directly from the surface.

Reducing the ontic experience of life to its abstract patterns has permitted him to show the ontological structures of a human transcendence on the very surface of the work. And our own perception of that abstract sensuous presentation incorporates the image of human disclosedness at its most intense projection: understanding what is expressed as a feeling, feeling what is understood in the expression, expressing what is felt in the understanding. That is how Dasein is there, in its world; and its interpretation is a circular process of expansion.

The poet once again has succeeded in making soul of light, and sound: his own, our own, and of all others who first show signs of life by an inarticulate crying and who, if they are wise, end that life of weariness by a sigh of renunciation. Geulincx's stoicism has never had a more compendious expression.

We have since the beginning of this book been following Beckett's personal voyage as a search for self-consciousness in two parallel directions. The avenue of poetry led to prose, and in the end to the "ping" of the articulation of sound; that of the drama led to the film and television, and, in the end, to a series of abstractions from the "pinging" sounds: to the inarticulate cries of a new born baby (through concrete articulation of the poetic voice), and finally to all that remains of the word when it has been reduced to its most essential condition – the breath of a sigh, where body and soul leave no difference to be perceived.

That has been a project – from *Whoroscope* to *Breath* – in which Samuel Beckett achieved his personal self-identity, a feat he was to duplicate by translating the works into still another language. But that is the subject of the following chapter.

CHAPTER XIII

SOUND, SENSE AND SOUND

One of the lessons Beckett had learned from his study of the sources of Joyce's works was enunciated by Giambattista Vico: before the concept was the metaphor, and before the articulation of metaphor was song. Whether or not this historical claim is true for the race of mankind, it rings true for the experience of every living man, who hears sounds of words long before he is capable of articulating the least of his thoughts. And this is as true of the mother tongue as it is of any adopted one. The poets amongst us needed no convincing, addicted already to the sense of sound as a means of conveying the sounds of sense; and Beckett has always been first and foremost a poet, an artist working with words.

But if the style of the literary artist is recognizable by the unique tensions created by the functional relationship between the aesthetic surfaces of his words and their resounding depth significations, how is this style to be translated into a language different from the original expression? And to whom does one entrust the task? Anyone can make a literal translation of another's thought as long as the concepts are not lacking in the one context or the other; it takes a poet to translate the sounds of another poet. For this reason, no doubt, Beckett has undertaken the task of composing his own translations. And the change has been slight: from Beckettian English to Beckettian French, or as the occasion demanded, the reverse. Always away from home, he could move in Italian, Spanish or Latin at ease; only Gaelic left him untempted. It was a stroke of chance that forced him to hear the barbaric sounds of Nazi German after he had chosen to live in France at war rather than in Ireland at peace. When the Gestapo had chased him from Paris to the Vaucluse in the unoccupied zone of France, he spent his days in agricultural work and his nights composing, among other things, *Watt*.

In the lean years before he had established his literary reputation with the success of *Godot*, Beckett made what meager living he did by translating the works of others. Like the penurious Irish poet, Ticklepenny, of *Murphy*, he was forced into other tasks; and like Murphy himself, he was human enough to give up his solipsism, the solitude of his poetical ways, to concern himself with those sick minds in other bodies who were to find solace in his short stories, novels, and plays. The only trick was to talk the Murphy of his split personality into exchanging places with the poet. That, as we know, happened, not without leaving the record by which we have come to know of the anguish of his decision.

Once the switch was made, the poet could spend his time investigating the springs of poetic expression in a new effort to discover what it is he had talked himself out of. But, then, if these investigations could themselves be given some form of poetic expression, as in *The Stories*, the trilogy, *Texts for Nothing, How It Is*, and in the dramaturgical plays, *Cascando, Embers, Words and Music*, he would have shown that nothing had been lost for having made the switch. Poems, novels, plays are after all things made – with words, that must be assembled in some order for the articulation of the least thought. Let us not forget the pathetic pleading of Croak, begging Words and Music to be friends. Unfriendly, they grate upon our ears or weary our intellects that might be more profitably engaged in other pursuits. Not even logicians are concerned with the content of a verbal expression; and, in the creative drive, music will often lead the way.

Whatever the art of literary creation, the art of translation is that of finding new orders in another musical system, a new surface for the old, while retaining as much of the lyrical tension and depth significance as linguistically possible. Since the aesthetic surface has to be changed in the effort, the experiential depth that is the soul of the piece must be retained. Images are repeatable from language to language, as are ideas; and the philosophical substance of a work may be found in an abstract form in the writings of others: a Dante, a Bruno, a Vico; a Hegel, Heidegger or Sartre. A Joyce, a Proust will show the way. With a hand or mouthful of abstractions, and a functioning pair of ears, the impossible becomes probable. The contingency is the translator's genius.

It may have been his love for "Anna Livia Plurabelle," or his great admiration for Joyce, or a desire, vaguely felt, to prove his brash words against the likes of Rebecca West, who had claimed that Joyce had written no story in *Finnegan's Wake* – at least nothing that could be understood – all this could be shown by finding the words that purveyed the multiple senses of his text. In that project he had the help of a number of collaborators, principally Alfred Péron, and including Joyce himself, who created the impression of being more interested in the sense of the rhythm than in the rhythm of the sense he had put into the original work.[1] The finished translation appeared in the *Nouvelle Revue Française* **36** (May), 1931; it was revived by Gallimard, in 1962. After *Finnegan's Wake* everything else must seem easy.[2]

But not all that easy. The very conception of the problem often obscures the issue at stake. Consider the cynical statement of that old bromide: translations are like women; when beautiful they are not faithful, and when faithful, not beautiful. Some translations, of course, are neither beautiful nor faithful, but they would hardly be acceptable to the author or to his readers. So, since *tertium quid non datur*, one has a choice to make, and the choice might be easier to make if it could be determined to what one is faithful, when one is, and what constitutes the beauty of the original expression when it is beautiful.

Beckett's own attitude seems clear. In his letters to Alan Schneider,[3] who directed *Godot* in Miami and *Endgame* and *Film*, he explained the author's responsibility in the initiation of the literary communication:

> Mon travail est un corps de sons fondamentaux (sans jeu de mots) produits aussi pleinement que possible et je n'accepte de responsabilité pour rien d'autre. . . .[4]

That's the poet speaking, accepting the "pensum" of producing the sounds that must be interpreted by another. If a sense is to be added to this basic body of sounds, then it must be another who supplies the interpretation. And of these, there may be as many as there are ways of attending to the sounds.

Yet between the ideas and the sounds there is another stratum of basic signification to be found in the words, in the very images they call forth. It is the relation between these images, and the relations between these relations that constitute the philosophical structure of the literary work's depth. If the images that form the first level of depth signification fail to fund into a coherent picture of the world as lived by everyone, so much the worse for naturalism as a literary genre. Recall Proust's rejection of the Goncourts,[5] and his insistence on creating a world structured on the principle of a regained lost time.

Like the Unnamable, the author lives in a world of literary significance – between the sounds of the words and the meanings they may intend. If his feelings are expressed, it is not the emotions of the natural world – recall the story by the Unnamable of the woman with the two dead husbands. The knowledge he has gained for writing his work is of the feelings expressed in the words, and that knowledge is not purely conceptual, not a matter of easy cross-reference in a bi-lingual dictionary. The narrator of *The End* had a son, whom his father had called "l'insupportable fils de putain."[6] The ascription is easily understood, given the description of this character – a brief-case toting, hat-lifting, kowtowing son of a bitch. And that is precisely how Beckett translated the expression: "an insufferable son of a bitch."[7] The older English "whoreson," which would have been more faithful to the original French, had, by lack of usage, become inadequate for conveying the living pragmatic equivalent of the current French expression. What was lost was slight; for where the original suggests the moral looseness of the son's mother, and the English loses that, the English maintains a negative connotation on the mother's character that is now evident in the character of the son. Thus, if the translator is to be true to his text, it is not the words alone, nor their most evident meanings that must be respected.

Nor, indeed, is it the rather idealistic (not to say "spooky" or "fuzzy") idea of the author's intent. That is the idea laid to rest by the current aesthetic concept of the "intentional fallacy," if not by the words Beckett addressed to his producer on the literary artist's sole responsibility, and his continual disdain for telling anyone under any circumstances what his works are supposed to mean. They can only mean what they say, and if their contexts are not sufficiently clear to the would-be interpreter, nothing could possibly make them any clearer.

The author, and the poet, if that venerable word is to retain its original meaning, makes a body of fundamental sounds to which he, among others, associates a body of sense. The working of the work is in that tense relationship. The translator's task is to substitute an equally tense body of secondary sounds. And that pattern of productive and reproductive activity – sounds, sense, sounds – repeats the basic rhythm built into the texture of the original creative word in a characteristic chiasm: silence, sound, silence. Before each sound there is a silence, and after as well; but that second silence is full of the sense of a conscious awareness. The space between the two silences, the one of a struggle with nothingness – the bare desire to make oneself heard – and the other a return thereto, in the appreciation of awakened consciousness, is filled by the expiration of the breath necessary to make the sound. In reading *How It Is*, one hears that breath creating the sounds as the story comes in bits and pieces. That is how the narrator of the story has himself heard the tale:

comment c'était je cite avant Pim avec Pim aprés Pim comment c'est trois parties je le dis comme je l'entends [8]

Instead of punctuation, the stopping of the breath: that, after all, is what punctuation intends in the first place. Our periods are but full stops, giving us time to breathe in before breathing out the sequel of our thought. Inspiration and expiration are two sides of the same process.

Showing us how it is in writing *How It Is* Beckett dispensed with the punctuation, and let the breathing develop its own rhythm until the very end:

bon bon fin de la troisième partie et dernière voilà comment c'était fin de la citation après Pim comment c'est[9]

And his translation follows the rhythm of his narrator's breathing, from the beginning to the end:

how it was I quote before Pim with Pim after Pim how it is three parts I say it as I hear it[10]

and

good good end at last of part three and last that's how it was end of quotation after Pim how it is[11]

Though the timbre of the languages differ, they can be made to mark the same time, in the same tempi, and in addition, present the same temporality, on the surface, alone, of the two versions of the "same" work. Perhaps it was for this reason that Joyce gave himself out to be more interested in the rhythm of his prose than in its overcharged sense.

But that too is an abstraction. For, if temporality may be presented on the surface of the sounded prose, a variation on human temporality may be represented in the meanings of the words. That, Beckett had found in his youthful study of Joyce, was the hallmark of a highly developed style. And the same phenomenon may be observed in his own work. The periods of time before, with, and after Pim were immense, purely relative determinations of a phenomenal seeming, in themselves as short or as long as they may seem to the consciousness that is its own temporality and expended in the task of solving the mysteries of its own existence. Consciously existing for itself, that temporality is finite, having begun with the first cry of the new born infant and having to end in the agony of death; but as long as there is life and breath, an on-going process of significance in the making, the development of that sense will take as much time as one is allowed to live – either waiting (inauthentically) or performing an act in resolute decision (authentically). As long as the breath continues to flow in and out, the round of personal decision will not end for the writer. Listen to the words, "après Pim comment c'est (.)" After Pim, how it is – the possibility of a new beginning (commencer). If that overtone of meaning has been lost in translation, it has already been established in the discussion of worldly justice in the third part of the story: there will always be a new beginning, of a new story, a new voyage, a new world to be created – by the sounds of the exhaled breath. Literature will endure.

And the reader, too, is a party to the enduring communication. His is the task of supplying the meanings. Whether or not a story happens to be about a consciousness in search of itself, the reader's consciousness, expending its temporality in phase with the time created in the work, is engaged. In the first instance, after all, the reader fills one of the functions already fulfilled by the author himself. Since the author's understanding does not precede his expression, nor his feeling his understanding, the reader expends his own temporal existence in phase with the author's own, as it is marked out within the structures of the created world.

Whether, within that world, soul is made light, character is made presence, or breath made the word, the reader's own consciousness is caught. Images may color it in different lights of varying values from black on white to white on white or to a uniform grey; the world in which it has been caught is structured by a fundamental philosophical idea and expressed in an orderly pattern of sound. Whence, in the beginning was the word, any organization of sounds that create images for all with enough imagination to sense them (a human power not tied to language), and these images may resonate to any degree of depth provided only that the ideas so generated are restricted to the bounds of sense (meanings) laid out in the original act of phenomenological bracketing.

If the author's task is to establish the original linguistic context, moving from sound to sense, the translator's is to perform his act in mirror image of the author's, moving from sense back to a new basis of sound. If Moran had had access to Molloy's pages, he might have given a translation of his own. The two activities would still have been performed in mirror image one of the other, and Moran's "report" would still point back to those original pages. And the content would likewise have remained the same: two consciousnesses moving together, then parting into that peculiar solitude experienced as a post chantal triste.

How many times have we heard it? Murphy-Endon, Watt and Sam, Mercier and Camier, Molloy and Moran, Malone and Macmann, the Unnamable and Mahood/Worm, Bom and Pim; the self can only be known in and through the other: Didi and Gogo in the foreground of Pozzo and Lucky; Hamm and Clov against the background of Nagg and Nell; all those lovers and would-be lovers driven out of their solitude only to return thereto, unenlightened for their union; and last but not at all least, Words and Music, conjoled to be friends under the beat of Croak's baton. If the message is missed in the depth, it is there to be grasped on the surface, a phenomenon under control of the Opener, who opens and closes his physical or physiological apparatus as the contingencies of the occasion demand.

In making his own translations, therefore, Beckett has become the complete sorcerer, performing his lying tricks by opening and closing one of his gadgets, and then opening and closing others to perform substantially the same trick.

Attending always to the effect of color, as controlled by the images his words evoke, substituting one pragmatic value for another, his significance, if not the verbal signification, remains the same. The problem becomes doubly compounded by the language of (Mrs.) Lousse's swearing parrot. We hear in the original French:

> Elle avait un perroquet, très joli, toutes les couleurs les plus appréciées. . . . Il disait de temps en temps, Putain de conasse de merde de chiaison. Il avait du appartenir à une personne française avant d'appartenir à Lousse. Les animaux changent souvent de propriétaire. Il ne disait pas grand'chose d'autre. Si, il disait aussi, Fuck! Ce n'était pourtant pas une personne française qui lui avait appris à dire, Fuck! Peut-être qu'il l'avait trouvé tout seul, ça ne m'étonnerait pas. Lousse essayait de lui faire dire, Pretty Polly! Je crois que c'était trop tard. Il écoutait, la tête de côté, réfléchissait, puis disait, Putain de conasse de merde de chiaison. On voyait qu'il faisait un effort.[12]

Pretty Polly had a genius for words that swear, a truly bilingual evil genius. His French is untranslatable into English, and the English loses all its force in French, since "Fuck" in French (*foutre*) has been used so much and in so many varying circumstances that it may mean an innocuous, but insistent "give," or "throw" or even the hearty American's "shit!"

Beckett solved the problem by merely inverting the order of the obscenities, and adding the ubiquitous "son of a bitch" not contained in the original, along with the suggestion that the forceful English had been learned from an American sailor. And throughout the changes, the tone, and the humor has been maintained:

> She had a parrot, very pretty, all the most approved colors. . . . He exclaimed from time to time, Fuck the son of a bitch, fuck the son of a bitch. He must have belonged to an American sailor, before he belonged to Lousse. Pets often change masters. He didn't say much else. No, I'm wrong, he also said, Putain de merde! He must have belonged to a French sailor before he belonged to the American sailor. Putain de merde! Unless he had hit on it alone, it wouldn't surprise me. Lousse tried to make him say, Pretty Polly! I think it was too late. He listened, his head on one side, pondered, then said, Fuck the son of a bitch. It was clear he was doing his best.[13]

In its own way, the parrot, under the training of his American and French sailor owners, gives the lesson for all would be translators: to find the expression that creates the proper impression.

As long as the proper impression is created, the translation will work. But

Beckett's addition of the supposed past ownership of the parrot, in the English context, is a privilege only the original author may assume when translating his own works. The second time around, substituting "a French sailor" for the respective "personne française," he was able to make more credible the cause of the parrot's language, and at the same stroke, to make Molloy's observations a bit more incisive. Numerous are the occasions when Beckett has availed himself of that privilege.

Where then is the original when the author has translated his own work? The German translator of Beckett's plays answered the problem with a bit of ingenuity on his own. His publishers[14] state boldly that under the circumstances indicated there exist two originals for each of the plays. That presumption allows the translator a still further alternative to the usual dilemma between faithless beauty and ugly fidelity, and allows the translator to intercalate a mean position on his own. Witness the *Missler* (the bungler) for Beckett's *Maunu* in the French and 'Woburn' in the English version of *Cascando*. The Italian title remained the same in all three versions, suggesting in both sound and sense the fate of the hapless *Maunu*/Woburn/*Missler*.

It is dangerous, too, for the bilingual reader of either of Beckett's original forms of expression to presume upon his own interpretations. If one is writing for an English speaking readership, it would certainly be preferable to search down Beckett's own translation. But this is not possible in every case, even when both texts are available. Chapter VI of this book was written from the reading of the French version of the *Texts for Nothing*. In the second Text, the narrator contemplates the question of where one could go, once one has become fixed for all eternity in the form of a literary essence. Where was he now? In French we read:

> Ce n'est pas la mort, ce n'est pas la tombe, loin de là, ca ne peut pas être la tombe, ce serait trop fort.[15]

Whether consciously, for a smoother transition, or unconsciously, by the jumping of an eye, that sentence has not been translated into the second text, as published in *No's Knife*. The English context suffices without the sentence to establish the narrator's condition – somewhere between life and death and feeling the necessity of groaning out the words and the impossibility of doing so satisfactorily, so that sentence may have been judged superfluous to the work's intent. But the author translating his own works has the privilege of second thought. He may have simply taken his narrator at his word and judged, on second reading, that the quoted words were indeed too strong, and merely deleted them in retelling the story. But they serve, in the original context to create the dilemma of the narrator's condition. The Gutenberg galaxy must be full of such black holes.

Other difficulties in the translation are found in the little losses or gains in the connotative power of the changed words. Joe, of *Eh Joe*, is noted for his use of the poetic expression to seduce his sexual quarries. The voice of the woman that haunts his conscience possessed the qualities of "du cristal de roche" in translation, whereas the original English read "Voice like flint glass." A gain, there, which another translator working from the French back into the English could establish by rendering the French as "rock crystal." Flint glass is more noted for its optical properties than the purity of its resonance. The clearness, the sharpness of the metaphorical comparison fits in either English version, and that allows the voice to play its ironic role in taunting its victim, but "flint glass" is too brittle to permit the lasting effect, and that brittleness can be heard in the two sharp syllables. Moreover, the hissing sibilance of the "glass" may fit the character of the woman who possessed the voice, in her role as taunting conscience, but the hiss is inconsistent with the purity of her voice.

The same kind of phenomenon is apparent in Beckett's translation of Joe's other poetic expression, made more poetic in the French than in the original English. The eyes of the girl who had committed suicide because of Joe's faithlessness veritably threw off light (as, I am told, do the eyes of most women in sexual embrace): nothing astonishingly new in that observation. In English Joe saw that pale green brilliance as "spirit made light." "Spirit" indeed, if the girl were all-fired up on her own; but then Joe would be less the seducer than the story calls for. Beckett improves the work once again in translation by removing that aura of self-determined activity and supplanting it with a deep welled pathos more fitting the relation between the seducer and the seduced: in French what Joe saw in those pale green, expectant eyes, was "de l'âme faite lumière," soul made light – a symbol so apt that it may be used to refer to the whole process of artistic creation in the photographic media.

And what is to be said of *Come and Go*? "Va et Vient" concretizes the sexual implications of the three women's situations that is lacking in the original English, or, if present, would suggest an abrupt ending of a sexual embrace and a quick parting of the ways: more fitting for the bordello than the minds of three young girls in a convent court yard. But that may be a critical aberration – one of the kind that motivated another change in the translation of *Godot*.

Didi and Gogo decide to make the time pass by insulting one another.

Vladimir begins the game with 'Misérable,' and Estragon acquiesces, "C'est ça, engueulons-nous." The repartee is as follows,

	Original	*Translation*
V:	Andouille!	Moron!
	E: Tordu!	Vermin!
V:	Crétin!	Abortion!
	E: Curé!	Morpion!
V:	Dégueulasse!	Sewer-rat!
	E: Micheton!	Curate!
V:	Ordure!	Cretin!
	E: Archi . . . tecte!	Crritic!
V:	Ohhh![16]	Oh![17]

The German translation is freer than Beckett's own, retaining the rapid-fire invective and trading on the pragmatic equivalences of the three languages. Thus, the original 'archi . . . tect,' and the "second" original 'crritic' became 'Ober . . . forstinspektor'![18]

On the whole, the intellectual Didi berates the more physical Gogo for his lack of intelligence, while Gogo retaliates by referring to Didi's false intellectuality. Each therefore touches the other where he hurts. But if the original French 'archi . . . tecte' is as pejorative as the English 'crritic,' the reason would be more difficult to grasp, at least by the members of that noble profession dedicated to building a better universe in which to live than for those others who, admittedly, do not always know what they are doing even when they are most insistent upon their right of doing it. At least the English actor, in the role of Estragon, can spit out the word 'crrritic' as if he were a snapping dog, and that would hurt Didi, who has a genuine affection for his suffering fellow. As the case stands, the English theatrical in-joke seems petty, a nonsensical replacement for an incomprehensible verbal jibe in the first instance. Be that as it may, the innumerable curates of the world may be flattered to envision themselves in the shoes of the architect. The German translator leaves the clergy in peace, but translates the final epithet as a reference to the highest form of German officiousness, not the working corps of forest rangers, but their superior officers. And perhaps that is as good a solution to the problem as might be found. Negative values are where one happens to find them.

Take the case of the narrator of 'First Love,' written in 1945 (Beckett's first French prose). That character found it useless (*oiseux*) to speak of the reasons for changing habitations from a bench in the open air to a closed stable, beyond the obvious one of the approaching cold winter, because those who would hear his words (the readers) are simple cullions (too bad that Elizabethan English expression has been lost, otherwise the current French *couillons* could be directly transcribed into the English without loss of the intensive irony). For, if a cullion is stupid in taking the essence of his existence to be located in his testicles, he is not, granting the fitness of the ascription, too stupid to recognize the significance of the warm, enclosed space of the stable as a simulacrum for the womb from which he was initially expelled into the world. The true *couillon* in the affair is the narrator who had to discover that significance as he wrote the name of the substitute womb that was his first love on the dried piles of bovine excrement found in the stable, licking his fingers after the achievement.[19]

In short, and in summary, the effectiveness of a translation depends upon the success of the translator to place the reader in a recognizably similar linguistic world, partially defined by the tension between the sounds of the words and their signfications – the existential spatiality of that world – and partially by the sonorous rhythms of the sounds as they are supported by the rhythmic occurrence of the coloring images – the existential temporality, or consciousness of inhabiting that space. And, on the whole, Beckett has succeeded in reproducing the conditions for experiencing the overlapping significances of the two worlds presented in his originals and their translated versions.

Perfection in this endeavor is an impossibility. Neither the timbre, nor the rhythms, nor the melodies of the two languages can be substituted point for point; a change in any context is a change in significance. The same narrator recounting the development of his first, and perhaps last, love gives two insights into the translator's predicament.

The first is his musing over the expression 'vase de nuit' (Chamber pot). Why should it remind him of Racine and Baudelaire? In the one case, perhaps, for the purity of its poetry; and in the other, for the intellectual synthesis of the symbolism; perhaps both together. At any rate, as his mind continues its reflection on the reverberations of those symbolic sounds, he is led to that place where the word stops – back to the silence on whose ground that figure has made its appearance. One would suppose that Dante, creating his own language, was not far away.[20]

But there is more. He who has discovered his first love has discovered a voice that sings. The meanings of the words are lost, but the music charms the savage breast. And his own words strike him, like those of all poets, with the same kind of charm. "Try now to throw me out," he tells his mistress who has allowed him to take root in her rooms. Both the sound and the sense of the words were grasped only after having been enunciated. The reason would be obvious to an ontological critic, for the narrator is those words and feels himself coming to life in their expression. Usually, upon reflection, his words were expressed in an "impeccable" grammatical form, and not lacking in meaning – there is always too much of that – but in a solid foundation for their statement. That, too, is obvious to the ontological critic, for the missing foundation is the substantial ego which never exists outside the expression of a human desire. Even though such meanings may be discovered after the expression, the speaker seems never to be unaware of the sounds emitted by the expression because that is where the poet lives, primarily in the universe of sounds.[21]

We hear them, and too often lose ourselves in the maze of meanings they suggest. But the narrator of 'First Love' shows us what the nature of that universe is: the sounds of his wife's lovers, her own pangs of childbirth, and the cries of their offspring drown out the sounds of his own breath, and expell him once again into the silence where he is united to himself once again, in solitude.

In this way, we may find textual verification of what Beckett wrote to Alan Schneider about the limits of an author's responsibilities, and the same explanation would be fitting his decision to undertake the translations of those two bodies of his own fundamental sounds. Beckett's decision to fulfill his own resolve for authenticity is visible in the solipsistic lover's refusal of the world.

The tense world of the writer is located between the sound his words emit and the meanings they suggest. The Unnamable knew no other world: that of sound and sense, but above all the tension of the pull between the figure of sound on a background of meaningful silence. The aesthetic surface of the writing controls everything that may be found in the experiential depth of the linguistic expression.

And, *mutatis mutandis*, the same thing is true for the dramatic and cinematographic media. Wherever he works, the true artist aims for the expression of more with less.

Beginning as a poet, Beckett moved out into the freer expression of prose, then into a foreign language, where he learned the true power of the word, which he then freely transposed from one language to another, fairly crying for all to hear: "Look Mom; no hands." And he blithely repeated the performance in at least two other media. What, then, are we to think of those critics who thought the Swedish Academy had covered itself with dishonor in honoring such a literary achievement?[22]

We may begin to formulate our judgment by concluding this work with a capsule statement of Beckett's literary and philosophical significance.

CHAPTER XIV

CLOSURE

The Principle. Linguistic contexts close by virtue of the saturation of a sign system. Opening the context, someone makes a gesture to articulate a meaning; and when another consciousness reacts to that gesture in a way similar to the opener, the meaning has been understood. Beckett's play *Cascando* dramatizes the activity of the "Opener," who, at will, opens and closes his boxes of music and voice to tell the story of Woburn (*Maunu*) falling at each new stage of his continuous voyage; and the play's dramaturgical structure gives several hints for the hermeneutical expansion of a general linguistic context.

First of all, whatever the signification embodied in that context, literary or more generally linguistic, the play is not in the Opener's head, as he says; it is nothing so abstract as an intention to be fulfilled. To argue that it is, is to defend the position of linguistic idealism, which has come to be denounced in contemporary criticism as the "fallacy of intentionalism."[1] If the artist had any intention at all, it was, presumably, to make a work of art, which may be constructed of meanings – of signs and symbols and their intentions – but which, as a whole, cannot be said to mean anything at all, if by 'meaning' is meant some reference to an extraliterary context. Such a reference would leave the signs of the original context unsaturated.

Secondly, the opener makes his work by controlling two related strands of significance, symbolized by the music, on the one hand, and by the words of the voice, on the other. Moreover, if all readers were sufficiently attuned to the music of the words' sensuous surface, a single box would suffice. For the words as signs are either heard or seen simultaneously with the active association of any image their syntactical union may suggest. Hence, where the sensuous surface is perceived, the depth experience that surface conveys begins to take shape in the imagination of the opener, as of his audience, in the activity of associating the conjured image with the field of sensuous stimulation. It is this splitting of the surface into two lines of significance that makes the word itself an apt image for the consciousness of man: both are inherently schizophrenic.

Normal to the linguistic context and endemic to the human animal's conscious evaluational experience, where it can become psychopathic if the creation of a personal world and a number of related personalities serve as an escape from the pressures of adjusting to the one universal world of nature, the phenomenon of schizophrenia is not an unreducible given, to be explained by the modern psychologist's classification of "abnormal," aberrant human behavioral traits, nor by the traditional metaphysical and religious interpretation of man as a composite entity of two interacting substances, the one a body and the other a mind. We are all too familiar with the history of modern philosophy and its seemingly interminable discussion of the relative values of percept and concept, sense and meaning, body and mind, in the development of human knowledge. Beckett's Opener is not concerned with giving *knowledge of* the external world, as were the modern philosophers, but with showing how a world is created with his schizophrenic words, in part a phenomenon of sense and in part, of meaning.

Thirdly, the first meanings created by his sounded words are of the images they represent. We "see" a man falling as we hear the words of the voice's box. But instead of attaching to the real world as our perceptual visual images must, if we are not to be deluded, these images attach only to the words that have created them. And they in turn may give rise to an idea – of a world of pure image, whose reality is the relation between the images as they unfold before our imagination. For these relations, created by the images, and hence in derivation by the words, there corresponds an idea of the one world created within the linguistic, literary context.

Such ideas may have been expressed before, in the philosophical descriptions of moralists (of an ideal and hence prescriptive model of human behavior), of epistemologists (of the way in which knowledge is gained), of metaphysicians (of the structures of the real world), or of ontologists (of the being of the entities discovered in human experience); or they may have been the personal discovery of the artist who has elaborated the images in the first place. In either way, the philosophical idea is said to provide the structure of the context's experiential depth, but it composes only one strand of the literary work's total linguistic texture.

Philosophically considered, the process is not as complicated as all that. The sounds of the surface present a series of sensuous Gestalten, which represent images formed by the imaginative consciousness, and these images of the imagination present in their turn, by virtue of their relationships, an idea of a single created, purely imaginary world.

We may therefore visualize the process as a functional relationship between two expansive fields of signification: that of the signs as perceived, running horizontally from opening phrase to closing period, and that of the vertically structured depth. Dante, it will be recalled, found four levels of vertically arranged depth structures: the literal (which corresponds to the first level of images conjured by the words), the moral (which corresponds to the evaluation of the imaginary events occurring in the life of any human subjects depicted in the images), the allegorical (corresponding to a substitute system of meanings associable with either one of the prior levels), and the anagogical (corresponding to the "spiritual" or mystical sense of the entire text of meanings).[2]

In more modern terminology, we refer to "allegorical" meanings as 'symbols,' and to the "anagogical," as the last and most basic philosophical idea by virtue of which the meaning context is brought to closure. And to understand the ideal of a unified total linguistic structure, this whole system of meanings must be perceived as the fulfilling intention of the Opener's words in a unique relationship to their surface "music."

Once the allegory of a symbolic reference has been established (raising a signified into a sign with a new signified), the context of expression has become intensified, the original word more potent; but if the context is to come to closure, there must be a single frame of reference with respect to which each sign is fulfilled in signification. The "anagogical" level of signification merely provides the conditions under which the context achieves its closure in the complete saturation of the original signs. The work comes to exist for any consciousness that is attentive to the structural relations of the context. So runs the structuralist theory, phenomenologically interpreted.

Only one additional caveat is in order; it is prompted by the inherent schizophrenic tendency of the words. An experience of the literary work is definable at any level. We may, like poets, be more interested in the sounds of the words than in their meanings, or in the images than in their interpretation(s), but what the title of the work names is the experience of the funding relationships between all the counters going to make up the texture of the context. To that phenomenon, I give the name 'total expressiveness' of the work. It may be visualized as the power of the word to express its systems of signifieds, stratified to whatever degree. To feel that power is to have related the horizontally defined system of sense. Thus, if it is true that in the beginning was the word, it is likewise true that in the end, too, it is the structured word that controls the significance felt.

In every poetic context, there is a contrapuntal relationship between the rhythm of the developing melodic line of the sense stratum, and the rhythms of the resounding depth levels of signification. The two may run together in harmony, or be contrasted for ironic effect. As we have seen, and as it has been amply pointed out by others,[3] Beckett is a master of the comic ironic effect; in the tone of his surfaces, in the tragi-comic reversals of meanings at the level of image and idea, and in the paradoxical play of surface tone against a first level meaning that itself constitutes meaning at the next higher level. Writing of clowns in a non-clownish manner, he has, for the most part, succeeded in showing the tragedy of human existence – a point to which I shall return.

Its Significance. Faced with a veritable maze of literary artistic effect, contemporary criticism has done Beckett's work a disservice – as is universally the case in such "critical" procedure – by attempting to classify the unique and unclassifiable. Such terms as 'avant garde' or 'absurdist' condemn the *oeuvre* by damming it with the meaningless epithet. The one term does little to hide an historical or traditional bias, and the other isolates a single strand of the expressive contexts and erects it into "the significance" of the literary achievement.

It was Marcel Achard, member of the *Académie Française*, who berated the Swedish Academy, in view of Beckett's 1969 Nobel laureateship, for its "literary snobbery."[4] But is it snobbery to have recognized that there is nothing new in the work of the avant-gardist who succeeded in bringing literary creation back to the fundamental properties of the expressive word? Or snobbery, to have understood that human meanings, if they are to be experienced, must be created by individuals suffering the conditions of human existence? And as for the "absurdity" of this or any other author, how can anything that is amenable to scientific or philosophical explanation be essentially absurd?[5] How one reacts to the absurdities of the human condition may be as rational or as irrational as the potentialities of human ingenuity permit. Literary works, like human beings, defy significant classification; and they may, within the bounds of their own significant closure, indicate the way in which the truly ingenuous creator transcends himself through his art. In this sense, Beckett's fictional creations display a double source of significance: they are about the act of creation at the same time they exhibit Beckett's own unique creativity, his own solutions to the universal problem of psychic stress and pain.

To state baldly that Samuel Beckett has created a literature of the absurd is to state a systematically misleading partial truth. His work is not unamenable to rational interpretation. If it at times depicts human schizophrenia and the author allows his very surface to illustrate the same phenomenon, schizophrenia is a natural event, dubbed "abnormal" only because of a groundless prejudice for adherence to statistical predictions. To cure such abnormal behavior the analyst must be led to discover how the suffering patient has undergone his personality disorganization; he can cure nothing with a name. And to make his analysis, he is forced to allow the patient to tell him "How It Is," how life, up there in the world of statistical light, was before it all went dark. That's what Beckett's characters do. It is not schizophrenia, the psychologist's syndrome, that explains the characters, but rather the characters that explain some forms of the syndrome.[6] And if we are to understand, we too must listen to how it is.

Personalities may disintegrate because they become convinced that there is no external foundation for their personal values; if so, we may all go mad, for there is no incontrovertible reasons to explain our or any one else's existence. That, we recall, was the "metaphysical absurdity" of the existentialists, made famous in the philosophy and literature of J.-P. Sartre; but he found the doctrine in Nietzsche's description of the spirit of seriousness and in Heidegger's phenomenological ontology.

Other personalities may disintegrate because of the awareness that the world of nature is indifferent to our every wish: the "moral absurdity" of Albert Camus, which he had discovered from his reading of the ancient Stoics.

Such facts as these are apparent to anyone who has read carefully the works of the various authors, and do not constitute grounds for comparing the systems of beliefs; for any such comparison would only establish that the literatures expressing such beliefs had this trait in common. And as for the other traits, with respect to which they differ, we would be left in total ignorance. We can suppose, however, that Beckett achieved his knowledge of philosophical ideas in the same way as the other philosophical literary artists of our time; he, too, had read his philosophy, and like one of his narrating characters had retained his "humanities."[7]

It was Geulincx who convinced him that man is free to determine his own fate; Descartes and Geulincx, that man is an apparent composite of body and mind capable of internally inconsistent impulses to action; Berkeley and Schopenhauer, that the world is a man's idea of supreme consistency. Indeed, the world makes itself felt as a regularly occurring order of phenomenal appearances, no matter what the philosophical explanation is for our relations to it, whether it be simply God's will or the "four-fold root of the principle of sufficient reason."[8]

What could be more natural, then, than the conception of two worlds, the one larger and external to the thinking subject to which one is forced by circumstances to adjust, and the other, the world of the mind divided against itself and incapable of adjusting, but free to make the attempt? To continue to desire something as a positive remedy to men's dual self-alienation (from the larger world to which he is attached by the body and from the self in its own internal schism) is to look for nothing that may succeed in relieving the problem of the pain's being felt. That is the pessimism largely celebrated in Schopenhauer's philosophy of human conduct, to which the stoical ethics of Geulincx had long before provided one practical solution: when one is nothing, one is wise to wish for nothing. In probing deeper into the causes of human unhappiness, Beckett is led to an aesthetic solution of this fundamental human predicament.

Beckett's characters have done nothing more nor less than renounce the impossible, and to suffer their fates – but not in silence; for that would make them something less than human. Man's consciousness, which is nothing substantial, no bearer of properties, is to be open to the effects of the larger world and projected into the search for a private, personal significance. Such is man's dignity, and the source of all his merit, as Hölderlin's poetic lines suggest:

> Voll verdienst, doch dichterisch
> Wohnet der Mensch auf dieser Erde.[9]

Full of merit, but poetically, man dwells on this earth. Beckett's life – or all of it relevant to his works – is found in the plight he had accepted in making those works, to live on this earth as he had chosen.

And, if Martin Heidegger is right, to dwell poetically means to construct a place in which to live, a poem, indeed, in whose tensions all of man's knowledge, whether it be feeling, understanding, or expression, comes to light. Living poetically, a man is always underway to language,[10] and his readers' consciousness must dwell along with him in his poems.

If, to cite Heidegger once again, the meaning of human being is its temporality, that temporality must disclose itself as it is in itself, spanning a past created by a gratuitous birth and a future, necessary death, but, in the present, always a passage in which man finds himself on the way. The past is no longer, and the future is not yet; the present is a flight in which I can escape everything except the knowledge that I am fleeing. We need only consult Beckett's *Film*. Falling from time to time, from authenticity into inauthenticity, I may elude myself temporarily, only to be called back to my true self-determination by the feeling of anxiety. The self fleeing its being seen cannot escape the gaze of its own inner seeing. It is nothing that produces that feeling of anxiety, but the experience of a call to conscience, to the assumption of my own responsibility for being myself. In this way, Heidegger's ontology is attachable to Schopenhauer's pessimism and within this scheme personal salvation would be the poet's own authenticity.

For Schopenhauer, who quotes Calderón, the only sin of man is to have been born; for in having been born, one has experienced the original Fall of all men – from the universal, metaphysical Will of nature itself, expressed ideally in the species, to the particular form of individuation one's psycho-physical substance is to become. Having been born, I have been expelled from union with the universal into a state of individual desire for which there is no cure except death, or the cessation of desire (whichever comes first).

But for Schopenhauer and Schopenhauereans like Proust,[11] one can escape the punishments of unfulfilled desire by achieving the status of a "pure will-less subject of knowing," in an act of contemplating the essences of Ideas into which the primeval Will has expressed itself in nature, and which in turn have found expression in successful works of art.

Music, in Schopenhauer's view, dispenses with the representational image, and gives a direct intuition of the metaphysical Will. But this form of intuition is lacking in Beckett's literary worlds. The music is there, but none of his narrators has been satisfied that this story has been successfully told. So the telling must suffice, and that is what Heidegger had indicated in his description of human consciousness as disclosedness: the openness to one's world, which is the being of Dasein's being there (*Da-sein: Das Sein des Da*), in its world by virtue of its feeling, understanding, and discourse.

In opening himself to such a world, laid out in the literary context, the reader undergoes a similar experience of existential openness, unless, of course, he or she artificially closes that context by any number of the false procedures that have plagued traditional criticism. The Opener opens and closes, and his audience must have its experience guided by the structures of the closed context.

Let us avoid, then, the easy but meaningless epithet. Let us return, with Beckett, to the sources of the creative word, and follow the path of human disclosedness on that continuous search for itself, that seemingly endless attempt of a consciousness to define itself in creative discourse. If we meet the absurd, it will only be one stage on the way to the epiphany of the human spirit.

Phenomenological Criticism. In his effort to avoid self-delusion with the easy, but misleading identification, the phenomenological critic proceeds by describing the objects of his own consciousness as he concretizes the individual works of his author, limiting his observations to the linguistic details of that particular context. The musical surface, significant in itself, expands and deepens. The process of isolating the horizontal development of the images, constituting the first level of depth, is of necessity slow; the effort, laborious. But if the images are to be seen as structured upon the principle of a philosophical idea, one must possess the total picture. There is no short cut or royal road to successful hermeneutics. Any obvious symbols or plays on words are aids in the interpretation, which itself is carried out by the method of hypothesis formation – a procedure only slightly different from that employed by the explanatory sciences.

In confirmation of the hypothesis, the literal signification of the words (the images called up) serve once again as the test for the hypothesis' workability. When, upon second reading, the hypothesis is seen to hold, the critic is armed with the knowledge that he has found *an interpretation* of the work, and his aesthetic analysis may begin. Each personal concretization by the reader may or may not be the one intended by the author. That is a detail of little importance; for as the Opener has said, the work is not in his head, but in the structures of the meanings built into the texture of that work as they relate to the "music" of the words.

The foregoing chapters, excepting the first, which was meta-critical, were all written in accordance with the prescriptions of the phenomenological methodology. Beckett had himself hit upon the relations between philosophy as a humanistic discipline and literature as a fine art in his studies of Joyce and Proust. For neither of these authors is the philosophical idea a truth to be grasped in the abstract, but an artistic convenience – or inconvenience – a mere device for structuring the literary world.

The philosophy is not the meaning of the work; for within the context there are others, and the total expressiveness of that work is the felt quality of the funding strands of signification, as these relate to the expressive surface. For this reason, it is useful to distinguish between 'significance' and 'signification.'

The signification of any sign is its reference to an object or idea outside itself, and there are as many levels of these as there are signs to be interpreted. For the whole work, there is no signification; for, the total system of signs and signifieds, being a closed, saturated system of linguistic relations, does not refer to anything external to itself. That is why the key to the treasure is the treasure itself. The saturation of that system is, however, felt; it is the quality of that system's wholeness, the significance that pervades the whole.

Even that fourth level of Dantean signification – the anagogical – must be thought of as being a reference back to the system of meanings from which it has been deduced. And if that level of signification is thought of as "summarizing" the depth meanings of the piece, it too only serves to organize the significance of the context. The hermeneutical, phenomenological critic maintains the strictness of the work's intent – that which had been isolated within consciousness in the first place by his voluntary decision to allow the work to speak for itself – by restricting his attention to the internal play of the work's linguistic structures. What is expressed is "there," in the texture of the linguistic relations; and what is there must come to expression in the critic's words, which describe the images and all the relations a thoughtful interpretation may find obtaining between them.

For the remainder of this chapter I shall revisit the progress made by the series of phenomenological renderings I have given the individual works in the Beckettian *oeuvre*. In attempting to describe the basic structure of these works in terms of a philosophical idea, I shall be doing nothing more than to practice on Beckett's work the same sort of criticism he has practiced on the work of others.

It will come as no surprise to anyone that, as Beckett's literary skill is perceived to have developed, the philosophical ideas he has used for the structure of the individual works have undergone a thoroughgoing change. Indeed, as the sequel will show, Beckett's philosophical development parallels that of his technical mastery, and is directed towards a new phenomenology of the human spirit; and it might be supposed that one reason he has had for telling the same basic story in so many different ways is the changing sense of adequacy in the philosophical structures employed. Some older ideas are retained in the residual works published long after their original formulation in longer pieces.[12]

The Aesthetic Inversion of Values. If the same overall feeling of pessimism hangs over this *oeuvre*, it should not be forgotten that there are at least as many reasons for pessimism as there are ways of envisaging the being of a consciousness in search for its own self-identity that in the end is achievable only in death. If we haven't read the philosophy of Heidegger, the same idea may be found as the depth structure of *Malone Dies*.

Every living failure is a continued reason for the maintenance of the pessimism, and only the individual engaged in the search for the meaning of his being is capable of describing the conditions under which the "successful" resolution of the project would have been achieved. Given the quasi-infinity of human desire, and the specificity of the drive to self-determination, failures should outnumber the successes by a statistically certain percentage. That much may be predicted.

But it is not pessimism that is expressed in these works; the pessimism, like any other counter of the context, has meaning only by virtue of its relationship to other counters, in particular the surface values of the literary construct. Beckett's wit and propensity toward irony already modify the pessimistic coloring of his fictional creations in which the humanity of the characters is displayed before us as forced to come to grips with their lives in their worlds, and with others, under the conditions presented to them. They all may have failed – out of ignorance, fearfulness, personal anxiety, false pride, social snobbery, and the like; but in their very failure they still serve to show what success in their ventures might have been. Within the depth structures of these works the negative value illuminates the positive in a way that a bald statement of that positive value can never itself manage. That too is an irony of human expressiveness; the human spirit is not less poignant for having failed to achieve a goal.

Beckett's "pessmism" reduces to an awareness of the power of human consciousness to negate, absolutely to refuse the sham, the false and the meretricious, be they the expression of an ideal, the values of everyman, or any of the various claims to transcendence by which mankind has heretofore beguiled itself into believing that it has found the truth, absolute and necessary for every human being. That kind of optimism should make pessimists of us all. As a positive value, the negation of these negations of the truth, such "pessimism" is liberating, and expresses the resolute decision of a consciousness to be itself; wielding No's knife is a natural defense against the overbearing positivity of the world, even if it means complete withdrawal into the smaller, inner world of the consciousness in flight. The province of the mind is in the negative; so Hegel has said, and so Beckett's characters have lived. His readers, in sympathetic reaction to all this negativity, are responsible for its transevaluation into a positive aesthetic experience.

Beckett's Tragic Vision: The Conflict of Time and Temporality. The only thing a consciousness can know for sure is the immanence of its own experiences, and this seems reason enough to begin our review of Beckett's works with the Cartesian poem, *Whoroscope*. For it was that philosopher who based all his claims to knowledge on the certainty of the act of cogitation.

That was a purely subjective certainty, to be sure; for no one could doubt that he was doubting, without establishing the doubt at least in the second instance, and the connection between the doubter and the world is left to the guarantee of the almighty, all-good, God (Camier's "omni-omni"). For the whole scheme to work, the existence of that God must be established by an idea of the consciousness whose negation entails a self-contradiction. But the "ontological" proof, as it is called, is logically defective, and Descartes' certainty rests upon nothing more substantial than his prior system of beliefs. But let us refer back to the phenomenologist's context. What philosophy do we find in those purely literary worlds?

In *Whoroscope*, the persona speaking is that of René Descartes. His body is at rest, and his mind is contemplating the events of his life, taking up time while waiting for his breakfast. The overriding irony of the composition is the demonstration of that mind's pettiness, including the superstition of its concealing the exact date and hour of its birth; for, if that were known, one could have projected a horoscope that would enable its possessor to predict the future of that "immortal soul."

Descartes, of course, had written his *Meditations* to prove the existence of God and the immorality of the soul, and the last image of Beckett's poem is a reference to the philosopher's future rebirth in the expressed hope that he be granted his "second/starless inscrutable hour."[13]

The poem presents Descartes' subjective, immanent experiences, his thoughts of his body, dying at the hands of an intellectual enemy, and his personal hope for immortality through the continued existence of the soul in a world where no stars may influence the perfection of that essential existence, and in which no one may be betrayed by enemies, real or imagined. Intertwined with this moment's reverie the reader can perceive the effects of the reality of the time spent in waiting for something that never comes (Descartes' breakfast), and the timelessness of the wish for deliverance from time, from mortality to immortality.

That poem, which was submitted as an exercise on the literary interpretation of time, was the winner of the competition, but a questionable literary success. It was written after Beckett's study of the philosophical structure of Joyce's *Work in Progress*, but before the publication of his *Proust*, dedicated to an explanation of that author's use of the notion of time to construct one of the twentieth century's masterpieces of philosophical fiction.

Descartes' purely subjective certainty leaves the thinker with the image of man as an intellectual solipsist – one holding the belief that the whole world is the idea of that person's mind. The guide here is Schopenhauer, whose *World as Will and Idea* opens with the telltale sentence, "The world is my idea."[14]

The solipsist in question is Belacqua Shua, who contemplates the falsity of the common expression of the "quick death" of a lobster being cooked alive. The expression is false in at least two interpretations, since the lobster does not die immediately upon entering the boiling water, and in the other sense of the term 'quick' no one and nothing dies quickly, since that expression is an oxymoron, false on the face of it, but true upon second glance, since everyone who is living is dying. But that is usually a slow death. Quicker in both senses of the term is the death of the solipsist who betrays his essence by falling in love; one lives, but only succeeds in committing intellectual and symbolic suicide by the speed with which the stimulated body reacts to sexual stimulation. Belacqua's life is a series of such symbolic suicides, and ends on the operating table by an oversight on the part of the attending surgeon. The gratuitous circumstances of this actual death recall Descartes' real historical death, and the continued dilemma of harmonizing mental with bodily existence in a finite spread of time.

The title of the novel, *More Pricks than Kicks*, is a reference to the ethical interpretation of the solipsist's life predicament: there are always more pains than pleasures, more pains than ways of combatting them. The hermeneutical expansion of the novel's depth structures follows the model of Dante.

Murphy is the narration of the life of another solipsist, tortured by the human scission into body and mind, plagued in the body by women, and haunted by others of both sexes seeking to intrude upon the privacy of his own inner world of the mind. Superstitious, like Descartes, Murphy acts only in accordance with his horoscope, and dies, accidentally, in the pursuit of his own predilection for privacy and comfort.

Murphy experiences the "Belacqua syndrome," and castigates his mistress for squandering her being into meaningless acts of doing. To please her he accepts a job that seems to him predicted in his horoscope, but he himself predicts the end – either of her, his body or his mind, the only three possessions he has. As male mental nurse, he recognizes his own failure in the success of his patients in creating their own personally significant albeit schizophrenic worlds. More consistent than Belacqua, his love is purely intellectual and is directed towards the products of his own mind. The Other? A hindrance to personal gratification, but omnipresent.

Watt incorporates the consciousness at odds with the world. The protagonist's body functions badly, and only upon the conscious direction of his mind, fully capable of inductive and deductive reasoning. But the "rational domain"[15] is still valid only for the reasoning consciousness that becomes progressively more alienated from the structures of the real world inhabited by other people. The ideal, and therefore unreal, world of numbers and logical matrices only serves to alienate the personality forced to live in the untidiness of the real world.

Watt enters the service of Knott, the Absolute Mind that needs nothing except a witness to its needing nothing, because it is the absolute center about which all things relative turn. Perennially changing, it contains all changes; and so, with respect to the individual relative changes (the perceptual phenomena of the world) itself does not change at all. Here the rationality is indisputably Hegelian.

Knott also represents the master by which the servant's consciousness receives its definition: the first admission of the necessity of the Other. To be self-conscious is to be conscious of oneself as opposed to that Other. That is the explantion of the prior solipsists' unhappiness, and the first modification of the solipsistic ideal. Watt discovers himself as the negation of the negation, and goes mad in the process, but succeeds in telling his story to "Sam," who relates it in naturalistic order to the reader. Watt's madness: a function of his worldly skepticism, and his tendency to accept as true both sides of a self-contradiction – a repetition of the theme of dual self-alienation, first from the world, and then from personality integration.

First Love marks Beckett's discovery of the first-person narrator, who recounts his first love as an alienation from self. The path of a life's development is from solitude to communion with the other, and from there to solitude once again. The revised, attenuated solipsism persists, but now exhibits the theme of the eternal return, which will eventually become the account of an individual life from the nothingness of its beginnings to the nothingness of its end: in between, life at odds with others, which is experienced within the social arena as a series of expulsions. The narrator and his love have replaced Watt and Knott as embodiments of the antitheses united in the dialectic of the Other.

Mercier et Camier is the story of a pseudo-couple, just as the prior tale contained the story of a false couple. But the falsity of this couple has a dual sense: they form a false unity because they are two at the first level of interpretation, and a false duality, at the second, since individually they are only partial functions of a single human personality. Both interpretations fit the context. The self-other-self relations between the two subtend another relative entity, the self, which contains its own other, and the unity and the duality both must end with the separation of the two. Thus, the basic dichotomy of the human being into mind and body receives its first transformation: consciousness and body are functions of a single entity, not two distinct substances.

The story is recounted by an omniscient narrator, with the female of the species still representing an obstacle to the union of the mental and corporeal functions of a man's transcendence.

Watt makes an appearance in the story to announce that Mercier and Camier must be friends, and that the one shall be born of them, who, having nothing, will desire nothing, if not to retain the nothing he has. Geulincx's stoicism is thus reaffirmed, in the return of consciousness to itself, but is surpassed by the skepticism of the couple's attitude on the voyage from their home territory and back again in a series of meaningless sorties.

But the story ends with the achievement of the ultimate development of Hegelian self-consciousness: the state of unhappiness, that would be reinforced by the failure of consciousness to achieve the enlightenment of the Spirit, whose existence the voyagers have denied. Themselves the living affirmation of self-contradictory functions, they embody the unhappy consciousness in their unity, and each must die in separation from the other. Indeed, their unhappiness comes to an end only in death.

Stories, the smaller trilogy, is a return to first-person narration. *The Expelled*, the life of a consciousness in itself, *The Calmative*, of the same consciousness as it is for itself, aware of what it was in itself; *The End*, of the original consciousness in its return to itself, now existing in and for itself as it recounts the event of its own suicide. This is the Hegelian dialectic in its broadest interpretation: life as the development of full self-consciousness.

The *Stories* culminate in the description of the dying narrator's experience of the "oceanic consciousness," under which condition alone solipsism may be a true description for the relationship between consciousness and the objects of its world. The solipsist's victory results from the ultimate act of refusal: suicide; but it is pyrrhic, occurring just before its final extinction. Perhaps this story was predicted by Watt's final defiance, in *Mercier et Camier*, as he excitedly shouted, "life down the drain."

Molloy represents a life in a return to its source, and a reflective narration of its essence, as having been lived. There is an appearance of a third "voice," that of conscience, uniting the active and reflective aspects of the human project, as the Hegelian dialectic of conscious experience is replaced by Heideggerean existential ontology. Molloy's mind is a region, illustrating an old theme first introduced in *Murphy*. But the region is traversed in time under the impulse of conscience felt as a call to authenticity, from a future self that does not yet exist to the present one lost in an activity that prolongs the effects of a badly remembered past. The meaning of the being of consciousness is the temporality it lives, as directed towards a future discovery of self. Philosophically, the theme has changed from the reality of time to the subject's inner experience of temporality.

At the level of symbolism, Moran's journey reflects in mirror image that of Molloy, the reader's that of the writer's. The reader's authenticity refers him to Molloy's own pages for the truth of Molloy's life. Simultaneous with the discovery of conscience, there is a denial of any external sanction for it: the abandonment of the individual to his own decisions for the determination of a life's project, ending as always with a return to self.

Malone Dies depicts the wholeness of a human transcedence: life as the anticipation of death, the present as the resolute anticipation of the future. If time is infinite, human temporality is not, and one's essence is achieved with the acceptance of human finitude, as a being-in-the-world. Malone's stories, the self-conscious narration of the imagined lives of others, concern Saposcat and Macmann, two universal symbols for all mankind, suffering, going mad, and placed under the care of the sado-masochistic "Lemuel." God's name is now 'Lem,' where before it was 'Sam.' He is the caretaker of the world of others, and theirs is the misfortune of having to suffer his justice.

The Unnamable depicts a human transcendence in its essential state, no longer transcending, but transcended by time and history: a description of Malone's continued existence as now embodied in the space and time of a literary, symbolic situation. With existence now itself transcended, the being of the narrator is purely essential, fixed in the symbolic relations created by the author, whose voice cannot be distinguished from the Unnamable's own or from those of the latter's creation, principally Mahood and Worm.

If consciousness in this state is unnamable, it is the fault of the pronouns; they must have an antecedent, but all their antecedents are insubstantial nothings: all are merely relative in their existence, and the author of all lends his own voice to their projects to hide his own dependence upon them.

The *Texts for Nothing* celebrate the nothingness of any possible narrator of a fictional character's existence. As in *The Unnamable*, the creator may only be defined by a relation to his creature, and neither exists outside of that relationship. Writing as an experience becomes the discovery of finding a way out, finding the unnamable words for the unnamable things. Conscience is now the voice of the silent self forced to deny itself by speaking and letting others speak. Thus: the three voices of Molloy in search of a way to continue.

How It Is reduces the consciousness of the narrator to its essential function, breathing to create the words that create the images of the literary world. It presents a life before, during, and after union with the Other. Before that union there is no effective conventional language, since that develops out of the union of the self and the Other; and after that union the narrator must await visitation by still another to reveal how it is to live as a creator of a fictional world. The body is now all but indistinct from the mud from which the characters have sprung, and the consciousness is a voice announcing the rebirth of the spirit, through the acquisition of language. The unhappiness of that consciousness stems from its final abandonment, or its alternative, the acceptance of the justice meted out by the supervising scribe, who listens to the narrator's tale.

Enough is a vignette, recounting the love of the old for the young; of a father or a friend for a younger man, left alone after years of silent communion. The love is misunderstood by the young man, but is made real by his feeling of regret of its having passed: the Self-Other-Self once again, with the second conscious self older and wiser for the communal past, but unhappy in the feeling of the regret.

Imagination Dead Imagine is all that remains from a larger, unpublished work. To imagine the death of the imagination is to assist at its rebirth (a Cartesian step in the proof of the existence of the self). Consciousness cannot deny itself, and will go on writing, as it does, by describing a world changing character from white and hot to black and cold. A man and a woman make their appearance circumscribed in a sphere, fixed in relation one to the other, but seeing each other only at slight intervals as their eyes open and close out of phase. Consciousness is now a series of images, as in Berkeleyan idealism; and the world is determined by the consciousness of the third-person narrator. Unhappiness is rendered as separation – from the other as from oneself, since to be is to be perceived, i.e. created as an image, and in the image of one's own creator.

Ping, likewise a residual piece of a longer work, returns to the act of creation. The world appears in a set of images, as the words are written, but without distinctive color. The characters appear as white figures on a white ground, as each new image is announced by the telltale "ping," the register of a sound for a new word to be heard, creating still another image. It is of a female character, eyes open, imploring liberation; but the narrator cannot see what the pings do not produce. That consciousness dominates this world, but it is reduced to a punctuated breathing that emits sounds creating the images.

This poetic image itself presents the poet's plight in full sublimity: if human life is reducible to a breath, give or take a few, it is with that smothering breath that the poet transcends his own condition; for as he breathes the words, they become images and the images, an idea of the creative spirit. That is one solution to the problem of pessimism. As long as there is breath, there is hope for creative transcendence.

From Sublimity to Tragedy. The dramatic pieces, for the most part, are experiments of the author to reproduce his poetic and prose achievements in related media. Where the value of the prose fiction slowly approaches and then overtakes the sublime, the sublimity of these fictional characters gives way to the highest form of contemporary tragedy – a development impossible to explain without reference to the artistic mechanisms by which the inversion of aesthetic values takes place for the reading consciousness.

Cascando, Words and Music, and *Embers* are dramaturgical plays; i.e. dramatic to the second degree, showing how an original first-order drama is created: out of the experiences of life (*Embers*); by the action of an Opener controlling the instruments of his craft (music and voice), for the harmonious narration of the life of the perpetual voyager, now called Woburn (*Cascando*); by finding, or forcing the harmony between the two strands of artistic creation, the surface and the stratified depth (*Words and Music*).

Waiting for Godot and *Endgame* reduce the dramatic characters to their very theatrical presence. The members of the various couples, Vladimir and Estragon, Pozzo and Lucky, Hamm and Clov, Nagg and Nell, are various impersonations of the master-slave in the dialectical relationship that announces the beginning of self-consciousness in the Hegelian account of human experience as the continual surpassing of an internal self-contradiction. But as time comes to a stop in the inaction of these plays, temporality comes to the fore as consciousness itself is shown in its enduring: the various couples wait, inauthentically, either for the event that does not occur or for the one that does. Waiting, after all, is an action that denies all action. But Godot never makes his appearance, and Clov finally leaves Hamm to his own nothingness. The artistic achievement is phenomenal: consciousness is temporality, is the presence of the characters, who merely await their ends. Their unhappiness is their abandonment; and left to themselves, they can do nothing, for their existence is inauthentic.

All That Fall makes explicit the transition from the earlier Schopenhauerean and Hegelian expositions of the human predicament to the later Heideggerean framework. The old couple have fallen into inauthenticity, and are too absorbed with themselves as determined by the structures of the phenomenal world and the existence of others to be able to discover themselves as free consciousnesses, whose freedom is to negate the power of the other to determine what it itself has to be. In the meantime, one waits – to die.

Krapp's Last Tape presents a consciousness lost in the record of its own past; and *Happy Days*, the play with ironic title, displays the unhappy condition of a feminine consciousness lost in the tangle of its present: two cases of an aborted love. Where Krapp's false ideal of his own identity and equally false estimation of his past lover's conscious self prevents their union – which persists in his regret, the loss of their child imprisons Winne and Willie into a present life situation that belies the cheerful suggestion of the connotations contained in their names. He cannot will and she cannot win. *Play* presents the same theme, but the loves form a triangle between adult consciousnesses shunted meaninglessly back into themselves, as the unfaithful husband, playing at love, succeeds in destroying all the angles and sides of the triangle, leaving only an unenclosed nothingness.

Beckett's dramatic experimentations begin to reach their heights in the abstractions of *Act Without Worlds I* and *II*, and in *Come and Go*. The abstraction is in the use of the medium to create a phenomenal world by turning the syntax of the medial relata into a newer, more powerful, form of semantics. The aim is to achieve more with less.

In the two *Acts Without Words*, the medium is mime, and the represented consciousness is its acts. Mind and body are therefore no longer separable. In the first of these mimes, the human being can find no way to die, unhappily deprived of the means to end its own unhappiness after two unsuccessful attempts; and in the second, two players and a sting complete a monotonous round of suprarational inauthenticity. But each is inauthentic in his own way: the one ignores time, and seeks deliverance by supplicating the divine, while the other is a victim of time, and is unaware of his own need for deliverance. Both sets of activities merely serve to pass the time of innumerable days each of which resembles the other in its monotony as the significance of self-transcending temporality goes abegging.

Come and Go presents a human mobile composed of the paths and patterns of the related consciousnesses, all female, moving from self to other and back to self, but suspended at a point insecurely anchored in the past, and coming to rest in the bodily configuration of the final tableau. The selves displayed in the mobile generate a series of chiasmatic figures, but have no future, since no one may recover a lost past. The attempt to recover any lost past locks the consciousness into a fixed, unopen present.

Film shows a consciousness in the vain attempt to escape from itself, and *Eh Joe* pictures a consciousness haunted by its conscience, speaking in the voice of a past, betrayed love. That consciousness, too, cannot escape from itself, and suffers the pangs of its never-ending guilt. And, finally, *Breath* reduces consciousness, on two levels, to a set of contrasts. Visually, from dark to medium light, where only a deserted world is seen; and audially, to the sounds of an inhaled and exhaled breath, punctuated by the cry of a baby announcing the "guilt" of its birth, and, in its repetition, the unhappiness of continued existence, until it too is extinguished by the return to the original silence of the dark.

In these final three "experiments," the writer has succeeded in attaining the artist's primary project: to make light and sound of soul, achieving the maximal aesthetic effect in the minimal surface artistic effect.

And for those of us who have not understood this literary and dramatic achievement, or who have understood it badly, the author has given us a second chance to grasp his display of the phenomenology of the human spirit in his various translations. If the English is not clear, try the French; and for those written in French, reverse the procedure.

The hope is that everyone may be led to see that Samuel Beckett is one of the most successful philosophical literary artists of our times. He has made his point by bringing his various fictional worlds to closure on the basis of a philosophical idea. In no *oeuvre* known to this writer has there been such a persistent attempt to come to grips with the understanding of personal existence, no work in which it has been shown how consciousness becomes fully aware of itself. That indeed was the project of the phenomenology of mind, which must itself define the meaning of "spirit," human if not absolute.

The consciousness of the writer constitutes the poet's plight and, at the same time, the paradigm example of human disclosedness: being in the world, feeling what it knows, but knowing only what it can successfully express in a world of its own making. It is for his readers to determine the degree of success, and they need only a method of procedure that guarantees the validity of its own results. The two procedures, of writing and of interpreting, are, like the life projects of Molloy and Moran, mirror image reflections of each other.

The unhappiness of both projects is felt as the uncertainty of their results. But all human projects must be brought to a close, and perhaps only a philosopher will attempt to achieve successful closure on the basis of a structural, philosophical principle. To write of the poetic plight of Samuel Beckett is to find oneself on the other side of that mirror of life.

What have we found there? First of all, irony; next, sublimity; and finally, tragedy, which must now be given its phenomenological interpretation.

Tragic art: the pessimist's deliverance. We are all familiar with the effects of a mirror image: the right side of an object reflected appears as the left side of the reflection, and vice-versa. We all know that the images on our retinas are registered in a similar inversion, the top of the object forming the bottom of the image, and vice-versa. Yet we see objects as they are, bottom side down and top side up. Although the mechanism of this seeing remains obscure, the mind of the perceiver has no difficulty inverting the order of its images. Something similar, I shall argue, happens in any percipient's experience of tragic values: within the context of an aesthetic expression the viewer is presented with a negative value – pain, suffering, frustration, defeat, and yet the experience as a whole is itself suffused with positive value – elation, joy, admiration, success. A phenomenologically constituted datum of the aesthetic consciousness, this inversion of values has become a problem for aestheticians, whether psychological or philosophical, from Aristotle to the present day.

Aristotle's solution, the catharsis of the "negative" emotions of pity and fear (pity for the protagonist and fear of a like fate for ourselves), held current until the eighteenth century, when in Lessing's *Hamburgische Dramaturgie*[16] it was raised into the status of dogma. The only meaningful alternatives to the Aristotelian interpretation were offered by two other German aestheticians. Schiller's *Aesthetical and Philosophical Essays*,[17] the first of these attempts, sought an explanation in the experience of the morally sublime, following the inspiration of Kant's aesthetics. The deeper the suffering or degradation of a protagonist consistent with the safe contemplation of such a fate, the greater the sense of moral uplift felt by the contemplator whose own nature is experienced under the guise of its freedom as it relates to the nature of the suffering hero under the guise of strict determination through the forces of nature. And Nietzsche, in the *Birth of Tragedy from the Spirit of Music*,[18] interprets this freedom as an expression of the will to power, through which primeval pain, acting as a catalyst, enhances the quality of the master's acceptance of the conditions of life. The spectacle – yea! even the experience of pain is never to be avoided provided that it may be turned into the continued experiences of a forceful life.

All three of these "explanations" are predicated upon the inversion of an originally negative experience: no surprise, since that is the phenomenon to be explained. Where Schiller's and Nietzsche's accounts are redolent with metaphysical suppositions, Aristotle's long tradition is either directly psychological or more derivatively intellectual, as when, in the hands of Schopenhauer, for example, the pleasure of aesthetic contemplation stems from the Idea contemplated in the tragic representation whose pain is transmuted into pleasure through the intellectual insight into the Will itself as causing such pain. But this too rests finally upon the metaphysical presuppositions of its author, as upon the "heroic" proportions of the classical tragic protagonist. If we are to avoid the extravagance of the metaphysical aestheticians and the limited world views of classic tragedians, then we must follow the phenomenological dictum of Husserl by returning to the things themselves, to those plays in which successful artists have solved "the problem" of tragedy by writing successful plays. We already know that Beckett's protagonists resemble nothing like a classical hero; for the creation of a contemporary tragedy, a passel of nondescript "anti-heroes" may do just as well.

Interested primarily in the avatars of human self-consciousness, Beckett, like the Opener of his *Cascando*, opens and closes the context of aesthetic significance in each of his works. He speaks the words that create the images, of ordinary or sub-human men creating or failing to create. The worlds of these created personages are structured according to still deeper constraints of human possibilities; the ontic experiences of his "heroes" or antiheroes (the experience of being in their worlds) thus reveal ontological structures (the basic principles by which everyday human experiences are understood to be possible in the first place). Such was the basis on which was constructed the analogy of ontological analysis and literary depth structures. But as in all purely literary worlds, the depth idea needs embodiment in a sensuous surface. The Opener controls two boxes, one for the voice and the other for the music.

If existing provides the ontic materials of the first level references in the Beckettian worlds, the inquiry into existence provides both the ontological questions – Where now? Who now? When now? – and the ontological answers. They are indeed the answers that make Beckett's dramatic art anything but absurd. The narrators of the novels are themselves full human beings: ontic-ontological projections into an unknown future. What that future had to be could only be determined by the choices of the characters; and they inevitably fail. The Unnamable can be known only in terms of his own creations, and the voices swarming around him; that is what he is, essentially.

When we turn to the dramas, however, the negative experiences of these "fallen" or inauthentic personalities merely appear as phenomenally dark images on the ground of full authenticity. They are all victims of time, locked in the past or the present without the saving grace of a liberating choice that would open up a significant future. Time obscures temporality as being lost in the world obscures our creative transcendence beyond it. And here is where the inversion of values takes place: in our understanding of this aborted transcendence. The darker the surface image, all that much clearer is the tension between the darkness of the image in its inauthenticity and the brightness of its structuring principle (that brightness, indeed, is an oft-used symbol of consciousness itself) – of what it is authentically for a human being to be, a projection towards ultimate significance. Not that the positive is expressed through the negative; No's knife is still Yes's wound.

The expressiveness of the aesthetic context results from the saturation of the linguistic symbols, and is felt as the tense opposition between the negative and the positive held necessarily together as they are forged into synthesis by the author's opening and closing words. In this way, Beckett's fundamental aesthetic project points the way toward a second solution to the problem of pessimism vs. optimism in the contemporary world. There is nothing particulary pessimistic about the world in which we live; the problem lies in us. However pessimistic a human life may be, it is always possible to invert that value in a successful work of art – provided that one is true to one's calling as a writer or reader, as a speaker or hearer of the spoken word, as long, that is, as one refuses to lie to oneself.

The road to Beckett's discovery was long: from Descartes, to Geulincx; through Berkeley and Schopenhauer to Hegel; and from Hegel to an extraordinarily vibrant version of *Existenzphilosophie*. Given his own doctrine according to which a work of literary art may be structured as a convenience to creation by the deep structure of a philosophical idea, his success as a writer may be measured by his mastery of those philosophical ideas, whether he has read them in any source other than his own personal experiences, or whether he has merely imagined them along with his characters. However come by, given his artistic mastery of these ideas, we can be sure they were aesthetically felt.

Beckett's personal optimism may be viewed in his perfectionism, in his honesty, in his unflagging pursuit of the significance to be found in the common reflective knowledge that living and dying are identical processes. That is perhaps the basic figure by which to conceive of the famous inversion of values: to show the life in the act of dying; but such an inversion is impossible outside of expression in irony, at times purely comic, at others directly of the sublime, suffering narrator, and at still others movingly tragic. Samuel Beckett's art moves us through the entire gamut of human experience, accepted for what it is and raised to the most intense form of human feeling under the control of artistic structure.

Perhaps the most optimistic of his beliefs is that his literary works may speak entirely for themselves. That and only that would mark the success of his own aesthetic project. And the uncertainty of that success makes his own project all the more insecure. Every Molloy needs his Moran, even if, like Mercier and Camier, they must part following a moment of communication. That is the fate of any self that seeks to relate itself to another.

Beckett's contribution to modern letters is not an insight into man's (or his time's) absurdity. The rational animal has always been capable of the most monstrous kinds of absurdity even when, on the surface, he is acting most rationally. That, we recall, was the tragedy of Watt. Beckett's contribution can be measured only by the degree to which man's deeper absurdity (having no cause other than one's own decision to act), with all its tragic consequences, may be transcended in living works of art. And for that judgment a reader may be as well placed as the author himself. To have to submit to that judgment is, in essence, the ultimate plight of Samuel Beckett.

REFERENCES

CHAPTER I

[1] *Beckett at Sixty* (London: Calder and Boyars, 1967).
[2] *Samuel Beckett, A Collection of Critical Essays* (Englewood Cliffs, N.J.: Prentice-Hall, 1965).
[3] *Op. cit.*, pp. 169–77.
[4] New Brunswick, N.J.: Rutgers University Press, 1962.
[5] In the collection, Configuration Critique, no. 8 (Paris: Minard, Lettres Modernes, 1964).
[6] Carbondale, Ill.: Southern Illinois University Press, 1962.
[7] *Ibid.*, p. 59.
[8] Edith Kern, 'Contents of Pockets,' in Friedman, *op. cit.*, pp. 69–81.
[9] Germaine Brée, in Friedman, *op. cit.*, pp. 83–97.
[10] Friedman, *op. cit.*, pp. 117–150.
[11] *Ibid.*, pp. 23–53.
[12] *Ibid.*, pp. 153–68.
[13] *Existential Thought and Fictional Technique: Kierkegaard, Sartre, Beckett* (New Haven and London: Yale University Press, 1970).
[14] New York: Dodd, Mead & Co., Inc., 1970.
[15] *Ibid.*, p. xi.
[16] *Ibid.*
[17] See Ch. II, *infra*.
[18] New York: Barnes and Noble, 1964.
[19] First published in Britain by Chatto & Windus, 1934, limited editions by Calder and Boyars, Ltd., 1966, 1967; reprinted by Calder and Boyars, 1970.
[20] Written in 1945 as *Nouvelles*, published by Les Editions de Minuit, Paris, 1958, along with *Textes pour Rien*, written in 1950. The two collections were published in English translation along with selected shorter prose pieces in *No's Knife, Collected Shorter Prose, 1945–1966*, (London: Calder and Boyars Ltd., 1967).
[21] *No's Knife*, pp. 165–68.
[22] See Hoffman, in Friedman, *op. cit.*, pp. 23–53.
[23] Berkeley and Los Angeles: University of California Press, 1968.
[24] Athens, Ga.: University of Georgia Press, 1968; University of Georgia Monographs, No. 15.
[25] New York: Scribner's Sons, 1969.
[26] See his *Ideas*, Boyce Gibson, trans. (New York: Collier Books, 1962), pp. 96–100, 115–170.
[27] *Cf.* Beckett on Marcel Proust's treatment of this theme, *Proust*, 7th printing (New York: Grove Press, Inc., n.d.), originally published, 1931.

[28] See Hegel, *Phenomenology of Mind*, trans with an Introduction by Sir James Baillie (London: George Allen and Unwin, Ltd., 1931), second, revised edition, pp. 251–267.

CHAPTER II

[1] Samuel Beckett, 'Dante . . . Bruno. Vico . . Joyce,' *transition*, 16–17 (June, 1929), 242.
[2] *Ibid.*
[3] Frederick Hoffman, *Samuel Beckett: The Language of Self*, p. 59.
[4] Beckett, 'Dante . . . Bruno. Vico . . Joyce,' p. 244.
[5] *Ibid.*, p. 248.
[6] *Ibid.*, p. 245.
[7] *Ibid.*, p. 248.
[8] *Ibid.*, p. 246.
[9] *Ibid.*, p. 247. The older Latin form for an oak tree is *ilex*, a correction noted in a later version of this essay, published as the lead article in *Our Exagmination Round His Factification for Incamination of Work in Progress*, Samuel Beckett *et al.* (London: Faber and Faber, 1972); originally published by Shakespeare and Co., Paris, 1929.
[10] See his *Language and Gesture* (New York: Harcourt, Brace and Co., 1952).
[11] Becket, *op. cit.*, pp. 248–49.
[12] *Ibid.*, p. 250.
[13] *Ibid.*, p. 252.
[14] *Ibid.*, p. 253.
[15] *Ibid.*
[16] *Ibid.*
[17] *Ibid.*, p. 244.
[18] New York: Grove Press, Inc., n.d.; original publication, 1931.
[19] *Proust*, p. 57.
[20] *Ibid.*, p. 1.
[21] *Ibid.*
[22] *Ibid.*, pp. 67–68.
[23] *Ibid.*, p. 69.
[24] *Ibid.*
[25] In the Little Library of Liberal Arts, T. E. Hulme trans. (New York: The Liberal Arts Press, 1949).
[26] *Proust*, p. 66.
[27] *Ibid.*, p. 60.
[28] *Ibid.*
[29] *Ibid.*, p. 67.
[30] *Ibid.*, pp. 37–38.
[31] *Ibid.*, p. 41.
[32] *Ibid.*, p. 3.
[33] *Ibid.*
[34] *Ibid.*
[35] *Ibid.*, p. 10.
[36] *Ibid.*
[37] *Ibid.*, p. 11.

CHAPTER III

[1] Samuel Beckett, *Molloy, Malone Dies, and The Unnamable, Three Novels* (New York: Grove Press, Inc., 1959), p. 483.
[2] *Ibid.*, p. 409.
[3] Beckett, *Poems in English* (New York: Grove Press, Inc., 1961), p. 16.
[4] *Ibid.*, p. 15.
[5] *Ibid.*
[6] *Ibid.*
[7] *Ibid.*, pp. 13–14.
[8] *Ibid.*, pp. 12–13.
[9] *Ibid.*, p. 13.
[10] *Ibid.*, p. 17.
[11] *More Pricks than Kicks* (London: Calder & Boyars, 1967), p. 39.
[12] *Ibid.*, p. 40.
[13] *Ibid.*, p. 17.
[14] *Ibid.*, p. 18.
[15] *Ibid.*, p. 21.
[16] *Ibid.*, p. 105.
[17] *Ibid.*
[18] *Ibid.*
[19] *Ibid.*
[20] *Ibid.*, p. 189.
[21] *Ibid.*, p. 25.
[22] *Ibid.*, p. 30.
[23] *Ibid.*, p. 189.
[24] *Ibid.*, p. 115.
[25] *Ibid.*
[26] *Ibid.*, p. 167.
[27] *Ibid.*, p. 204.
[28] *Ibid.*, p. 184.
[29] *Ibid.*, p. 186.
[30] *Ibid.*, p. 189.
[31] *Ibid.*
[32] *Murphy*, tenth printing (New York: Grove Press, Inc., 1957), p. 108.
[33] *Ibid.*, p. 109.
[34] *Ibid.*
[35] *Ibid.*
[36] *Ibid.*, p. 275.
[37] *Ibid.*, p. 2.
[38] *Ibid.*, pp. 77–8.
[39] *Ibid.*, p. 82.
[40] *Ibid.*, p. 62.
[41] *Ibid.*, p. 32.
[42] *Ibid.*, p. 33.
[43] *Ibid.*, p. 48.
[44] *Ibid.*, p. 178.

45 *Ibid.*
46 *Ibid.*, p. 179.
47 *Ibid.*
48 *Ibid.*, p. 183.
49 *Ibid.*, p. 81.
50 *Watt*, tenth printing (New York: Grove Press, Inc., 1959), p. 168.
51 *Ibid.*, p. 215.
52 *Ibid.*, p. 39.
53 *Ibid.*, p. 43.
54 *Ibid.*, p. 254.
55 *Ibid.*, p. 129.
56 *Ibid.*, p. 62.
57 *Ibid.*, p. 44.
58 See Ruby Cohn, 'Philosophical Fragments in the Works of Samuel Beckett,' in M. Esslin, ed., *Samuel Beckett* (Englewood Cliffs, N.J.: Prentice-Hall, 1965), p. 174.
59 *Watt*, p. 44.
60 *Ibid.*, p. 30.
61 *Ibid.*, p. 73.
62 *Ibid.*
63 *Ibid.*
64 *Ibid.*, p. 75.
65 See *infra*, Chapter V, pp. 87–88, 90–106.
66 *Watt*, p. 81.
67 *Ibid.*, p. 77.
68 *Ibid.*, p. 78.
69 *Ibid.*, pp. 203–04.
70 *Ibid.*, pp. 202–03.
71 *Ibid.*, p. 159.
72 *Ibid.*, p. 253.
73 *Nouvelles et Textes pour rien* (Paris: Editions de Minuit, 1958).

CHAPTER IV

1 See Gérard Durozoi, *Beckett*, in the collection *Présence Littéraire* (Paris and Montreal: Bordas, 1972), p. 26.
2 This image is used as a title for the collection of stories and texts, written in 1945 and 1951 respectively, and first published in French; see the English publication, *No's Knife* (London: Calder and Boyars, Ltd., 1967).
3 "Poor papa, he must have been quite vexed that day, if he could see me, see us; vexed for me, I mean. Unless, in his great wisdom as disembodied, he saw further than his son, whose cadaver was not yet quite to the point of perfection." Trans. mine. – Beckett, *Premier amour* (Paris: Les Editions de Minuit, 1970), p. 17.
4 "What is called love is exile, with a postal card from the homeland from time to time; that's my feeling this fine evening." Trans. mine. – *Ibid.*, p. 22.
5 *Ibid.*, p. 21.
6 The anecdote is recounted by Durozoi, *op. cit.*, pp. 28–9.

[7] *Mercier et Camier* (Paris: Les Editions de Minuit, 1970), p. 17.
[8] *Ibid.*, p. 10.
[9] *Ibid.*, p. 32.
[10] *Ibid.*, pp. 91–92.
[11] *Ibid.*, p. 104.
[12] *Ibid.*, p. 108.
[13] *Ibid.*, p. 198.
[14] *The Unnamable,* in *Three Novels by Samuel Beckett* (New York: Grove Press, 1959), p. 409.
[15] *Mercier et Camier*, p. 39.
[16] *Ibid.*, p. 38.
[17] *Ibid.*, p. 39.
[18] *Ibid.*, p. 125.
[19] See that novel, New York: Grove Press, Inc., 1964, pp. 128–29.
[20] *No's Knife*, p. 13.
[21] *Ibid.*, p. 16.
[22] *Ibid.*
[23] *Ibid.*, p. 67.
[24] *Ibid.*
[25] See Ludovic Janvier, *Beckett par lui-même*, in the collection *Écrivains de toujours* (Paris: aux Editions du Seuil, 1969), *Chronologie et parcours*, 1930. Nonpaginated.

CHAPTER V

[1] I owe this chapter title to the suggestion of L. Janvier; see his *Beckett par lui-même* (Paris: Editions du Seuil, 1969), p. 128.
[2] *Three Novels by Samuel Beckett* (New York: Grove Press, Inc., 1959), pp. 5, 17, 29.
[3] *Ibid.*, p. 433.
[4] See Heidegger, *Being and Time*, Macquarrie and Robinson trans. (London: SCM Press Ltd., 1962), pp. 307 ff.
[5] *Ibid.*, p. 322.
[6] *Three Novels*, p. 562.
[7] *Being and Nothingness*, Hazel Barnes, trans. (New York: Philosophical Library, 1956), pp. 102–05.
[8] *Three Novels*, p. 113.
[9] *Ibid.*, p. 121.
[10] See Heidegger, *op. cit.*, pp. 91–148, and 149–224.
[11] *Three Novels*, p. 18.
[12] *Ibid.*, pp. 48–49.
[13] *Ibid.*, pp. 37–38.
[14] *Ibid.*, pp. 178–79.
[15] *Ibid.*, p. 178.
[16] *Ibid.*, p. 202.
[17] *Ibid.*, p. 228.
[18] *Ibid.*, p. 231.
[19] *Ibid.*, pp. 231–32.
[20] *Ibid.*, p. 240.

[21] *Ibid.*
[22] *Ibid.*, p. 31.
[23] *Ibid.*, p. 300.
[24] *Ibid.*, p. 391.
[25] See Heidegger, *op. cit.*, pp. 303–11.
[26] *Ibid.*, pp. 307–11.
[27] *Three Novels*, p. 243.
[28] *Ibid.*, p. 249.
[29] *Ibid.*, p. 285.
[30] *Ibid.*, p. 267.
[31] *Ibid.*, p. 253.
[32] *Ibid.*, p. 260.
[33] *Ibid.*
[34] *Ibid.*
[35] *Ibid.*, p. 397.
[36] *Ibid.*, p. 262.
[37] *Ibid.*, pp. 397–98.
[38] *Ibid.*, pp. 474–75.
[39] For the most concise statement of the distinction between essence and existence, see Heidegger, *op. cit.*, p. 67.
[40] *Three Novels*, p. 577.
[41] Sartre, 'No Exit,' in *No Exit and Three Other Plays*, no trans. (New York: Vintage Books, 1958), p. 47.
[42] *Three Novels*, p. 401.
[43] Arthur Koestler, *Darkness at Noon*, a Signet Book (New York: The New American Library, 1955), pp. 171–189.
[44] *Three Novels*, p. 513.
[45] *Ibid.*
[46] *Ibid.*, p. 439.
[47] See his *Monadology* in Religion of Science Library, Philosophical Classics, No. 52, *Discourse on Metaphysics*, G. R. Montgomery, trans. (La Salle, Ill.: Open Court, 1947), pp. 251–72.
[48] *Three Novels*, pp. 446–47.
[49] *Ibid.*, p. 447.
[50] *Ibid.*, p. 517. The image reoccurs from *More Pricks than Kicks.*
[51] *Ibid.*, p. 464.
[52] *Ibid.*, p. 465.
[53] *Ibid.*, p. 470.
[54] *Ibid.*
[55] *Ibid.*, p. 480.
[56] *Ibid.*
[57] *Ibid.*, p. 527.
[58] *Ibid.*, p. 298.
[59] *Ibid.*
[60] *Ibid.*, p. 487.
[61] *Ibid.*, p. 500.
[62] *Ibid.*, p. 509.

[63] *Ibid.*, p. 510.
[64] *Ibid.*, p. 516.
[65] *Ibid.*, p. 532.
[66] *Op. cit.*, pp. 211–14, 318, 196 ff.

CHAPTER VI

[1] *No's Knife, Collected Shorter Prose, 1945–1966* (London: Calder and Boyars, 1967), p. 71.
[2] *Ibid.*, p. 72.
[3] *Ibid.*
[4] *Ibid.*, p. 75.
[5] *Ibid.*
[6] *Ibid.*, p. 130.
[7] *L'imaginaire* (Paris: Gallimard, 1940), p. 233.
[8] This observation is by G. Durozoi, *Beckett* (Paris and Montreal: Bordas, 1972), p. 108.
[9] "It's not death, it's not the tomb, far from it; that cannot be the tomb, the expression would be too strong." Trans. mine. – *Nouvelles et Textes pour rien* (Paris: Editions de Minuit, 1958), p. 140. This passage was left untranslated by Beckett.
[10] *No's Knife*, p. 78.
[11] *Ibid.*, p. 80.
[12] *Ibid.*, p. 81.
[13] *Ibid.*, p. 83.
[14] *Ibid.*, p. 86.
[15] *Ibid.*, p. 87.
[16] *Ibid.*, p. 88.
[17] *Ibid.*
[18] *Ibid.*, p. 89.
[19] *Ibid.*, pp. 89–90, italics mine. Cf. Heidegger, *Being and Time*, p. 416.
[20] *No's Knife*, p. 90.
[21] *Waiting for Godot*; see *infra*, Ch. VIII.
[22] *No's Knife*, p. 93.
[23] *Ibid.*, p. 95.
[24] *Ibid.*, p. 100.
[25] *Ibid.*, p. 107.
[26] *Ibid.*, pp. 107–08.
[27] *Ibid.*, p. 108.
[28] *Ibid.*, p. 109.
[29] *Ibid.*, p. 111.
[30] *Ibid.*, p. 114.
[31] *Ibid.*, p. 117.
[32] The original French for this sentence fragment, my own translation, was " . . . ne connaissant personne, de personne connu. . . . " It was rendered by Beckett, *ibid.*, p. 127, as "knowing none, known of none . . . ," which destroys the symmetry of the French. For that reason, and because this chapter had been written from the original French source before Beckett's English translations were available to me, I have retained my own translation of the fragment. *Nouvelles et Textes*, p. 210.

[33] *No's Knife*, p. 123.
[34] *Ibid.*
[35] *Ibid.*, p. 130.
[36] *Ibid.*, p. 135.
[37] *Ibid.*
[38] *Ibid.*
[39] *Ibid.*
[40] Cf. Maurice Merleau-Ponty, *Le visible et l'invisible* (Paris: Gallimard, 1964), pp. 268, 316–18.

CHAPTER VII

[1] *How It Is* (New York: Grove Press, Inc., 1964), p. 7.
[2] Frederick J. Hoffman, *Samuel Beckett: The Language of Self* (Carbondale, Ill.: Southern Illinois University Press, 1962), p. 59.
[3] *Samuel Beckett: The Comic Gamut* (New Brunswick, N. J.: Rutgers University Press, 1962), pp. 167–68.
[4] *How It Is*, p. 7.
[5] See his *Being and Time*, pp. 188–95.
[6] *How It Is*, p. 54.
[7] *Ibid.*, p. 40.
[8] *Ibid.*, p. 7.
[9] *Ibid.*, p. 9.
[10] *Ibid.*, p. 13.
[11] *Ibid.*
[12] *Ibid.*, p. 17.
[13] *Ibid.*, p. 24.
[14] *Ibid.*, p. 42.
[15] *Ibid.*, p. 59.
[16] *Ibid.*, p. 60.
[17] *Ibid.*, p. 69.
[18] *Ibid.*, p. 78.
[19] *Ibid.*, p. 97.
[20] *Ibid.*, p. 109.
[21] *Ibid.*, pp. 114–15.
[22] *Ibid.*, p. 118.
[23] *Ibid.*, p. 128.
[24] *Ibid.*, p. 139.
[25] *Ibid.*, p. 144.
[26] See L. Janvier, *Beckett par lui-même*, (Paris: Editions du Seuil, 1969), unpaginated "Chronologie," 1961.
[27] Published in *No's Knife*, pp. 153–68.
[28] *No's Knife*, p. 159.
[29] This image is borrowed from a title of the later Heideggerian work, *Unterwegs zur Sprache*, which draws out some of the linguistic implications of the hermeneutics in *Being and Time. Unterwegs* (Pfullingen; Neske, 1959).

30 *No's Knife*, p. 156.
31 *Ibid.*, p. 166.
32 *Ibid.*, p. 167.
33 *Ibid.*, p. 168.

CHAPTER VIII

1 See the account of the Miami fiasco in John Lahr, 'The Rise and Fall of Beckett's Bum: Bert Lahr in Godot,' *Evergreen* 13 (Sept., 1969), pp. 29–32, 79–86. Citation, p. 32.
2 All citations of the plays published before 1964 are taken from the trilingual version published in Germany. *Dramatische Dichtungen*, Vol. I (Frankfurt am Main: Suhrkamp Verlag, 1963), p. 365. The English translations are Beckett's own.
3 *Ibid.*
4 *Ibid.*, p. 365.
5 *Ibid.*
6 *Ibid.*
7 *Ibid.*, pp. 399–400.
8 *Ibid.*, p. 400.
9 *Ibid.*, p. 401.
10 *Ibid.*, p. 402.
11 *Ibid.*, p. 405.
12 *Ibid.*, p. 410.
13 See Stanley Cavell, 'Ending the Waiting Game,' in *Must We Mean What We Say*? (New York: Scribner's Sons, 1969).
14 'Waiting for Godot,' in *Dramatische Dichtungen*, Vol. I, p. 448.
15 *Dramatische Dichtungen*, Vol. I, p. 473.
16 *Ibid.*, p. 493.
17 *Ibid.*, p. 501.
18 *Ibid.*
19 *Ibid.*
20 *Ibid.*, p. 502.
21 *Ibid.*, p. 523.
22 *Ibid.*, p. 527.
23 *Ibid.*, pp. 529–30.
24 See Heidegger, *Being and Time*, pp. 396–400.

CHAPTER IX

1 Heidegger, *Being and Time*, p. 171.
2 *Ibid.*, p. 220.
3 *Dramatische Dichtungen*, Vol. II, pp. 8–10.
4 *Ibid.*, p. 12.
5 *Ibid.*
6 *Ibid.*

[7] *Ibid.*, p. 20.
[8] *Ibid.*, p. 144.
[9] *Ibid.*
[10] *Ibid.*, p. 66.
[11] *Ibid.*, p. 114.
[12] *Ibid.*, p. 118.
[13] *Ibid.*
[14] *Ibid.*, p. 130.
[15] *Ibid.*, p. 134.
[16] *Ibid.*, p. 136.
[17] *Ibid.*
[18] *Ibid.*, p. 144.
[19] *Ibid.*, p. 272.
[20] *Ibid.*
[21] See my 'Method and Methodology in Literary Criticism,' *Art and Existence* (Lewisburg, Pa.: Bucknell University Press, 1970), pp. 99 ff, for an elucidation of the critical concepts employed here.
[22] *Dramatische Dichtungen*, Vol. II, pp. 276–78.
[23] *Ibid.*, p. 278.
[24] *Ibid.*
[25] *Ibid.*, p. 282.
[26] *Ibid.*, p. 288.
[27] *Ibid.*, p. 290.

CHAPTER X

[1] *Being and Nothingness*, pp. 361 ff.
[2] *Dramatische Dichtungen*, Vol. II, p. 88.
[3] No prediction seems to be intended. For Krapp to have recorded his voice thirty years ago on magnetic tapes (in 1958, the year *Krapp's Last Tape* was written), they would have to have existed in 1928. Beckett seems therefore to have noted the setting of the play in the future to avoid anachronism. Given the essential characteristics of the ontological determination of the play, however, a prediction would not be out of order.
[4] *Dramatische Dichtungen*, II, pp. 92–94.
[5] *Ibid.*, p. 96.
[6] *Ibid.*
[7] *Ibid.*
[8] *Ibid.*
[9] *Ibid.*, pp. 100–02.
[10] *Ibid.*, p. 106.
[11] *Ibid.*
[12] *Ibid.*, p. 150.
[13] *Being and Nothingness*, pp. 47 ff.
[14] *Dramatische Dichtungen*, II, p. 150.
[15] *Ibid.*, p. 156.
[16] *Ibid.*, p. 158.

17 *Ibid.*
18 *Ibid.*, pp. 160–62.
19 *Ibid.*, p. 164.
20 *Ibid.*
21 *Ibid.*, p. 166.
22 *Ibid.*, p. 194.
23 *Ibid.*, p. 200.
24 *Ibid.*
25 *Ibid.*, p. 208.
26 *Ibid.*
27 *Ibid.*, p. 214.
28 *Ibid.*, p. 224.
29 *Ibid.*
30 *Ibid.*
31 *Ibid.*, p. 226.
32 *Ibid.*, p. 230–32.
33 *Ibid.*, p. 232.
34 *Ibid.*
35 *Murphy* (New York: Grove Press, Inc., 1957), p. 4.
36 *Dramatische Dichtungen*, II, p. 242.
37 *Ibid.*, pp. 236, 238, 240.

CHAPTER XI

1 See the account of same, *supra*, Ch. VIII.
2 See Heidegger, *Being and Time*, pp. 91 ff.
3 See *Ibid.*, pp. 96–102.
4 Heidegger has made this commentary on his own principal work in criticism of Sartre's use of it. See his *Über den Humanismus* (Frankfurt am Main: Klosterman, 1946) pp. 15–16. See also *Being and Time*, p. 171.
5 *Dramatische Dichtungen*, II, p. 515.
6 *Ibid.*
7 See Heidegger, *Being and Time*, pp. 171–72.
8 'Come and Go,' in *Breath and Other Shorts* (London: Faber and Faber, 1971), p. 19.
9 That was in the French version; in the original English, the expression was toned down to the startled 'Oh!' – *Ibid.*
10 In English, 'God grant not.' – *Ibid.*
11 *Ibid.*, p. 21; the characterizations of these seven positions as determinations of the developing drama are my own.
12 See the original version, *Le Visible et l'invisible* (Paris: Gallimard, 1964), pp. 252, 268, 316, 317–18, 321.

CHAPTER XII

1 *Evergreen*, 13 (Sept., 1969), 29–32, 79–86.

[2] *Eh Joe and Other Writings* (London: Faber and Faber, 1967), p. 31.
[3] *Ibid.*
[4] *Ibid.*
[5] *Ibid.*
[6] See his *Being and Nothingness*, pp. 252 ff.
[7] *Ibid.*, p. 70.
[8] *Ibid.*, for Sartre's source on this concept, cf. Heidegger, *Being and Time*, pp. 304–05. Cf. also, Sartre, *op. cit*, p. 149.
[9] In the French version of the play, *Dis Joe* in *Comédie et actes divers* (Paris: Editions de Minuit, 1966), p. 84; the English, in *Eh Joe* (London: Faber and Faber, 1967), p. 16, is "The best's to come." In translating from the English to French Beckett has made the meaning more explicit.
[10] Another improvement in French translation; the original English was "spirit made light." See *Eh Joe*, p. 19.
[11] Following the French translation, *Comedie et actes divers*, p. 91.
[12] See Sartre, *Being and Nothingness*, p. 21 ff.
[13] Heidegger, *Being and Time*, pp. 153 ff.
[14] See Hegel, *Phenomenology of Mind*, Baillie trans. (London: George Allen and Unwin, Ltd., 1961) pp. 234 ff.
[15] See Heidegger, *op. cit.*, pp. 51 ff.

CHAPTER XIII

[1] The fact is reported by Gérard Durozoi, in *Beckett, Présence Littéraire* (Paris: Bordas, 1972), p. 21.
[2] *Cf.* Samuel Beckett, et al., *Our Exagmination Round His Factification for Incamination of Work in Progress* (London: Faber and Faber, 1972), originally published in Paris: Shakespeare and Co., 1929.
[3] Published as 'Fourteen Letters from Samuel Beckett to Alan Schneider,' in the *Village Voice*, 19 March 1958.
[4] See L. Janvier, *Beckett par lui-même*, p. 158.
[5] Projected into the life of his character, Marcel, *Le temps retrouvé* in vol. III, *A la recherche du temps perdu* (Paris: Bibliotheque de la Pléiade, Gallimard, 1954), pp. 708 ff, 718–21, 779, 855, 833 ff.
[6] *Nouvelles et Textes pour rien*, p. 97.
[7] *No's Knife*, p. 53.
[8] *Comment c'est* (Paris: Editions de Minuit, 1961), p. 9.
[9] *Ibid.*, p. 177.
[10] *How It Is*, p. 7.
[11] *Ibid.*, p. 147.
[12] Molloy (Paris: Editions de Minuit, 1951), p. 55.
[13] *Three Novels*, p. 46.
[14] *Dramatische Dichtungen*, two vols. (Frankfurt am Main: Suhrkamp Verlag, 1963–64) inside flap, dustcover, vol. I.
[15] *Nouvelles et Textes*, p. 140.
[16] *Dramatische Dichtungen*, I, p. 158.

[17] *Ibid.*, p. 432.
[18] *Ibid.*, p. 159.
[19] *Premier amour* (Paris: Editions de Minuit, 1970), pp. 26–29.
[20] *Ibid.*, p. 44.
[21] *Ibid.*, p. 46.
[22] Durozoi, *op. cit.*, p. 9.

CHAPTER XIV

[1] See W. K. Wimsatt, Jr., and Monroe Beardsley, 'The Intentional Fallacy,' *Sewanee Review*, LIV (Summer, 1946), 468–88.
[2] Dante, *Il Convivio*, critical edition *a cura di* Maria Simonelli (Bologna: Casa Editrice Patron, 1966), pp. 30–33.
[3] See Ms. Ruby Cohn, *Samuel Beckett: The Comic Gamut* (New Brunswick, N. J.: Rutgers University Press, 1962).
[4] See G. Durozoi, *Beckett, Présence Littéraire*, (Paris and Montreal: Bordas, 1972), p. 9.
[5] See Martin Esslin, *The Theatre of the Absurd*, an Anchor Book (Garden City, N.Y.: Doubleday, 1961).
[6] See G. D. Barnard, *Samuel Beckett: A New Approach* (New York: Dodd, Mead, & Co., Inc., 1970).
[7] *How It Is*, p. 42.
[8] The title of Schopenhauer's doctoral dissertation, 1813; trans. by Mme. Karl Hillebrand, revised ed. (London: A. Bell, 1891).
[9] See his *Erläuterungen zu Hölderlins Dichtung* (Frankfurt am Main: Klostermann, 1963), pp. 39 ff.
[10] The theme is already pronounced in *Being and Time*, and projected as a title to a collection of his essays in *Unterwegs zur Sprache* (Pfullingen: Neske, 1959).
[11] See Beckett, *Proust* (New York: Grove Press, Inc., 1931).
[12] See his 'Residua,' *No's Knife*, pp. 153–68.
[13] *Poems in English by Samuel Beckett*, p. 15.
[14] Vol. I (London: Routledge & Kegan Paul, Ltd., 1959), p. 3.
[15] See Hugh Kenner, *Samuel Beckett: A Critical Study* (Berkeley and Los Angeles: University of California press, 1969).
[16] Georg Witkowski, ed., *Lessings Werke* (Leipzig und Wien: Bibliographisches Institut, n.d.), Vols. IV and V.
[17] Nathan H. Dole, ed., *Complete Works* (Boston: The Aldine Publishing Co., 1910), two vols., no translator indicated.
[18] Tr. by Clifton Fadiman. In *The Philosophy of Nietzsche* (New York: Random House. Inc., n.d.), pp. 949–1088.

BIBLIOGRAPHY

PRIMARY

Beckett, Samuel, 'Dante . . . Bruno. Vico . . Joyce,' *transition* 16–17 (1929), 242–53.

Beckett, Samuel, *et al.*, *Our Exagmination Round his Factification for Incamination of Work in Progress*, London: Faber and Faber, 1972. Original, Paris: Shakespeare & Co., 1929.

Beckett, Samuel, *Poems in English by Samuel Beckett*, New York: Grove Press, Inc., 1961. Original version of 'Whoroscope,' Paris: The Hours Press, 1930.

Beckett, Samuel, *Proust*, New York: Grove Press, Inc., n.d. Original publication, London, 1931.

Beckett, Samuel, *More Pricks than Kicks*, London: Calder and Boyars, 1966. Originally, Chatto and Windus, 1934.

Beckett, Samuel, *Murphy*, New York: Grove Press, Inc., 1957. Originally, London: Routledge, 1938.

Beckett, Samuel, *Watt*, New York: Grove Press, Inc., 1959. Originally, Paris: Olympia Press, 1953. Written, 1942–44.

Beckett, Samuel, *Premier amour*, Paris: Editions de Minuit, 1970. Written, 1945.

Beckett, Samuel, *Mercier et Camier*, Paris: Editions de Minuit, 1970. Written, 1946.

Beckett, Samuel, *Nouvelles et Textes pour rien*, Paris: Editions de Minuit, 1958. *Nouvelles* written, 1945; *Textes*, 1950.

Beckett, Samuel, *No's Knife, Collected Shorter Prose, 1945–66*, London: Calder and Boyars, 1967.

Beckett, Samuel, *Molloy*, Paris: Editions de Minuit, 1951. Written, 1948.

Beckett, Samuel, *Malone meurt*, Paris: Editions de Minuit, 1951. Written, 1948.

Beckett, Samuel, *L'Innommable*, Paris: Editions de Minuit, 1953. Written, 1949.

Beckett, Samuel, *Molloy, Malone Dies, The Unnamable, Three Novels by Samuel Beckett*, New York: Grove Press, Inc., 1959.

Beckett, Samuel, *Comment c'est*, Paris: Editions de Minuit, 1961.

Beckett, Samuel, *How It Is*, New York: Grove Press, Inc., 1964.

Beckett, Samuel, *Têtes mortes*, augmented edition; Paris: Editions de Minuit, 1972.

Beckett, Samuel, *From an Abandoned Work*, London: Faber and Faber, 1958.

Beckett, Samuel, *Dramatische Dichtungen in drei Sprachen*, Frankfurt-am-Main: Suhrkamp Verlag, 1963/4. 2 vols.

Beckett, Samuel, *Comédie et actes divers*, second, augmented edition; Paris: Editions de Minuit, 1972.

Beckett, Samuel, *Eh Joe and Other Writings*, London: Faber and Faber, 1967.

Beckett, Samuel, *Breath and Other Shorts*, London: Faber and Faber, 1971.

SECONDARY

Barnard, G. C., *Samuel Beckett: A New Approach*, New York: Dodd, Mead and Co., Inc., 1970.

Calder, John, ed., *Beckett at Sixty* [A Festschrift], London: Calder and Boyars, 1967.

Cavell, Stanley, 'Ending the Waiting Game,' in *Must We Mean What We Say*? New York: Scribner's Sons, 1969.

Cohn, Ruby, *Samuel Beckett: The Comic Gamut*, New Brunswick, N. J.: Rutgers University Press, 1962.

Durozoi, G., *Beckett, Présence Littéraire*, Paris and Montreal: Bordas, 1972.

Esslin, Martin, *The Theatre of the Absurd*, An Anchor Book; Garden City, N.Y.: Doubleday & Co., Inc., 1961.

Esslin, Martin, ed., *Samuel Beckett, A Collection of Critical Essays*, Englewood Cliffs, N.J.: Prentice-Hall, 1965.

Fletcher, J., *The Novels of Samuel Beckett*, New York: Barnes and Noble, 1964.

Friedman, M. J., ed., *Samuel Beckett*, Configuration Critique No. 8; Paris: Minard (Lettres Modernes), 1964.

Harrison, R., *Samuel Beckett's Murphy, A Critical Excursion*, University of Georgia Monographs, No. 15; Athens, Ga.: University of Georgia Press, 1968.

Hoffman, F. J., *Samuel Beckett: The Language of Self*, Carbondale, Ill.: Southern Illinois University Press, 1962.

Heidegger, M., *Being and Time*, J. Macquarrie and E. Robinson trans., London: SCM Press, Ltd., 1962.

Heidegger, M., *Über den Humanismus*, Frankfurt-am-Main: Klostermann, 1947.

Heidegger, M., *Erläuterungen zu Hölderlins Dichtung*, Frankfurt-am-Main: Klostermann, 1963.

Heidegger, M., *Unterwegs zur Sprache*, 3rd edition; Pfullingen: Neske, 1965.

Hegel, G. W. F., *The Phenomenology of Mind*, J. B. Baillie trans., New York: Macmillan, 1961.

Husserl, E., *Ideas*, Boyce Gibson, trans., New York: Collier Books, 1962.

Janvier, L., *Samuel Beckett par lui-même*, Paris: Aux Editions du Seuil, 1969.

Kaelin, E. F., *Art and Existence*, Lewisburg, Pa.: Bucknell University Press, 1970.

Kenner, Hugh, *Samuel Beckett: A Critical Study*, New edition, with supplementary chapter; Berkeley and Los Angeles: University of California Press, 1968.

Leibnitz, G. W., *Discourse on Metaphysics*, G. R. Montgomery and A. R. Chandler trans., La Salle, Ill.: Open Court, 1947.

Merleau-Ponty, M., *Le visible et l'invisible*, Paris: Gallimard, 1964.

Sartre, J. -P., *L'Imaginaire*, Paris: Gallimard, 1940.

Sartre, J. -P., *Being and Nothingness*, Hazel Barnes trans., New York: Philosophical Library, 1956.

Sartre, J. -P. *No Exit and Three Other Plays*, New York Vintage Books, 1958.

Schopenhauer, A., *The Fourfold Root of the Principle of Sufficient Reason*, Mme. K. Hillebrand trans., revised edition; London: G. Bell, 1891.

Schopenhauer, A., *The World as Will and Idea*, R. B. Haldane and J. Kemp trans., London: Routledge, Kegan Paul, Ltd., 1950, 3 vols.

INDEX

ANALECTA HUSSERLIANA

The Yearbook of Phenomenological Research

I. *Analecta Husserliana.* 1971.

II. *The Later Husserl and the Idea of Phenomenology. Idealism-Realism, Historicity and Nature.* 1972.

III. *The Phenomenological Realism of the Possible Worlds. The 'A Priori', Activity and Passivity of Consciousness, Phenomenology and Nature.* 1974.

IV. *Ingardenia. A Spectrum of Specialised Studies Establishing the Field of Research.* 1976.

V. *The Crisis of Culture. Steps to Re-Open the Phenomenological Investigation of Man.* 1976.

VI. *The Self and the Other. The Irreducible Element in Man, Part I: The 'Crisis of Man'.* 1977.

VII. *The Human Being in Action. The Irreducible Element in Man, Part II: Investigations at the Intersection of Philosophy and Psychiatry.* 1978.

VIII. Yoshihiro Nitta and Hirotaka Tatematsu (eds.), *Japanese Phenomenology. Phenomenology as the Trans-Cultural Philosophical Approach.* 1979.

IX. *The Teleologies in Husserlian Phenomenology. The Irreducible Element in Man, Part III: 'Telos' as the Pivotal Factor of Contextual Phenomenology.* 1979.

X. Karol Wojtyła, *The Acting Person.* 1979.

XI. Angela Ales Bello (ed.), *The Great Chain of Being and Italian Phenomenology.* 1981.

XII. *The Philosophical Reflection of Man in Literature.* 1982.

XIII. Eugene F. Kaelin, *The Unhappy Consciousness. The Poetic Plight of Samuel Beckett, An Inquiry at the Intersection of Phenomenology and Literature.* 1981.

GPSR Compliance

The European Union's (EU) General Product Safety Regulation (GPSR) is a set of rules that requires consumer products to be safe and our obligations to ensure this.

If you have any concerns about our products, you can contact us on ProductSafety@springernature.com

In case Publisher is established outside the EU, the EU authorized representative is:

Springer Nature Customer Service Center GmbH
Europaplatz 3
69115 Heidelberg, Germany

Batch number: 10761599

Printed by Printforce, the Netherlands